EARLY STAGES IN THE EVOLUTION OF MESOPOTAMIAN CIVILIZATION

EARLY STAGES IN THE EVOLUTION OF MESOPOTAMIAN CIVILIZATION

Soviet Excavations in Northern Iraq

Edited by

Norman Yoffee and Jeffery J. Clark

The University of Arizona Press / Tucson

The University of Arizona Press
www.uapress.arizona.edu

First paperback edition 2016

Printed in the United States of America
21 20 19 18 17 16 7 6 5 4 3 2

ISBN-13: 978-0-8165-1393-2 (cloth)
ISBN-13: 978-0-8165-3281-0 (paper)

Front cover: Front view of a clay anthropomorphic vessel from Yarim Tepe II, lower Halaf level; *Back cover:* Yarim Tepe I. Building plan of level 5, upper Hassuna level. Images courtesy of R. M. Munchaev.

Library of Congress Cataloging-in-Publication Data
Early stages in the evolution of Mesopotamian civilization : Soviet excavations in northern Iraq / edited by Norman Yoffee and Jeffery J. Clark.
p. cm.
Translations of papers originally published in Soviet journals 1973–1989.
Includes bibliographical references and index.
ISBN 0-8165-1393-7 (cl : alk. paper)
1. Neolithic period—Iraq—Sinj¬ār Region. 2. Excavations (Archaeology)—Iraq—Sinjār Rgion. 3. Sinjār Region (Iraq)—Antiquities. 4. Iraq—Antiquities. 5. Sovetskaia arkheologicheskaia ekspeditsiia v Irake (1969–). I. Yoffee, Norman. II. Clark, Jeffery J.
GN776.32.173E27 1993
935—dc20 93-3129
CIP

British Library Cataloguing-in-Publication Data
A catalogue record for this book is available from the British Library.

♾ This paper meets the requirements of ANSI/NISO Z39.48-1992 (Permanence of Paper).

Contents

List of Illustrations vii

Acknowledgments xiii

Notes to Transliterations xv

1 Introduction 1
N. Yoffee

2 Tell Maghzaliyah: An Early Neolithic Site in Northern Iraq 7
N. O. Bader
- a. 1977–1978 Results (From *Sovetskaya Arxeologiya* [*SA*] 1979)
- b. 1979–1980 Results (From *Sumer* 1984)

3 The Early Agricultural Settlement of Tell Sotto 41
N. O. Bader
(From *SA* 1975)

4 Results of the Excavations at the Early Agricultural Site of Kültepe in Northern Iraq 55
N. O. Bader
(From *Arxeologiya Starogo i Novogo Svete* [*ASNS*] 1982)

5 Summary of *The Earliest Agriculturalists of Northern Mesopotamia* (1989) 63
N. O. Bader

6 Yarim Tepe I 73
N. Ya. Merpert and R.M. Munchaev
- a. The Upper Hassuna Levels (From *Iraq* 1973)
- b. The Lower Hassuna Levels (From *Iraq* 1987)

7 The Archaic Phase of the Hassuna Culture 115
N. Ya. Merpert
(From *ASNS* 1982)

8 Yarim Tepe II: The Halaf Levels 129
N. Ya. Merpert and R. M. Munchaev
a. The Upper Halaf Levels (From *Iraq* 1973)
b. The Lower Halaf Levels (From *Iraq* 1987)

9 Yarim Tepe III: The Halaf Levels 163
N. Ya. Merpert and R. M. Munchaev
a. Results of the 1978–1979 Field Season (From *Sumer* 1984)
b. Results of the 1980 Field Season (From *Sumer* 1984)

10 Burial Practices of the Halaf Culture 207
N. Ya. Merpert and R. M. Munchaev
(From *ASNS* 1982)

11 Yarim Tepe III: The Ubaid Levels 225
N. Ya. Merpert and R. M. Munchaev
(From *SA* 1982)

12 The Earliest Evidence For Metallurgy in Ancient Mesopotamia 241
N. Ya. Merpert and R. M. Munchaev
(From *Rannezemledel'cheskie Poseleniya Severnoj Mesopotamii* 1981)

13 Some Problems in the Archaeology of Mesopotamia in Light of Recent Research by the Soviet Expedition to Iraq 249
R. M. Munchaev
(From *Kratkie Soobshcheniya Instituta Arxeologii* [KSIA] 1984)

14 Mesopotamian Interaction Spheres 257
N. Yoffee

Select Bibliography 271
Index 281

Illustrations

Figures

Key Russian terms in the figures: печь = oven; яма = pit

2.1	General view of the site from the west.	8
2.2	Section of the south side of squares L-8 and K-8.	10
2.3	Plan of Level 13 in squares K-6, L-6, M-6, K-7, L-7, and M-7.	11
2.4	Plan of Level 13 in squares J-7, K-7, L-7, J-8, K-8, and L-8.	11
2.5	Part of house 3.	12
2.6	Clay storage bin with opening at top.	12
2.7:1,2	Obsidian cores and flint celts.	13
2.8	Flint projectile points and geometric microliths.	14
2.9	Obsidian sickle blades.	16
2.10	Bone polishers.	17
2.11	Grinding stones.	18
2.12	Clay, bone, and stone figurines, pieces of malachite, copper awl, ceramic sherds.	19
2.13	Fragments of marble and plaster vessels.	20
2.14	Groundstone mortars.	21
2.15	Stone pendants.	22
2.16	Clay female figurine.	22
2.17	Upper excavations and the defensive wall.	26
2.18	Defensive wall in Level 4.	27
2.19	Plan of the dwelling in Level 2.	29
2.20	Stone foundation of the Level 2 dwelling.	30
2.21	Part of the Level 3 dwelling.	31
2.22	Different examples of masonry foundations.	33
2.23:1,2	Different examples of masonry foundations.	34
2.24:1–6	Basketry impressions on gypsum and bitumen.	35
2.25:1–30	Stone and obsidian beads, pendants, and bracelets.	36
2.26:1,2	Burials in squares P-3 and P-4.	38

3.1	1973 field season. Section 8 through squares 10 and 11.	42
3.2	1974 field season. Ground plan of structures in Level 7.	44
3.3	Dwelling 29 with storage area for grain and cooking pot.	46
3.4	Dwelling 78 with an oven, large jar, and a bench near the wall; excavations of 1975.	47
3.5	Crude vessels from the lower levels.	48
3.6	Various appliqué designs on crude vessels from the lower levels.	49
3.7:1–6	Photographs of appliqué designs from the lower levels.	50
3.8	Typical pottery forms.	51
3.9	Painted fine wares.	51
3.10:1–5	Stone bowls and clay objects.	52
3.11:1–5	Clay female figurines from the lower levels (#5 from the surface).	53
3.12:1–9	Stone implements from the lower levels.	53
4.1	Cross-section 2 along northern balk of square 55.	56
4.2	Level 1, floor plan of large house.	57
4.3	Coarse ware cooking vessels from lower levels.	58
4.4	Painted fine ware pottery.	58
4.5	Level 1, ceramic bull head and black stone slab with cross-hatch incisions.	59
4.6	Level 1, yellow marble bowl.	60
4.7:1,2	Level 1, marble vessels.	60
6.1	Level 4 in the central part of the settlement with a 10 m. grid.	74
6.2:1–3	Levels 4–7.	77
6.2:1	Levels 4–5.	77
6.2:2	Round oven, Level 4.	77
6.2:3	Two-stage pottery kiln, level 7.	77
6.3	Building plan of Level 5.	78
6.4	Building plan of Level 6 in the central part of the settlement.	81
6.5:1	Level 6 from the south.	82
6.5:2	Levels 7–9 from the south.	83
6.6:1–6	Yarim Tepe I ceramic vessels Levels 3–6, scale 2:5.	86
6.7:1–3	The pottery from Level 5.	88
6.7:1	Samarra vessel.	88
6.7:2	Samarra vessel.	88
6.7:3	Depiction of a scorpion on a Standard Hassuna vessel.	88
6.8:1–7	Yarim Tepe I and II Pottery: 1–4,6,7 Yarim Tepe I: 1–4 Samarra pottery; 7, Halaf vessel from grave; 5, Yarim Tepe II Halaf vessel.	89
6.9:1–8	The finds. 1–5, 7–8 Yarim Tepe I (1–4 from Halaf graves); 6 Yarim Tepe II. 1–4 stone, 5–6 clay, 7–8 bone objects.	90
6.10:1–12	Yarim Tepe I (1–11) and II (12). Clay anthropomorphic figurines (1–10, 12) and one zoomorphic object (11).	92
6.11	Plans of building levels. I. Pits and ovens. II. Level 12.	94
6.12	Plans of building levels. I. Level 10. II. Level 11.	96
6.13	Plans of building levels. I. Level 9. II Level 8.	97
6.14:1	Excavation of the lower levels: from the south.	98
6.14:2	Excavation of the lower levels: from the north.	99

6.15:1	Buildings of Level 12: 31.	100
6.15:2	Buildings of Level 12: 35.	100
6.15:3	Buildings of Level 12: Tholos 319.	100
6.16:1	Buildings of the lower levels: Level 9.	101
6.16:2	Buildings of the lower levels: Level 10.	101
6.17:1–4	Burials from the lower levels. 1. No. 102, building Level 8. 2. No. 103, building Level 8. 3. No. 102, building Level 8 (different angle). 4. No. 132, building Level 9b.	102
6.18:a–d	Vessel types and decorative styles. a. Type I. b. 1–8 appliqué, c. Type II. d. 1–5, 9–11 painted. 6–7 appliqué, 8 incised.	106
6.19:a,b	Vessel types and decorative styles. a. Type III. b. 1–11 painted, 12–13 incised (a-inner surface, b-outer surface).	107
6.20:a,b	Vessel types and decorative styles. a. Type IV. b. 1–18 painted, 19–20 incised, 21 painted and incised, 22–26 appliqué.	108
6.21	Vessel types. Type V.	109
6.22	Types of vessels and decorative styles from the lower levels. 1–4 from pit 71 in the original ground surface; 5–8 from Level 10.	109
6.23:1–5	Vessels from Levels 12 (1,3,5) and 11 (2,4).	110
6.24:1–5	Vessels and example of applied decoration from Levels 10 and 9.	110
6.25:1–6	Vessels from Levels 9 (1–3,5) and 8 (4,6).	111
6.26:1–6	Obsidian tools and arrows (1–6) from lower levels.	112
6.27	Flint and stone tools: axes and projectile points.	112
6.28	Bead necklace from building Level 12.	113
6.29	Stone seals from lower levels (1–2 Level 8, 3 Level 9).	114
8.1:1–9	Halaf pottery. scale 2:3.	133
8.2:1–4	Levels 3–4. Halaf pottery 1–3 scale 1:4; 4 scale 2:5.	134
8.3:1–6	Levels 3–4. Painted Halaf pottery with naturalistic motifs, scale 3:4.	135
8.4	Level 3. Painted Halaf sherd showing cheetah, scale 1:2.	135
8.5	Photographs from the first excavation season (1969).	136
8.6	Plan of the grid over the site and excavated area.	137
8.7	North-south section.	138
8.8	Plan of the earliest level (9).	140
8.9	Remains of domestic auxiliary structures.	141
8.9:1	Remains of tholos 73.	141
8.9:2	Remains of tholos 61.	141
8.9:3	Remains of tholos 67.	142
8.10:1–7	Objects from tholos 67.	142
8.11	Pottery kiln (structure 345).	143
8.12	Pottery kiln 345 plan and sections.	143
8.13:1–3	Views of a clay anthropomorphic vessel.	146
8.14:1–2	Photographs of both sides of a clay zoomorphic vessel.	146
8.15	Drawings of five views of a clay zoomorphic vessel.	147
8.16	Flint tools.	148
8.17	Flint tools.	149
8.18	Obsidian tools.	150
8.19:1–6	Stone vessels.	151

8.20:1–11	Stone seal-pendants.	151
8.21	Clay scrapers.	152
8.22:1–8	Photographs of coarse and thin-walled vessels.	153
8.23:1–6	Photographs of painted bowls.	154
8.24:1–5	Photographs of painted pots.	155
8.25:1–11	Photographs of painted sherds.	156
8.26:1–3	Drawings of painted bowls.	157
8.27:1–4	Drawings of painted bowls.	157
8.28:1–9	Drawings of painted bowls.	158
8.29:1–12	Drawings of shallow bowls and platters.	159
8.30:1–10	Drawings of painted sherds.	160
8.31:1–8	Drawings of painted sherds.	160
8.32:1–9	Drawings of clay anthropomorphic figurines.	161
9.1	Plan of Halaf Level 1.	168
9.2:1	Grain storage structures 92 and 94.	169
9.2:2	View of complex 9.	169
9.2:3	Rooms 102 and 104.	169
9.3	Plan of Halaf Level 2.	170
9.4:1	View of complex 10 and tholos 118.	171
9.4:2	Tholos 118.	172
9.4:3	Tholos 133.	172
9.5:1,2	Halaf Level 3 structures.	173
9.5:1	Room 132.	173
9.5:2	Complex 9-D.	173
9.6	Coarse ware ceramic vessel.	176
9.7:1–3	Ceramic vessels from the upper Halaf levels.	176
9.8:1–4	Ceramic vessels from the upper Halaf levels.	177
9.9:1,2	Ceramic vessels from burials 26/1 and 27/2.	178
9.10:1–6	Painted ceramics from the upper Halaf levels.	179
9.11:1–6	Painted ceramics from the upper Halaf levels.	180
9.12:1–7	Zoomorphic decorations on Halaf pottery.	181
9.13:1–5	Stone tools from the upper Halaf levels.	182
9.14:1–3	Stone palette, tools, seals, pendants, and beads from the upper Halaf levels.	183
9.15:1,2	Bone implements and clay rings from the upper Halaf levels.	184
9.16:1	Clay beads.	185
9.16:2	Clay model of a hoe.	185
9.16:3	Clay zoomorphic figurines.	185
9.17:1–4	Anthropomorphic figurine fragments.	186
9.18	The architectural plan of Halaf Level 3.	188
9.19	View of tholos 137.	189
9.20	View of structure 139.	189
9.21	Ovens and hearths for Halaf Level 3.	193
9.22:1–9	Lithic and bone artifacts from Halaf Level 3.	194
9.23	Ornaments from Halaf Level 3.	195
9.24	Bone and horn artifacts from Halaf Level 3.	195

9.25:1–3	Ceramic artifacts from Halaf Level 3.	196
9.26	Ceramic spindle whorls from Halaf Level 3.	196
9.27:1–2	Clay objects from Halaf Level 3.	197
9.28	Ceramic rings from Halaf Level 3.	198
9.29	Ceramic spoons from Halaf Level 3.	198
9.30	Ceramic seal pendant from Halaf Level 3.	198
9.31	Ceramic vessels from Halaf Level 3.	198
9.32:1–2	Painted ceramic vessels from Halaf Level 3.	199
9.33:1–2	Painted "tree of life," animal designs on pottery from Halaf Level 3.	199
9.34:1–2	Painted ceramic vessels from Halaf Level 3.	200
9.35:1–6	Miscellaneous ceramic vessels from Halaf Level 3.	200
9.36:1–3	Bichrome painted animal designs on pottery (1–2), and spouted vessel (3) from Halaf Level 3.	201
9.37	Painted ceramic vessel depicting owls.	201
9.38:1–2	Naturalistic clay anthropomorphic figurines from Halaf Level 3.	202
9.39:1–6	Stylized clay anthropomorphic figurines from Halaf Level 3.	203
9.40:1–5	Stylized clay anthropomorphic figurines from Halaf Level 3.	203
9.41:1–2	Photographs of clay zoomorphic figurines from Halaf Level 3.	204
9.42:1–3	Drawings of clay zoomorphic figurines from Halaf Level 3.	204
10.1	Burial 61, Yarim Tepe II: Two adults.	209
10.2:1	Burial 44, Yarim Tepe II.	210
10.2:2	Burial 46, Yarim Tepe II.	210
10.3:1	Burial chamber, Yarim Tepe II.	211
10.3:2	Outline of burial 45, Yarim Tepe II.	211
10.4:1–4	Plans of cremation burials 50, 54, 51, 53 (1–4) at Yarim Tepe II.	212
10.5	Cremation burial 40, Yarim Tepe II.	213
10.6:1–5	Mortuary offerings in cremation burial 40 at Yarim Tepe II.	214
10.7:1–4	Mortuary offerings in cremation burial 40 at Yarim Tepe II.	215
10.8:1,2	Mortuary offerings in cremation burial 43 at Yarim Tepe II.	216
10.9	Main section at Yarim Tepe I showing burial chambers.	218
10.10	Burial 45, Yarim Tepe I.	220
10.11	Burial 47, Yarim Tepe I.	220
10.12:1	Ceramic vessel from Yarim Tepe I burial 56.	222
10.12:2	Ceramic vessel from Yarim Tepe I burial 52.	222
10.12:3	Ceramic vessel from Yarim Tepe I burial 45.	222
10.13:1–3	Stone finds from Yarim Tepe I burials.	222
10.13:1	Vessel from burial 63.	222
10.13:2	Vessel from burial 45.	222
10.13:3	Macehead from burial 56.	222
11.1	Plan of structures in Levels III and IV. Rooms and features are numbered.	226
11.2	Plan of structures in Levels I and II. Rooms and features are numbered.	227
11.3:1–4	Structures of the Ubaid levels:	
11.3:1	Complex 1, view from the north.	228
11.3:2	Complex 6, view from the east.	228
11.3:3	Complex 5, view from the south.	229
11.3:4	Complex 4, view from the south.	229

11.4:1–5	Structures and a burial from the Ubaid levels.	230
11.4:1	Complex 1, view from the northeast.	230
11.4:2	Passageway between complexes 1 and 2.	230
11.4:3	Entryway in wall of room 1, view from the west.	231
11.4:4	Oven in room 4 of complex 3, view from the west.	231
11.4:5	Wall between structures 63 and 64 in complex 5, view from the east.	231
11.5:1–12	Ceramic vessels I.	234
11.6:1–6	Ceramic vessels II.	236
11.7:1	Vessel with zoomorphic decorations.	237
11.7:2	Anthropomorphic figurine.	237
11.8:1–31	Stone, bone, and clay tools.	238
11.9:1–21	Miscellaneous clay objects.	239
11.10:1–18	Yarim Tepe III: Ornamental objects and small artifacts.	240
14.1	Extent of Halaf, Samarra, and Hassuna material culture.	264

Maps

1.1	Neolithic sites in northern Iraq (redrawn from Munchaev and Merpert 1981:13).	3
1.2	Yarim Tepe (Munchaev and Merpert 1981:19).	3
2.1	Topographical plan of Tell Maghzaliyah.	24
9.1	Topographic map of Yarim Tepe III and excavated areas.	164

Tables

12.1	Chemical Analysis of Metal Finds at Yarim Tepe I and II.	247

Acknowledgments

I am grateful, naturally, to the members of the Soviet expedition to Iraq for providing the materials that comprise this book. While I was in Moscow, I met with authors to consider which essays would be most appropriate for translation and all authors submitted the illustrative material, much of it original photos and drawings, for inclusion here. The members of the expedition also gave freely of their time: Prof. Merpert discussed with me the rise and progress of archaeological theory in the Soviet Union; Nikolai Bader guided me through the materials that were being analyzed in Moscow; Valeri Gulaev, professor and international archaeologist, invited me to address the Sector for Archaeological Theory and facilitated intellectual communication within the Institute; a newer member of the team, Shaxmardan Amirov, not only described his thesis on Halaf pottery but was a friendly informant on life in Moscow and in Dagestan; Arsen Faradzhev was my vade mecum in matters of daily existence, and his friendship is much appreciated. This volume is but a token of my esteem for these scholars and gentlemen.

The International Research and Exchange Board supported my Moscow research and also provided a grant that funded the translation of these essays by Dr. Stephen and Mrs. Ethel Dunn of the Highgate Road Social Science Research Station in Berkeley. IREX has sponsored from its inception the archaeological exchange program that has brought dozens of Soviet and American colleagues together, promoted international collaboration and collegiality, and not least, international friendship. Here's a toast—*na zdorov'e*—to Wesley Fisher and IREX.

I am grateful to Robert McC. Adams, Robert Dyson, and especially to C. C. Lamberg-Karlovsky, the three organizers of the first two USSR-USA archaeological exchanges, for including me in their midst and to Carl Lamberg-Karlovsky also for his support of the project that has resulted in this volume.

This project was brought to its final stage thanks to a grant from the Vice President for Research Administration, University of Arizona, Charles Geoffrion. This grant funded Jeffery J. Clark, a Ph.D. student in the Department of Anthropology who has worked in a variety of sites in West Asia, as well as in Peru and in the American Southwest, to translate the English translation of the Russian text into "archaeologese." Clark regrouped the various essays on sites into the present organization of chapters according to the chronology of the sites. He further correlated the illustrative material with the text, and entered the text into the computer.

The last chapter was written and the last version of the book was composed while I was NEH Resident Scholar at the School of American Research in Santa Fe (1991–92). Philip Kohl, Wellesley College, provided exemplary comments on the shape of the book and good advice on particular matters of translation. Gil Stein, Northwestern University, located some obscure references for me. Reinhard Bernbeck, Institut für Vorderasiatische Altertumskunde, Berlin, read the entire manuscript and improved the archaeological terminology throughout. I am grateful to the SAR (Douglas W. Schwartz, President) for its fellowship and for the help from my friends.

Further thanks are due to Doris Sample, who expertly machine-processed the entire manuscript through its many drafts, and to Kathryn Keith, for proofreading, xeroxing, and general advice on the manuscript.

Grateful acknowledgment is also due to the Department of Anthropology, College of Social and Behavioral Sciences, University of Arizona, and to the Vice President for Research, University of Michigan, for grants that have subvened the costs of producing this volume.

Finally, I wish to thank the folk at the University of Arizona Press for their enthusiastic commitment to this publication: Stephen Cox, Director, Gregory McNamee, (then) Editor-in-Chief, and especially Dr. Christine Szuter, social science acquisitions editor. They are not only skilled book producers, they are academic colleagues.

Notes to Transliterations

Insofar as possible, the following table of transliterations for Russian characters without clear English phonetic equivalents has been used in this book. Due to published alternate spellings, however, some inconsistencies in phonetic transcriptions are unavoidable.

е = e or ye

ж = zh

й = j

х = x

ц = ts

щ = shch

ы = y

я = ya

ь = ‘

Arabic names are given in conventional forms as found in the archaeological literature and this has resulted in a few inconsistencies, e.g., “Hassuna” but “Maghzaliyah.”

EARLY STAGES IN THE EVOLUTION OF MESOPOTAMIAN CIVILIZATION

1

INTRODUCTION

Norman Yoffee

In 1981 the first USA-USSR symposium on the archaeology of the Near East, Central Asia, and Ancient India was held at Harvard University, 9–14 November. The second symposium was held in 1983 in Samarkand, Uzbekistan, 5–10 September. During these symposia the American delegation was treated to a series of papers reporting the latest findings and interpretations made by the Soviet expedition to northern Iraq. Although we had known about this work, since an article had been published in English in the journal *Iraq* (chapters 6a and 8a of this volume) and occasional reports had appeared in the journal *Sumer,* we soon learned that much research had been done that was simply unknown to Western scholars — and that our understanding of the early stages of Mesopotamian culture would have to be revised because of it.

My own paper, on "Mesopotamian Interaction Spheres," delivered in Samarkand in 1983 and published in abbreviated form in Russian (Yoffee 1986), occasioned long conversations, especially with V. A. Bashilov, in Samarkand and Moscow. Since that paper (see chapter 14 in this volume) referred to work done by the Soviet expedition on early agricultural settlements in Mesopotamia, it became clear that if I were to continue with this study, I needed to know more about the latest, important views held by the Soviet team. In 1986, with support from the International Research and Exchanges Board (IREX, which had sponsored the first two archaeological symposia), I spent five weeks in the Soviet Union looking at materials and illustrations and especially discussing with my colleagues their analytical methods and their interpretations of social organization and social change in the Neolithic of northern Iraq.

During this research visit I also learned of recent publications that were unknown to archaeologists outside the USSR. With the help of R. M. Munchaev and V. I. Gulaev I gathered a representative group of these articles and formulated a plan to publish English translations of them. Through the generosity of Prof. Munchaev, I was also able to collect the original plans, illustrations, and photos that were published in these articles.

Original Russian References to the Articles Published in This Volume

Chapter 2

a. "Tell Maghzaliyah: Ranneneoliticheskii pamyatnik na severe Irake." In *Sovetskaya Arxeologiya* 1979:117–32. ("Tell Maghzaliyah:

an early Neolithic site in northern Iraq." In *Soviet Archaeology,* hereafter *SA*).
b. This section is reprinted with the permission of the authors and from the journal *Sumer* 43 (1984):45–53.

Chapter 3

"Rannezemledel'skoe poselennie Tell' Sotto (po raskopkam 1971, 1973–74)." In *SA* 1975: 99–111. ("The early agricultural settlement of Tell Sotto [according to excavations of 1971, 1973–74]").

Chapter 4

"Nekatorye rezul'taty rabot na rannezemledel'skom poselenii Kyultepe v severnom Irake." In *Arxeologiya Starogo i Novogo Svete,* edited by V. I. Gulaev (1982):50–57. Moskva: "Nauka." ("Some results of the work on the early agricultural settlement of Kultepe in northern Iraq." In *Archaeology of the Old and New World,* hereafter *ASNS*).

Chapter 5

This chapter is reprinted (with minor editorial revisions) from *Drevnejshie Zemledel'tsy Severnoj Mesopotamii* (*The Earliest Farmers of Northern Mesopotamia*) by N. O. Bader, Moscow: "Nauka" (1989): 346–56.

Chapter 6

a. This section of the chapter is reprinted by permission of the authors and from the journal *Iraq* (1973) 35:94–108, published by the British School of Archaeology in Iraq.
b. This section of the chapter is published by permission of the authors and from the journal *Iraq* (1987) 49:1–21.

Chapter 7

"K voprosu ob arxajcheskoj etape xalafskoj kul'tury." In *ASNS*:11–27.

Chapter 8

a. This section of the chapter is reprinted by permission of the authors and from the journal *Iraq* 35 (1973):108–13.
b. This section of the chapter is reprinted by permission of the authors and from the journal *Iraq* (1987) 49:21–36.

Chapter 9

a. This chapter is reprinted by permission of the authors and from the journal *Sumer* 43 (1984):54–68.
b. This chapter is reprinted by permission of the authors and from the journal *Sumer* 43 (1984):32–45.

Chapter 10

"Pogrebal'nyj obryad plemyen xalafskoj kul'tury (Mesopotamiya)." In *ASNS*:28–49.

Chapter 11

"Poselenie ubejdskoj kyl'tury Yarym-tepe III v severnoj Mesopotamii." *Sovetskaya Arxeologiya* 4 (1982):133–49 ("The Ubaid settlement of Yarim Tepe III in northern Mesopotamia").

Chapter 12

"Drevnejshaya metallurgiya Mesopotamii." Prilozhenie 3 in *Rannezemledel'cheskie Poseleniya Severnoj Mesopotamii,* by R. M. Munchaev and N. Ya. Merpert (1981):307–16. ("The most ancient metallurgy in Mesopotamia." Appendix 3 in *The Early Agricultural Settlements of Northern Mesopotamia*).

Chapter 13

"Nekatorye problemi arxeologii Mesopotamii v svete novejshix issledovanij (iz itogov rabot Sovetskoj ekspeditsii v Irake)." In *Kratkie Soobshcheniya Instituta Arxeologii,* no. 180 (1984):38–45. ("Some problems in the archaeology of Mesopotamia in light of the latest discoveries [from results of Soviet expedition in Iraq]." In *Brief Reports of the Institute of Archaeology,* hereafter *KSIA*).

Chapter 14

A first version of this paper was presented at the second USA-USSR archaeological symposium in Samarkand. An abbreviated version was published as part of the proceedings of the symposium (Yoffee 1986). In December 1986 I read a revised version of this paper at

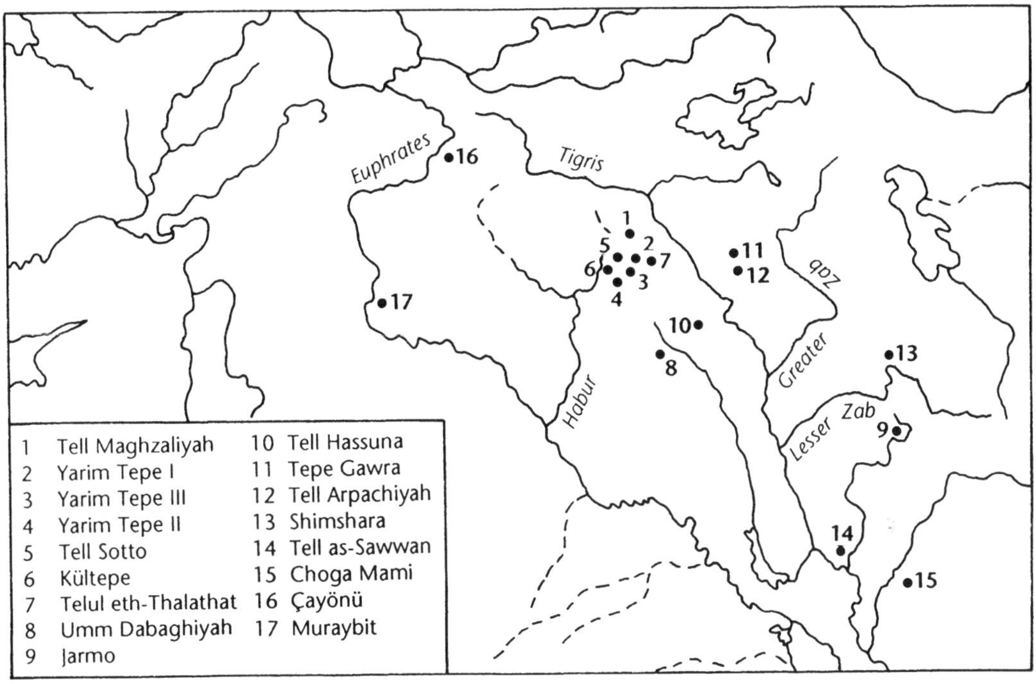

Map 1.1 Neolithic sites in northern Iraq (redrawn from Munchaev and Merpert 1981:13).

the invitation of V. I. Gulaev in the Sector for Archaeological Theory, Institute of Archaeology in Moscow, at which time Profs. Merpert and Gulaev and other archaeologists discussed the *doklad*.

Note that the order of the chapters does not reflect the order of publication but rather the relative chronology of the sites excavated.

The Soviet Expedition to the Sinjar Plain

The Soviet archaeological expedition worked from 1969 until 1980, twelve seasons in all, in the Sinjar Plain, which spreads southward from the Jebel Sinjar (a range rising to a height of 1250 m), and extends to the Mosul region to the east (see Map 1.1). The main work was conducted near the town of Tell Afar on a group of mounds situated 10 km southwest and called Yarim Tepe, since one of the mounds had been cut in half by a local brook. (Yarim Tepe means "half-mound" in Turkish; see Map 1.2).

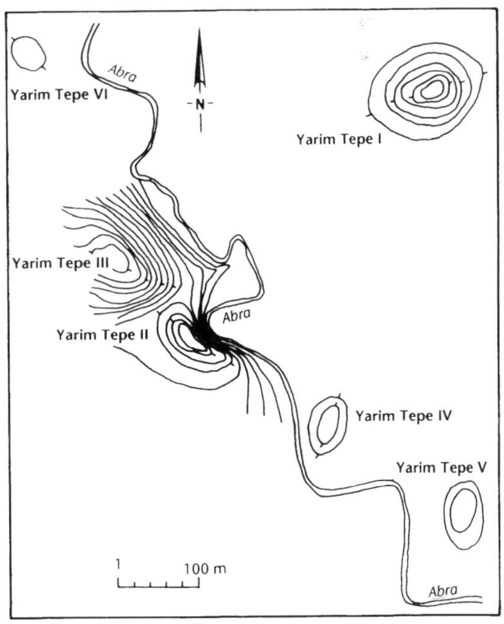

Map 1.2 Yarim Tepe (Munchaev and Merpert 1981:19).

The leaders of the expedition, Profs. R. M. Munchaev and N. Ya. Merpert, distinguished prehistorians of the Soviet Union, were invited by the Iraqi government, which had close political ties with the USSR at the time, to launch excavations in Iraq. David and Joan Oates, British prehistorians who had long worked in the region, provided invaluable aid and advice to the newly formed expedition, which Merpert and Munchaev have always acknowledged.

Previously, Rauf Magomedovich Munchaev had done extensive work in his native Caucasus region (e.g., see R. M. Munchaev 1975) and was at the time of the expedition vice-director of the Institute of Archaeology, Moscow Branch, Academy of Sciences of the USSR. Nikolai Yakovlevich Merpert, a veteran of many seasons' work, notably in the Balkan region and Russia (e.g., see N. Ya. Merpert 1978), was professor in the Institute of Archaeology, Moscow. These two distinguished prehistorians have brought a wealth of archaeological field skills and careful, nonpoliticized interpretive method to the project in northern Iraq. In particular, their experience with questions of ethnicity in the archaeological record, especially Merpert's long interest in the Indo-European question, have led to certain inferences about the culture change in the Neolithic of Mesopotamia (see especially chap. 7).

Joining Munchaev and Merpert was Nikolai Ottovich Bader, son of a famous Soviet archaeologist and himself a skilled prehistoric archaeologist. Bader was placed in charge of the three earliest sites that were investigated, Maghzaliyah, Sotto, and Kültepe. His interpretations of the sequence of change in these sites and his views of cultural continuity in Mesopotamia from the earliest village farming communities in the region through the Hassuna occupations at Yarim Tepe I are represented in chapters 2–5.

Finally, two other principal members of the field team were V. I. Gulaev and V. A. Bashilov. These archaeologists are specialists in the prehistory of, respectively, the Maya and Peru. Although they have assiduously followed the scholarly literature in these two areas and have written analyses of various ancient New World societies, neither has been able to undertake fieldwork in his area of primary expertise. As members of the Institute of Archaeology in Moscow, however, both were experienced excavators in the USSR and elsewhere and were invaluable members of the Soviet team.

In the history of the archaeology of Mesopotamia, it is hard to find a long-term regional project comparable to that conducted by the Soviet team, certainly not in modern times, and never on prehistoric sites. This volume presents some of the major findings of the expedition that are changing the way we currently perceive the cultural history of the Neolithic in Mesopotamia. It does not, however, translate all the publications of the team and does not cover the most recent work done by the expedition in Iraq and Syria.

Notes on the Contributions of the Soviet Expedition

As Munchaev reports in chapter 13, little work had been done in the Sinjar Plain before the Soviet expedition, in spite of the many tells that dot the horizon. The northern Mesopotamian prehistoric cultures called "Hassuna" and "Halaf" were known either from important (but not long-term) excavations undertaken decades ago (like Tell Hassuna and Arpachiyah), or were inadequately published, or were from small-scale work designed mainly to explore stylistic changes in the ceramic assemblage.

It was one task of the Soviet team, then, to expose sufficient areas of sites to understand residential structures and to place in context the artifactual inventory of people whose ceramics had defined the Hassuna and Halaf periods. Furthermore, the sequence of cultural life from sites like Jarmo, in the Zagros region, to Hassuna was little known, and a site like Umm Dabaghiyah, that preceded Hassuna, seemed anomalous. In one synthesis of the prehistory of the area (Mellaart 1975), each site seemed to be its own "culture."

The Soviet objectives were admirably fulfilled. Yarim Tepe I provides the longest sequence, best residential plan, and largest artifactual assemblage for the Hassuna. Although the mound contains Hassuna ceramics almost exclusively, some Samarra ware was found in the upper levels, and thus it provides evidence of social interaction (see chapter 14) between northern and middle Mesopotamia. Yarim Tepe II is the best-attested Halaf site and hence allows clarification of long-standing problems (chaps. 8 and 9 show, for example, that *tholoi* are domestic structures) while necessitating that new questions about the period be posed. Yarim Tepe III documents the transition between the Halaf and Ubaid (chaps. 9 and 11). Maghzaliyah, Sotto, and Kültepe provide a link from Jarmo to the Hassuna culture (chaps. 2–4).

While some of the finds of Yarim Tepe I and II have been published in English (chaps. 6 and 8) and in a monograph in Russian (Munchaev and Merpert 1981), less has been known in the West about Maghzaliyah, Sotto, and Kültepe. In his chapters in this volume (chaps. 2–5), Bader presents some basic results of the work on these sites. From Maghzaliyah, an aceramic site (there are other uses of clay at the site, however), the assemblage shows many similarities to the lower levels of Jarmo and Shimshara, and perhaps a connection to Çayönü. The defensive wall and impressive residential structure at the site, which was discovered in 1977, are also described (chaps. 2 and 5).

The assemblage at Maghzaliyah precedes but is connected to that at Tell Sotto, which was excavated from 1971–75. Sotto is described by Bader as following Jarmo (and Maghzaliyah) and leading to the earliest stages of Tell Hassuna (the "archaic" levels) and the site of Kültepe (excavated in 1975–76). The expedition's work shows clearly that the early levels of Tell Hassuna, described as a campsite, do not reflect the earliest settlement of northern Mesopotamia.

The interpretation of the pre-Hassuna materials is debated by Bader (in the above cited chapters) and Merpert (chap. 7). While Bader prefers to see a cultural continuity in the occupation of northern Mesopotamia, at least through the Hassuna period, Merpert contends that cultural change is not internal, but occasioned by movements of peoples. His argument of ethnic groups in prehistory, specifically in the relations between Hassuna, Samarra, Halaf, and Ubaid, is taken up by me in chapter 14. Bader's case of continuity is strengthened, it seems to me, with the discovery of new sites in northern Mesopotamia, Qermez Dere, near Tell Afar and Yarim Tepe (Watkins and Baird 1987; *Iraq* 49 [1987]:244; *Iraq* 51 [1989]:260–261 Watkins [1992]; Watkins, Baird, and Betts [1990]), and Nemrik, on the Tigris north of Eski Mosul (Kozlowski, *Iraq* 49 [1987]:242; *Iraq* 51[1989]:258–259; Kozlowski [1990a, 1990b]) that are earlier than Maghzaliyah and provide a continuous sequence from preagricultural settlements to the pre-Hassuna sequence documented by the Soviet team.

Chapter 10 presents evidence of burial practices, including that of the Halaf cemetery that was founded atop Yarim Tepe I (presumably by people from Yarim Tepe II) after Yarim Tepe I was abandoned. This cemetery, which contains the remains of adults (while children and infants seemed to have been buried under house floors and in other domestic contexts), supports the argument that Yarim Tepe II was founded by a group of a different ethnicity than that of the group which occupied Yarim Tepe I (this argument is additionally supported by changes in house plans and in ceramic types; see chapter 14).

Chapter 12 summarizes the rich finds of metals by the Soviet team and provides new evidence for early metallurgical practices in West Asia. This chapter, however, does not attempt to survey all the data on the early use of metals in West Asia, omitting the discoveries in Israel, for example; it was written as an appendix to the Yarim Tepe monograph that appeared in 1981 and so does not include the metal finds from Maghzaliyah.

Although these chapters do not present all the finds of the Soviet expedition (one must see Bader's book, for example, for a full description of the defensive wall at Maghzaliyah; that book, however, appeared well after the initial translations were completed and major excerpts from it could not be included here), they provide data from many years of research and analysis. Information about the economy of these sites (the domesticated and wild species of plants and animals that were utilized), the lithic assemblage, the ceramics, both figurines and pottery, the domestic and ritual architecture, storage facilities, and plazas are documented in these chapters. Finally, the many plans, photos, and illustrations that accompany the chapters show what the sites and materials look like. The detailed presentation of these data will doubtless lead to further studies on social life in Mesopotamia in the periods investigated by the Soviet team.

Missing from these chapters, but recounted to a certain extent in the two monographs by expedition members (Munchaev and Merpert 1981; Bader 1989), are discussions of excavation and retrieval methods. Several other omissions that may trouble Western archaeologists are the essentially noninterdisciplinary nature of the fieldwork, the lack of quantitative methods in data analysis, and the relative paucity of interpretations of sociocultural change. A consideration of these topics is not germane to this volume, since one would have to discuss the historical circumstances in which archaeology was practiced in the former USSR, the financial difficulties of assembling library resources and of launching an expedition from Moscow to Iraq, and the constraints placed on a variety of possible theoretical avenues of social interpretation. The IREX-sponsored and other initiatives to bring Soviet and Western scholars together (and which have made possible the publication of this volume) have afforded significant opportunities not only for the dissemination of information, but also for dialogues on archaeological method and theory.

Perhaps a more glaring omission in this book is a period-by-period reconstruction of basic subsistence economies and attempted correlations with social organization and social change. Whereas this synthesis might be regarded as an editorial duty, I have thought it presumptuous to speak for my Russian colleagues, who have provided several chapters of synthesis themselves and who know the archaeology of these sites and periods better than I do. Since I assume that Near Eastern archaeologists can and will build inferences of their own using the material and bibliographic resources presented here, chapter 14 is offered, not as a summary of the volume, but as an indication of what new interpetive possibilities are opened by the Russian work.

2

TELL MAGHZALIYAH: AN EARLY NEOLITHIC SITE IN NORTHERN IRAQ

N. O. Bader

a. 1977–1978 Results

Aceramic Neolithic sites represent the earliest stage in the development of food-producing economies in the Near East and are therefore of great interest. Unfortunately, excavations of these sites are few in number and generally of low quality. In Iraqi Kurdistan, Robert Braidwood traced the origins of the food-producing economy from the archaeological remains at the sites of Jarmo and Karim Shahir (Braidwood and Howe 1960). In the same region, similar remains were recovered by P. Mortensen at Shimshara; the lower strata of this site (14–16) were apparently contemporaneous with the very end of the preceramic phase at Jarmo (Mortensen 1970, 46). Preceramic early agricultural sites are also known in Luristan and Khuzistan, Iran (Hole, Flannery, and Neely 1969; Hole 1977). However, the archaeological materials recovered to date from these regions yield only an incomplete picture of the early Neolithic. Additionally, no consensus exists on the origin of the Hassuna, the first fully agricultural society of the northern Mesopotamian plain.

Braidwood believes that the Hassuna culture follows Jarmo in an evolutionary sense, but at the same time there exists a certain chronological and cultural disparity between them. Braidwood saw a transitional complex at the poorly studied site of Ali Aga (Braidwood and Howe 1960, 162). V. M. Masson (1964, 51) concurred with Braidwood on this issue. P. Mortensen (1970, 120) posits that at Shimshara and Jarmo, pottery was adopted from the Hassuna culture and that the Hassuna and Jarmo may have been contemporaneous for a significant period of time. David and Joan Oates (1976, 59) hold similar points of view. D. Kirkbride (1972, 1973a, 1973b, 1974, 1975), excavator of Umm Dabaghiyah, a site of pre-Hassuna date with early Hassuna features, came to the conclusion that there was no genetic connection between this site and the Hassuna. Thus, she postulates that the material culture at Umm Dabaghiyah was derived from a western source. This is also the opinion of J. Mellaart (1975, 135–41, 149).

The excavation of sites with assemblages similar to Tell Sotto (see chapters 3 and 5, this volume) has revealed that these settlements, like those firmly within the Hassuna, had fully adopted an agricultural economy. Thus, to find the origin of food production, one must look further back in time. Tell Maghzaliyah, a newly discovered aceramic Neolithic site, has yielded an archaeological assem-

Figure 2.1. General view of the site from the west

blage that predates Tell Sotto. This site was discovered by the present author in the spring of 1977.

Tell Maghzaliyah is located 7.5 km northwest of Yarim Tepe, at the northern edge of the village of Abra Najar, approximately 1 km north of the village of Abrahi Kabir, which lies along the old road from Tell Afar to Beled Sinjar. The site is near the right bank of the Abra River, on one of the hilly flanks of the Jebel Sinjar, rising 50–70 m above the plain (Fig. 2.1). The boundary between the mountains and the plain is approximately 1 km to the south with each of the regions representing different ecological zones. Tell Maghzaliyah is situated among numerous natural hillocks along the slopes of mountains. The soil here is quite rocky, and limestone outcrops emerge from the surface in many places. The site is also at the boundary between land currently under cultivation and that used for animal pasturage. In general, the rocky soils are unsuitable for cultivation. Farmland comprises only a part of the area and alternates with pastures which stretch further up the slopes of the mountains. The proximity of the Abra River was undoubtedly a key factor in the settlement of the site. Here the river runs through a deep picturesque ravine which continues 1.5 km upstream. On the whole, this area is better suited to hunting and gathering than to agriculture, and the remains from the site confirm this.

The eastern half of the tell has been destroyed by the river, and the northern edge has been cut by a road. The surviving portion of the site encompasses an area of about 4500 square meters. The depth of the cultural deposit exceeds 8 m. This deposit is comprised largely of a number of stone buildings that in many places are visible on the surface. For this reason, the tell has very steep sides. The site has not been disturbed by modern plowing, building, or digging except in its southern half, which is occupied by a modern cemetery.

The site is bounded on the north by a flat expanse of land of which at least a hectare is contained within a bend of the steep-banked river. Here, on modern cultivated land, a large quantity of flint flakes and debitage and a number of tools similar in type to those from the site itself were found. The main difference between the site and this outlying area is that obsidian is lacking in the latter, whereas on the tell itself it is abundant. To determine whether this region is a temporary camp for Mesolithic hunters or an outlying area of the main settlement that was not built up further requires a more intensive study of the area.

During the survey, the expedition received permission from the Department of Antiquities of Iraq to conduct excavations at Maghzaliyah. A 100 square meter area of the tell was excavated down to sterile soil. As a result, a fairly complete reconstruction of the settlement history of the site was obtained, including the tell's stratigraphic sequence, its architecture, and material culture assemblage over the entire period of its occupation. The preliminary soundings of the site were facilitated by the fact that the tell is cut to its base by an erosional channel, extending more than 100 m in a north-south direction. Excavation of the site began on the northern side of the tell away from the modern cemetery. Here, two small excavations were conducted at the edge of the gully, one at the base of the tell and the other at its summit, for a total area of about 30 square meters. These excavations were linked together stratigraphically by the channel which has been cleared and faced to obtain a vertical section.

At least in the excavated area, the tell is composed entirely of the remains of structures constructed of pisé walls erected on stone foundations (Fig. 2.2). They are visible throughout the the channel section and on the surface of the undisturbed western slopes of the tell, from summit to base. The lowest stratum in the excavated area is not architecturally distinct in any way from the upper deposits. However, the initial settlement of the site may lie elsewhere in the mound than the central area where the excavation was conducted.

The material culture assemblage of Tell Maghzaliyah differs significantly from the sites investigated by the expedition on the loess plain of the Sinjar. Tell Maghzaliyah is situated at a higher elevation than the loess soil zone. The cultural stratum at the site is saturated with a limestone loam, differing from the clay used for the structures, which was brought in from other areas. Therefore, structure walls were usually clearly distinguishable by color and density. Separate strata were easily defined by stone foundations, plaster floors, and ash accumulations. The stone foundations were usually 40–60 cm in height with a thickness of 60–70 cm. The high visibility of the architectural features separating strata allowed us to excavate in 5 m squares with 50 cm balks rather than in the usual 2.5 m squares done at other sites on the plain. Thus, the notes, sections, floorplans, and recording of material culture were based on a 5 m square grid. The excavation grid was oriented along the cardinal directions. Stratigraphic sections and floorplans were drafted using a scale of 1:20 (Figs. 2.2–2.4). The highest point of the mound was arbitrarily chosen as the primary datum.

A 15 m step trench extending down to the bottom of the cultural deposit gives an idea of the stratigraphy of the site. The top level of the excavated area starts 0.60 meters below datum (hereafter mbd), and sterile soil is reached at a depth of 8.2 m. At least one more cultural stratum is present at the summit of the mound and is not included in the excavated section. Not including this uppermost area, there are fifteen building levels present with an average thickness of 50–60 cm.

There are no obvious breaks in the stratigraphic sequence (Fig. 2.2). In general, the

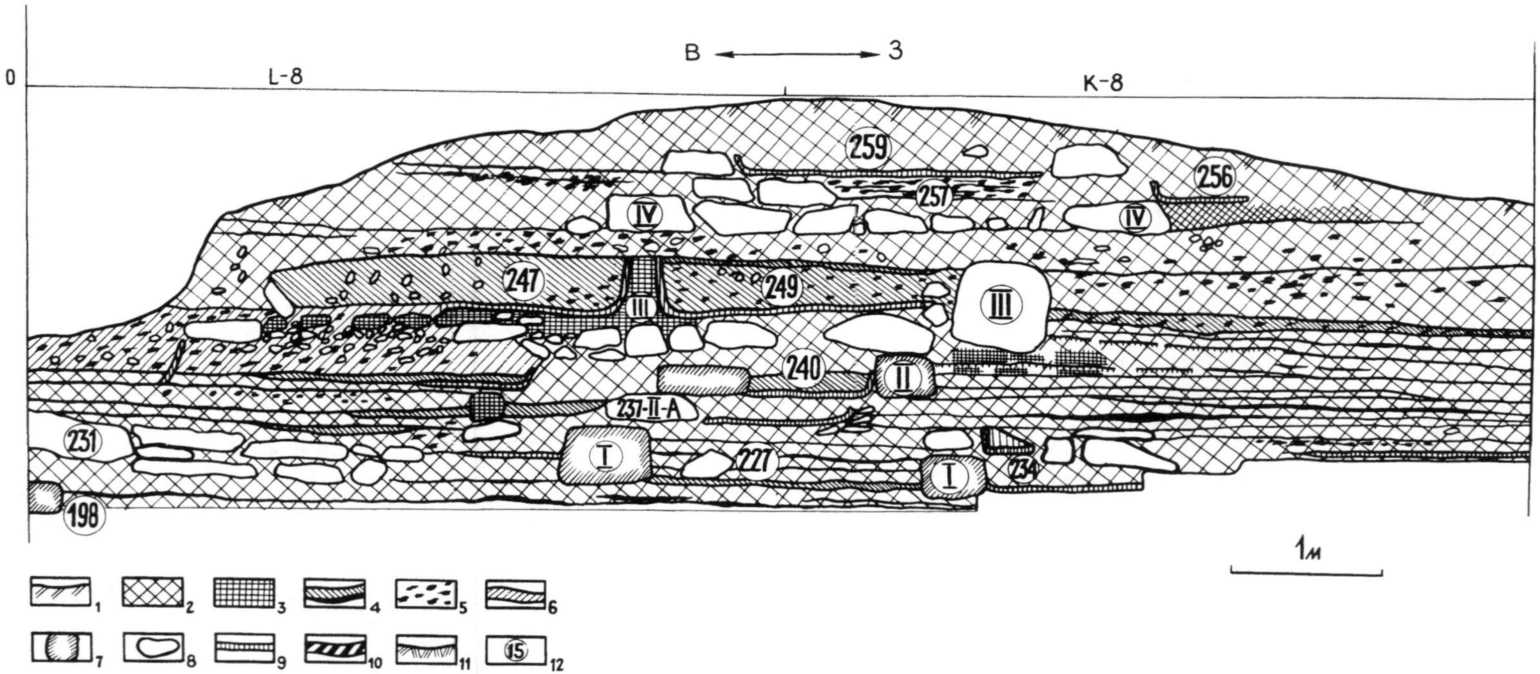

Figure 2.2. Section of the south side of squares L-8 and K-8. Key: *1,* Surface; *2,* Loam; *3,* Compact clay; *4,* Charcoal layers; *5,* Ash layers; *6,* Humus; *7,* Tauf-constructed walls; *8,* Stone features; *9,* Plaster floors; *10,* Base of fired kiln; *11,* Virgin soil; *12,* Feature numbers.

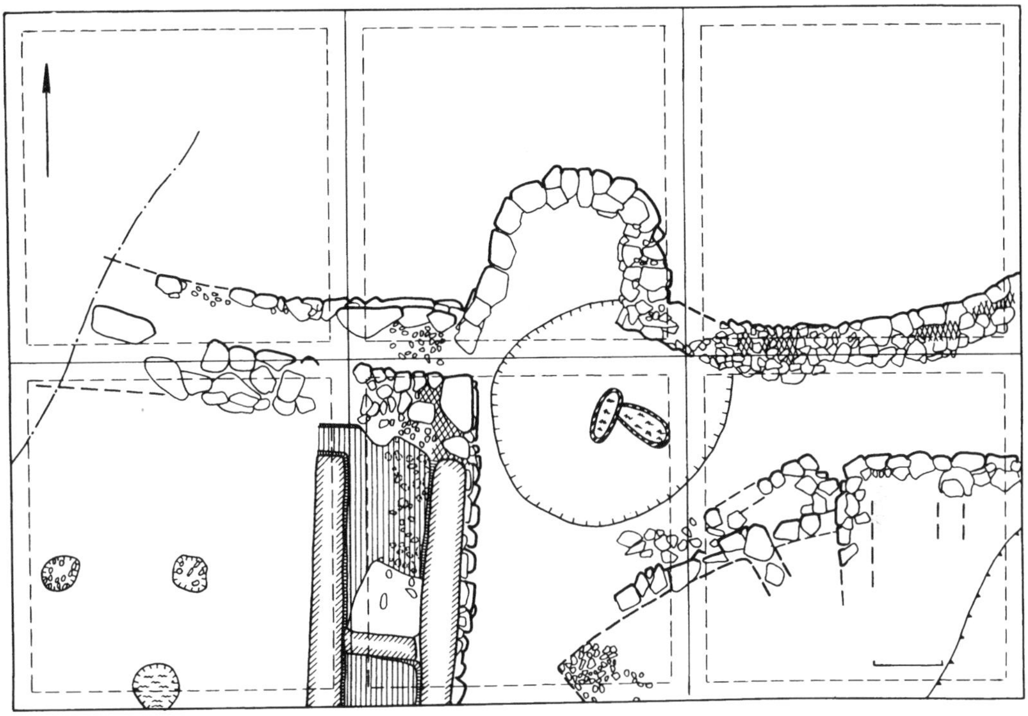

Figure 2.3. Plan of Level 13 in squares K-6, L-6, M-6, K-7, L-7, and M-7

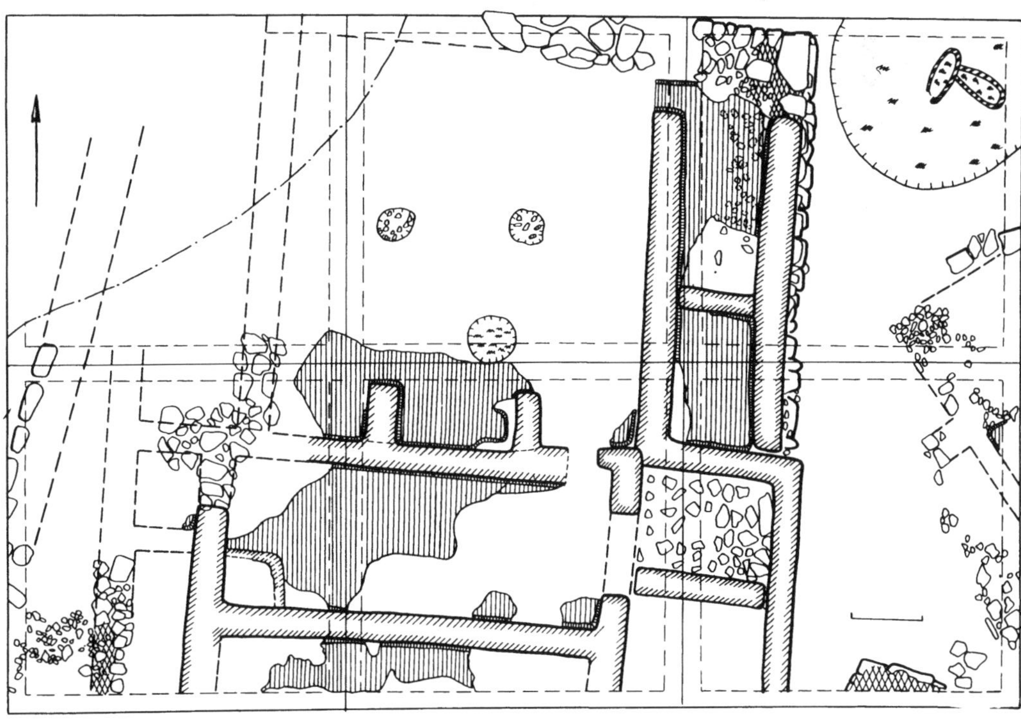

Figure 2.4. Plan of Level 13 in squares J-7, K-7, L-7, J-8, K-8, and L-8

Figure 2.5. Part of house 3

Figure 2.6. Clay storage bin with opening at top

settlement was densely packed with architectural features. Three buildings, one built directly atop the other and preserved only in floorplan, were commonly found. The material culture remained relatively uniform throughout the settlement history of the site. All strata are characterized by a preponderance of obsidian. The number of flint tools increases somewhat only in the uppermost stratum, but the material culture remains essentially unchanged throughout, suggesting that the settlement was occupied over a relatively short period of time.

All the strata of Tell Maghzaliyah are characterized by rectangular structures. Houses were raised on thick stone foundations (Fig. 2.5), with pisé walls. This technique is well known from Hassuna sites. The thickness of the layers forming the average wall is approximately 6–7 cm and the space between them is usually 1–2 cm. In one case, the stone foundation was laid in a specially prepared trench that is clearly visible in section, but the typical stone foundation is aboveground. The width of the foundation can reach 60–70 cm and its height up to 60 cm. Room floors are also usually laid on stone slabs, smoothed over with clay, and then covered with plaster. The plaster coating, as a rule, shows indications of being reapplied many times. In some cases, the plaster coating is preserved on the internal faces of structure walls. A fair quantity of bitumen was also found throughout the cultural deposit. On the floor of one of the rooms a large piece of bitumen bearing the impression of a reed mat was preserved. In general, during the entire settlement history of the site, building techniques underwent no significant changes. It is still premature to talk about houseplans and the settlement layout as a whole.

An oven of unusual shape was found in the lowest cultural stratum. The base of the oven is a rectangle with sides measuring 0.80 m × 1.8 m, and dug 20 cm into sterile soil. The chamber and the mouth of the oven have not been preserved. The oven was filled with a large quantity of charcoal, ash, and fire-cracked rocks, and its floor and walls show evidence of repeated exposure to high temperatures. The base of the oven wall is about 20 cm thick and is reinforced with stones of the type used in residential architecture. A layer of gravel and other small rubble was placed under the stove and then smoothed over with a layer of clay, 2–3 cm in thickness. It is interesting that the same construction method is found throughout the lowest stratum of Tell Sotto. The central partition wall of the oven has been clearly preserved. It abuts the oven's external wall, thus, it probably was a later addition. The partition shows evidence of contact with fire only on the side that faces the main chamber.

A grain storage bin of unbaked packed clay (Fig. 2.6) was found in the same stratum as the rectangular oven. Its shape is almost that of a cylindrical barrel, widening slightly at its flat base, with a height of 65 cm and a diameter of 45 cm. The side wall thickness is 2–3 cm and in its lower portion, a circular opening about 10 cm in diameter has been clearly preserved. To this day, similar grain storage bins are used by the local inhabitants. Their most characteristic feature is the opening in the lower part through which the grain is poured. It is this opening that clearly indicates that the bin was an aboveground structure rather than a subterranean feature.

Let us take a closer look at artifact assemblage recovered from the site. Obsidian predominates throughout the sequence, and the quantity of flint increases slightly only in the uppermost stratum. Debitage is considerably more plentiful in the flint portion of the lithic remains. Thus, it is obvious that obsidian was brought to the site from a considerable distance in the form of prepared flakes and cores.

The knapping technique obviously harks back to the Mesolithic tradition (Figs. 2.7 and 2.8). Flint and obsidian cores can be divided into two types: small wedge-shaped cores with one oblique platform; and conical cores with a straight platform. Knife-like flakes (long, thin, and of regular shape) predominate over other types of flakes.

Figure 2.7:1,2. Obsidian cores and flint celts

As mentioned above, Mesolithic knapping techniques are still well represented in the lithic assemblage, but geometric microliths are rarely encountered. We have found only two flint microlith inserts to date: a segmented blade and a trapezoid. The length of the former is approximately 3 cm and its

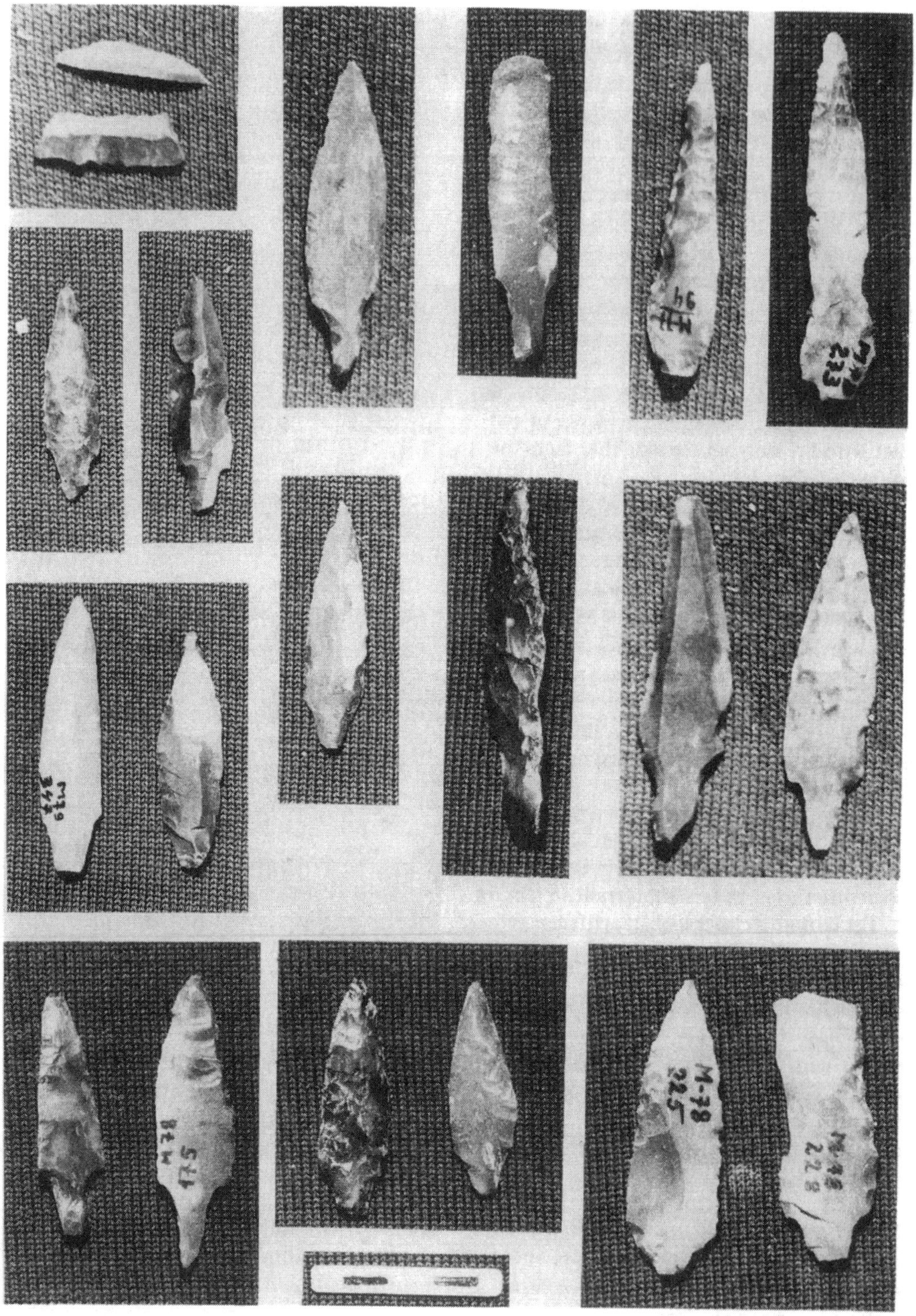

Figure 2.8. Flint projectile points and geometric microliths

width less than 1 cm. The severely worn ends of the cutting edge exhibit careful back side retouch whereas the inset side of the tool remains unworked. The trapezoid has a length of 3 cm and a height of 1 cm. Both side edges and the inset edge of the microlith have been retouched. The retouch is quite fine and steeply angled. The inset side of the tool is approximately two-thirds the length of the cutting edge on the opposite side. Both inserts were made from knifelike flint flakes.

Among the other lithic tools are scrapers, cutting tools, projectile points made from knifelike flakes, and reaping-knives, with scrapers the most frequently recovered tool type. End-scrapers made from large flakes predominate whereas round scrapers made from smaller flakes are rarer. Massive scrapers made from tabular flint are also present in the assemblage. Such tools are characteristic of early Hassuna sites. Many scrapers have retained traces of wear, including flat microchipping on the working edge, a characteristic wear pattern indicating that the tool was also used as a chisel. The ventral side of obsidian tools often show a broad matte finish which could be formed through contact with soft materials.

Uniformly shaped cutting tools are also well represented in the lithic assemblage. Angled cutting tools made from knifelike flakes predominate. Despite the great simplicity and uniformity seen in this tool type, the working edge has been carefully prepared, with one or more long cutting flakes removed, and the tool is often jagged from use. Sometimes, both corners of one or both ends of the flake are formed by removing small flakes.

"Drills" are represented by a small but distinctive series. All tools are of one type, prepared from knifelike flint flakes by dorsal retouch. In only a few cases is the retouch from the opposite side. Breakage resulting from drilling is usually clearly visible on the working edge. Furthermore, the right edge of the butt of the drill is always well worn, indicating that the drill was rotated clockwise. Large and shallow flake scars from use indicate that the tool was used for drilling hard materials.

Projectile points made from knifelike flakes also make up a significant part of the assemblage. All the points are made from flint. Tanged, leaf-shaped points are also commonly found. The length of the point varies from 5–7 cm. Both types have a fine lanceolate blade and a short, massive, hafted end. The ends of the points usually exhibit flattening retouch applied perpendicularly from the butt end. In most cases, the blade is unretouched or at most has very lightly retouched surfaces. On tanged points, the tanged end is created by steeply angled retouch from the backside.

In general, knifelike flakes and tools made from them predominate in the assemblage. Most of these, particularly the small obsidian flakes, were used without further modification. Carefully retouched flakes are rare. Steeply angled dorsal retouch and flattening retouch proceeding from the butt end are equally common and are encountered in various combinations. Steep lamellar retouch gives the entire assemblage a distinctive character. Obsidian flakes worked in this way have a distinctive long, curved, concave profile. The same methods of retouch are commonly found in the lithic assemblage from the lowest stratum of Shimshara. Reaping-knives were most frequently manufactured in this manner.

Reaping-knives are represented by numerous examples and are one of the most diagnostic classes of the entire assemblage. These are made from large flint or obsidian flakes and may have been used without a handle or as part of a straight composite tool. Sturdier flint flakes were usually not retouched and can be identified by their well-smoothed working edges. The most interesting examples are the carefully worked obsidian reaping-knives. Their purpose is indicated by the characteristic signs of wear on the working edge, which has a matte, whitish surface with oblique strokes. Obsidian flakes are usually

Figurc 2.9. Obsidian sickle blades

bifacially worked with oblique denticulate retouch from the back or from the front (Fig 2.9). Many reaping-knives have an unretouched projection on one of the ends and may, because of this characteristic trait, be called spiked reaping-knives. Frequently, the spike was left on the bulbar end of the flake. However, in most cases, it bears traces of use and therefore cannot have served as a haft for attachment to the handle. Probably the projection was left on the end of the working edge of the knife to make it easier to grasp stalks during cutting. An intact tool of this kind was found at Shimshara. It was a composite tool comprised of two obsidian flakes glued together with bitumen and thus did not have a wooden or bone frame. One of the ends had a spike of the kind described above. P. Mortensen (1970:38, 39, Figs. 30–31) has called this tool a dagger. However, like the numerous flakes analogous to it from Shimshara and Maghzaliyah, it bears the traces of wear characteristic of sickles. Finally, bifacially worked obsidian flakes exhibiting carefully executed steep-angled dorsal retouch can be placed in a separate category. These flakes do not bear traces of wear, and it is difficult to determine their function.

The technique for working bone was also rather advanced. In addition to polishing (Fig. 2.10), bone carving was widespread. In one case, incised on a fragment of hollow bone was a decoration in the form of a rhombus outlined with parallel lines. Several fragments of very fine bone needles have also been found. As in later sites, awls made from hollow bones predominate at Maghzaliyah.

Grain grinders, milling stones, and pestles make up a distinctive category of tools from the site (Fig. 2.11). In most cases, as is the case at Tell Hassuna, they are made from basalt. Milling stones are oval-shaped, with a flat or slightly convex working surface. In contrast to this the grain grinders have a concave working surface. A single pestle made from a cobble has been found that is of the usual elongated proportions, with two working surfaces. Several spherical stones bear traces of having been beaten and may have been used as anvils.

The cultural deposit contains many small cobbles and pebbles which certainly were selected and brought to the site by human action. Two general types predominate: flattened, oval, or pear- shaped pebbles; and cobbles in the form of small pestles. The latter are often stained with ochre on the worn end and in the shape of elongated cones. They also have bases that are polished and covered with scratches. The flat pebbles, as we will see below, were made into pendants and other ornaments.

The inhabitants of Tell Maghzaliyah were well acquainted with the properties of clay, used it widely, and were certainly on the verge of making ceramic vessels. They used clay to make houses, ovens, containers for grain, and to mold figurines (Fig. 2.12). Many lumps of kneaded unfired clay were found throughout the cultural deposit. Apparently, clay vessels were already being molded, simi-

Figure 2.10. Bone polishers

Figure 2.11. Grinding stones

Figure 2.12. Clay, bone, and stone figurines, pieces of malachite, copper awl, ceramic sherds.

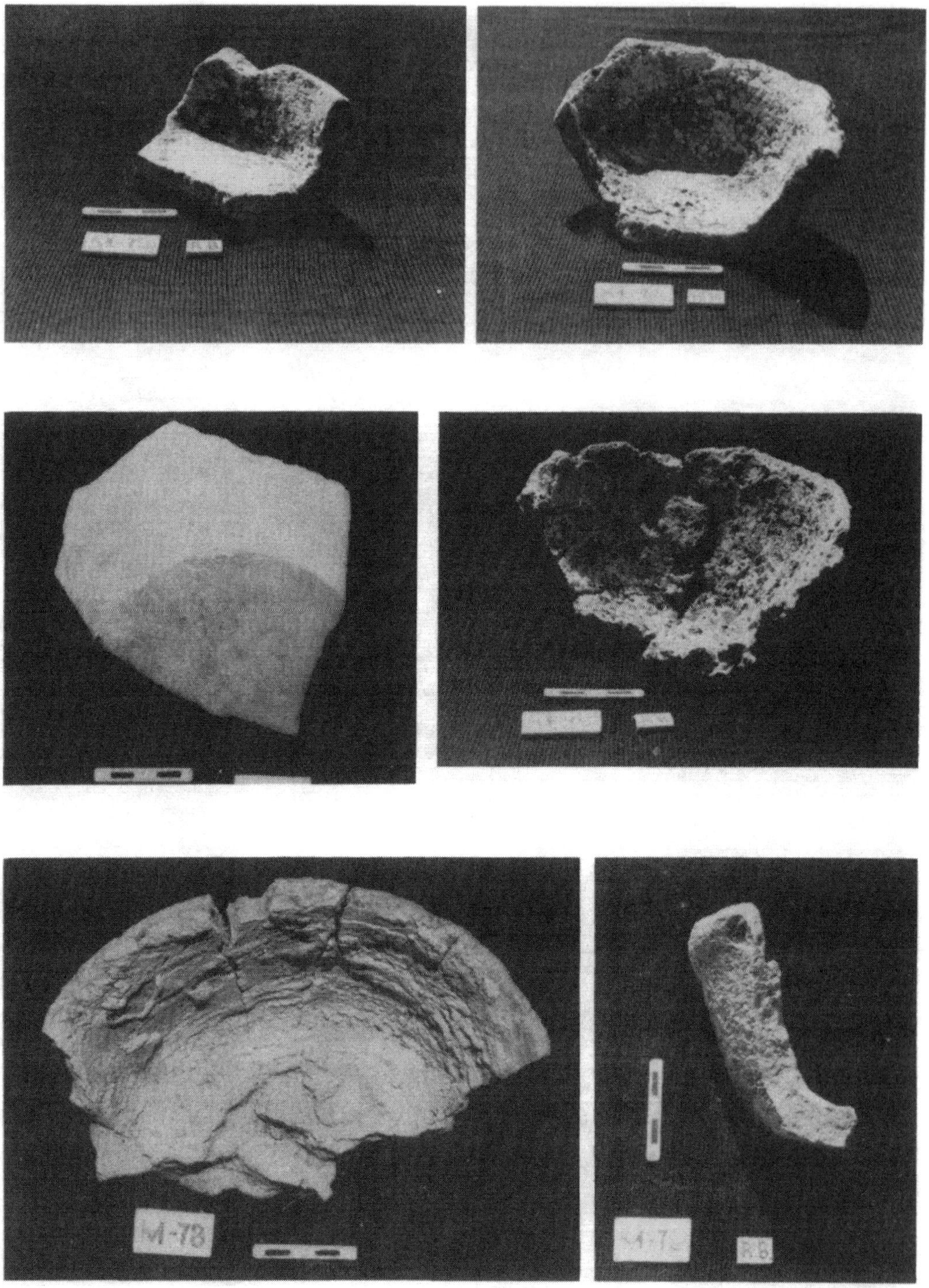

Figure 2.13. Fragments of marble and plaster vessels

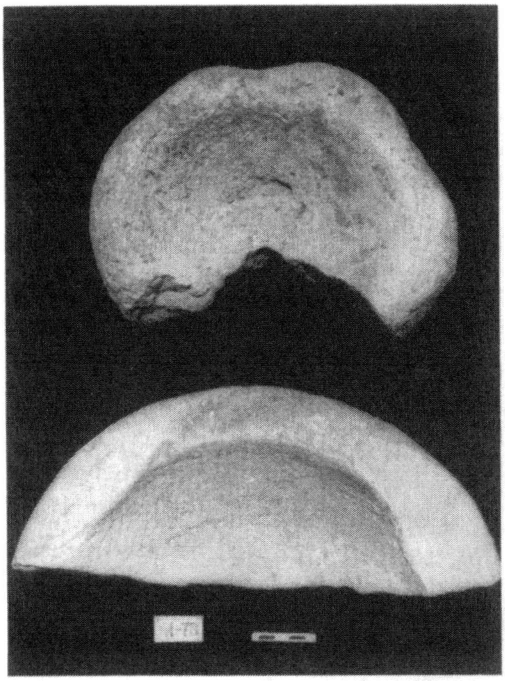

Figure 2.14. Groundstone mortars

lar in shape to their stone counterparts, but not yet in wide circulation. The only sherd recovered from the site was found at a depth of 4.70 mbd in square M9 (Fig. 2.12:bottom center). This is part of a small, poorly fired, carinated cup or pot, and fully conforms to our idea of archaic pottery, but its stratigraphic position is not certain.

One of the interesting features of the Tell Maghzaliyah assemblage is stone vessel fragments. The manufacture of stone vessels required great expenditures of labor, but these vessels were significantly superior in quality to clay ones. Apparently, only the transition and commitment to agriculture that created the need for mass quantities of storage vessels that were convenient and easy to prepare caused the older stoneware tradition to be abandoned. The transition to ceramic vessels may also have been connected with the emergence of people onto the northern Mesopotamian plain where stone sources are rare, but clay is abundant.

To date, ten fragments of stone vessels have been found at Maghzaliyah (Figs. 2.13 and 2.14). As observed at early Hassuna sites, these vessels were made from marble or marblelike stone, usually white or white-veined with red. A half-finished vessel has been found, permitting us to reconstruct the manufacturing technology. Marble is a soft material which is easily smoothed, but does not fracture readily. For this reason, the vessel's original form is produced by hammering. It is clear that the hollowing out of the internal cavity was begun by gouging, and that only in the later stages of work was it drilled and polished. Only fragments of stone vessels are represented in the collection recovered thus far. From these fragments it is difficult to reconstruct the form precisely. Apparently, a spherical pot with a broad neck, flat bottom, and plain rim is the typical form. A plaster dish with a flat bottom, vertical sides, and a diameter of about 35 cm has also been found. To date, no ribbed-wall stone vessels have been recovered.

Advanced techniques for stoneworking are indicated by the various stone ornaments. A fragment of a bracelet made of white marble has a cross-sectional diameter of more than 2 cm and formed a circle with a diameter of about 9 cm. Flat pebbles were a favorite material for making pendants (Fig. 2.15). We can distinguish two types that are most commonly found. The first class is composed of oval- and pear-shaped pendants with a 2–3 mm diameter hole drilled from both sides near the edge of the pendant. Besides the perforation, the pebble was not further worked. The second group includes disks made from pebbles with a hole drilled in the center. The diameter of these artifacts is usually about 5 cm, and the cylindrical hole is about 5 mm across. Several disks were cut from soft stone. Three identical disks were found next to each other and were apparently strung together. On one of these, a circle was carefully etched around the inner diameter. A well-polished fragment of a drill-perforated, rectangular marble

Figure 2.15. Stone pendants

Figure 2.16. Clay female figurine

plaque is also worth mentioning. Thus far, not a single polished, wedge-shaped axe has been found at Maghzaliyah. However, a crude discoidal chopping tool, hammered around the perimeter as indicated by large chips removed from both sides, was used for working wood. The discoid has a diameter of about 15 cm. Finally, two spherical stones with holes drilled in the center may have been used as weights.

The anthropomorphic sculpture is also of great interest. To date, two fragments of clay figurines, a stone statuette and a stylized head on the end of a marble pestle, have been recovered. The leg of a seated female statuette made of clay was found in square M-10 at a depth of 7.50 mbd (Figs. 2.12:center, and 2.16). Its dimensions are 4.3 cm × 3.4 cm × 2 cm. The convexity of the thigh is clearly visible, whereas the lower part is flattened. The surviving joint shows that the statuette was molded part by part. A fragment of a cylindrical object of poorly fired clay with a widened base was found in the same general stratigraphic provenience. Its height is 6 cm and the diameter of the upper portion is 3.8 cm. It is difficult to interpret this figurine from the surviving fragment. Perhaps it has a certain similarity to the stylized cylindrical figurines from Jarmo, although it significantly exceeds those in size. An anthropomorphic figure of soft white stone was found on the floor of a structure in the uppermost stratum of square M-8. Only the head and bust have been preserved with the lower torso missing. This figurine follows a cruciform pattern and lacks any detailed work. A small drilled hole is visible at the breast. The height of the figure is 6.5 cm, the width of the breast 7.5 cm, and the thickness is 2.5 cm. The style of this figurine is apparently unique since we have no knowledge of similar forms from other sites.

A small conical pestle with an anthropomorphic, stylized head was found in square M-10 at a depth of 5.60 mbd (Fig. 2.12: bottom left). The artifact is made of white marble with red veins, a favorite material used in stone vessel manufacture. The pestle is conical in form with the flat side used as the working surface. It is characteristic of this group of tools. The top of the cone is slightly flattened on the sides, and in profile, is reminiscent of the head of a statuette. The details of the face are not depicted.

In general, Tell Maghzaliyah presents us with the remains of a culture possessing an advanced lithic technology and a highly developed architectural tradition but is still apparently aceramic. The subsistence economy is characterized by a mix of hunting and gathering with agriculture. Hunting is indicated by the large numbers of tanged projectile points made from knifelike flakes and the bones of large, and apparently wild, cattle. However, the sedentary character of the overlying settlements, the large number of milling stones, reaping-knives, blade inserts for composite sickles, grain storage containers, and other lines of evidence permit the assumption that agriculture was already an important element in the site's subsistence strategy.

Until recently, the pre-Hassuna Neolithic of northern Iraq was only known from several scattered settlements, among which the best known was Jarmo. These newly discovered sites may be classified into two temporally successive groups: preceramic and early ceramic sites. The first group includes the lowest stratum of Jarmo, the lowest stratum of Shimshara, and Tell Maghzaliyah.

The latter group includes the lower strata of Tell Sotto, Kültepe, Umm Dabaghiyah, and Telul eth-Thalathat, and the upper stratum of Jarmo and Ali Aga. Close to this latter group, although perhaps somewhat later in time, are the lower strata of Yarim Tepe I and Tell Hassuna 1a. The lowest stratum of Tell es-Sawwan, where crude ribbed pottery is encountered along with stone vessels, has much in common with strata containing analogous vessels known from Kültepe and Umm Dabaghiyah and can be placed close to this group in time.

Tell Maghzaliyah, Jarmo, and Shimshara preserve the microlithic tradition characteristic of the preceding Mesolithic period. The large quantity of imported obsidian places these sites within the same general time period. Stone vessels, small conical pestles of a uniform type, and cobble tools for grinding ochre are characteristic of all of these sites. Close parallels to the stone discoids of Maghzaliyah are found in the preceramic stratum of Shimshara. The previously mentioned fragment of a perforated rectangular marble plaque found at Maghzaliyah is reminiscent of a similar artifact from Zawi Chemi Shanidar. The reaping-knives of Maghzaliyah and Shimshara show an especially marked similarity. They both have the characteristic spike, and are worked by oblique denticulate retouch.

Maghzaliyah, Jarmo, and Shimshara are apparently contemporaneous since definite material culture ties are found among them. Additionally, the exchange of obsidian among the three sites is certainly plausible. However, the assemblages of the three sites are not identical since flint projectile points are unknown at Jarmo and Shimshara, and no clay figurines were found at Shimshara. The tanged projectile points which are characteristic of the Tell Maghzaliyah assemblage has parallels in the west, in the aceramic strata of Beidha. In the Sinjar region, these projectile points continue to be found in the early Hassuna strata at Tell Sotto and Umm Dabaghiyah.

The character of the Maghzaliyah settlement is also somewhat different from that found at Jarmo and Shimshara. Maghzaliyah is an intensively built-up tell with clearly defined boundaries. Shimshara is situated on a natural hill and Jarmo on a high plateau bounded by precipices on three sides. Despite the 7 m cultural deposit at Jarmo, its boundaries are vague, which may be explained by the more random layout of the site and its less dense construction.

The nature of the ties between Tell Magh-

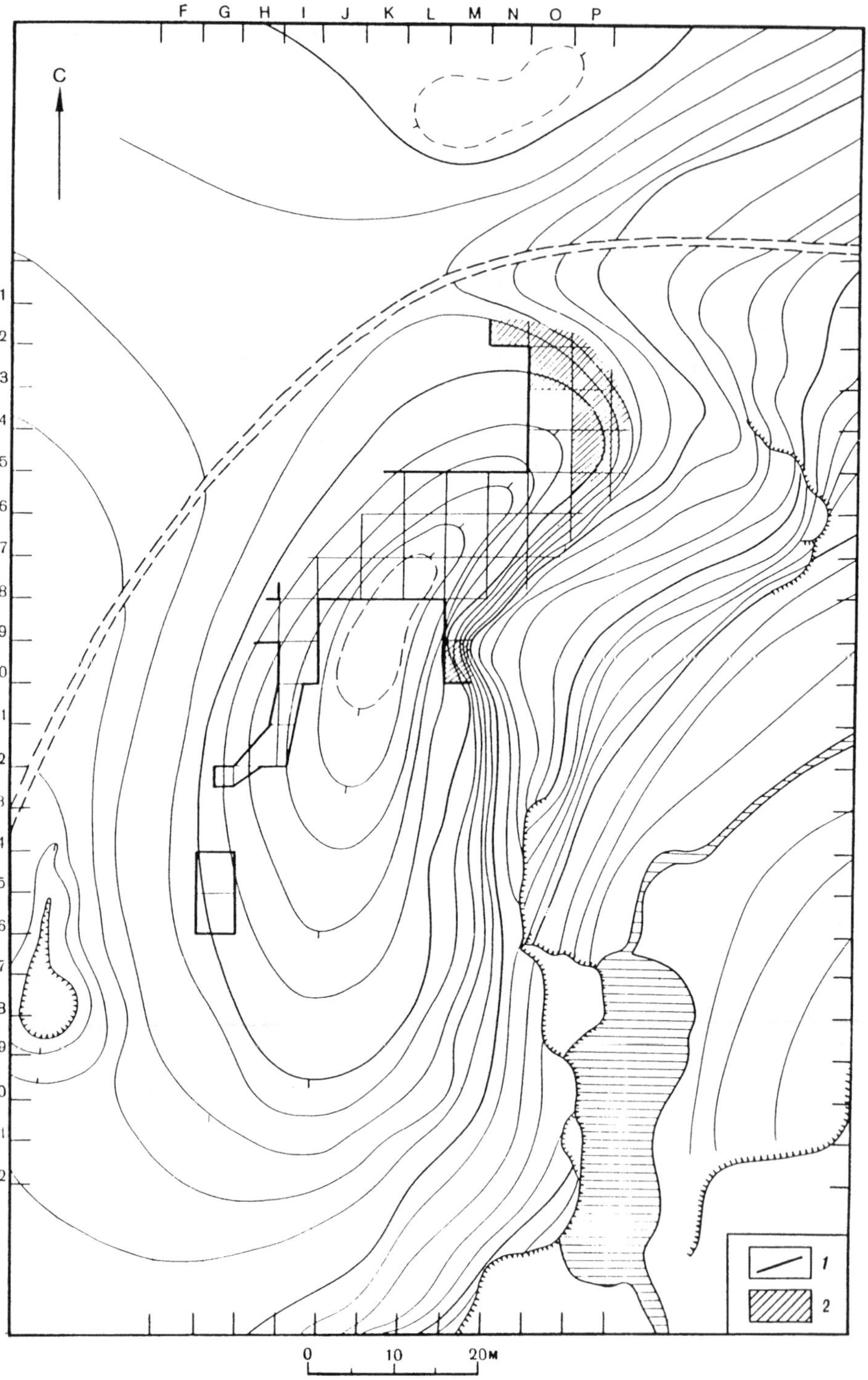

Map. 2.1. Topographical plans of Tell Maghzaliyah.

zaliyah and Tell Sotto remains an open question. We have already spoken of a certain similarity between the culture of Tell Sotto and the ceramic phase of Jarmo (Bader 1975, chapter 3 this volume). Differences in the level of lithic technology and in the types of flint and obsidian tools indicate that a significant chronological gap exists between Tell Sotto and Maghzaliyah. In both assemblages, stone vessels, stone bracelets, and hafted arrowheads are found, but these artifacts existed over a long period of time and have a wide distribution and thus are not discerning chronological markers. It is interesting that the lowest stratum of Telul eth-Thalathat, which according to its material culture assemblage belongs among the sites of the Sotto type, yields a large and varied series of statuettes similar to those found at Jarmo. Unfortunately, the stone sculpture assemblage of Tell Maghzaliyah and Tell Sotto is, to date, extremely fragmentary. New materials recovered from the sites of the region reduce the chronological gap between the aceramic and early ceramic sites, but the exact nature of the ties between them is still to be defined.

The limited material recovered from Tell Maghzaliyah does not yet give a definitive answer to the questions posed, but significantly expands the scope of research. As we have noted, in the area between Tell Maghzaliyah and Yarim Tepe, the Soviet expedition has found possibly ten sites of the Sotto type. The majority of them are located in the region of the plain directly adjacent to the mountains and some of these are only 1–2 km distant from Maghzaliyah. However, not a single site of this type has yet been found on the slopes of the mountains, just as no aceramic site has yet been found on the plain.

Apparently, the final transition to a food-producing economy and the settlement of the Sinjar Plain by farmers took place within the relatively short period of time which divides Maghzaliyah from sites of the Sotto type. Whether or not there is a direct genetic connection between them, and how this transition in fact proceeded requires further work at Tell Maghzaliyah and a search for new early agricultural sites in the surrounding region.

b. 1979–1980 Results

Results from the 1977 and 1978 field seasons at Tell Maghzaliyah have already been published (Bader 1979; chapter 2a of this volume; Bader, Merpert, and Munchaev 1981; Merpert, Munchaev, and Bader 1981; and Munchaev, Merpert, and Bader 1979). In 1979 and 1980 the total excavated area of the site reached 625 square meters. Originally the tell occupied an area of roughly 0.65 hectares. However, after the final abandonment of the settlement, the eastern part of the mound was eroded away by the nearby river, and on its northern slope an area has been washed away by water following the path of the road. The total area affected by these natural and cultural processes is over 2000 square meters or approximately one-third of the mound. Hence, the four years of excavation at the site have exposed roughly one-tenth of the orginal tell. The largest excavation is located at the top of the mound and covers 375 square meters. Comparing the position of the defensive wall that was traced in Levels 3 and 4 with the contour of the unexcavated part of the tell at the same height, one can calculate rather precisely the size of the settlement within the walls, an area of at least 1000 square meters. Thus, to date, the excavations in this part of the tell have covered more than one-third of the walled settlement area of this period.

The main objective of the 1979 field season was to determine the intrasite settlement plan at the top of the mound (Map 2.1). To achieve this goal, we continued down in squares K-7, L-7, M-7, K-8, and L-8 in the upper excavation area, and opened up new squares (I-8, J-8, M-8, I-9, and I-10) on the western slope of the tell.

Figure 2.17. Upper excavations and the defensive wall

In several squares at the top of the mound, depths of 2.90–3.80 meters below datum (hereafter mbd) have been reached. Several dwellings and the defensive wall surrounding them were exposed in Levels 3 and 4 along with other information pertaining to the settlement behind the defensive wall. Work also continued in the excavations at the base of the mound in squares N-2, O-3, O-4, P-4, O-5, and P-5. The excavation of squares O-4 and O-5 has progressed to a depth of 5.50 mbd while the other squares in this area have reached sterile soil. The expedition has sampled the entire cultural deposit and determined the character of its material culture in each level. Additionally, the excavation has fully exposed a dwelling in square P-4.

In 1980, V. A. Bashilov continued exposing the structures at the bottom of Level 4 in square K-7 and the defensive wall in squares J-8 through J-13, and N-10 in the upper excavation area (Fig. 2.17). A. V. Kuza and O. G. Bol'shakov continued down in squares M-7, M-8, M-9, N-7, N-8, O-5, O-6, O-7 and P-6, obtaining a more accurate stratigraphic correlation between the upper and the lower excavations. N. O. Bader did not take part in this year's excavation season.

From an examination of the natural stratigraphic section found in the slope eroded by the river along the M and N line, fifteen building levels were identified within the entire cultural deposit. As a result of the excavations in the northern half of the tell, the settlement history of the site has been tentatively subdivided into at least four larger periods which represent substantial rebuilding episodes.

The original settlement emerged along the bank of the river. The lower levels of Tell Maghzaliyah have been more fully exposed in squares O-3 through O-6, P-4, P-5, and P-6 below the upper (northern) slope of the tell. The squares along the O line clearly demonstrate the north-to-south downward slope of

Figure 2.18. Defensive wall in Level 4

the natural strata in a direction opposite to the modern surface of the mound. These strata apparently follow the contour of one of the earliest settlements at the site. If this is the case, then the expansion of the original settlement proceeded along the river in a southerly direction down the slope of the original mound. Its eastern flank was destroyed by the river much later, and finally, water erosion following the course of the modern road gave the mound its current profile. On the basis of this profile, the western boundary of the settlement can be defined.

During the 1980 field season, the contour of one of the earliest settlement mounds was delineated in the south. In squares N-6, N-7, N-8, O-6, O-7, and O-8, a south-to-north downward slope was discovered. Ancient canals that crosscut the boundary between the northern and southern mounds make it difficult to establish the stratigraphic relationship between them with any degree of confidence. The southern mound continues under the central and, probably, the extreme southern part of the tell. Given the limited area of the excavations in this part of the site to date, it is difficult to determine the extent of its boundaries in the south. We may assume, though, that the earliest settlement consisted of a row of contemporaneous dwellings along the bank of the river. The excavated area of the northern mound includes the earliest building levels (9–15) comprising Period 1, each of which is defined by minor rebuilding events. Since the stratigraphic association between the northern and southern mounds has not yet determined, the evidence from each will be treated separately.

A new settlement (Levels 5–8) was built in Period 2 on top the the mound formed by the ruins of Period 1 buildings. An extensive rebuilding of the site at this time is indicated by the wall debris found in the section. The settlement of Levels 6–8 has remained unexcavated except for isolated structures exposed in squares O-6, O-7, N-7, and N-8. A more complete settlement plan has been obtained for Level 5 of which an area of more than 128 square meters has been uncovered in squares L-7, L-8, M-7, M- 8, N-7, and N-8. At this time, the dwellings on the northern slope of the mound were enclosed by a massive wall which was probably part of the first defensive wall that encircled the entire settlement.

The third period (Levels 3 and 4) is defined by a new defensive wall which has been traced over a large area on the northern and western slopes of the tell. The lower stratigraphic boundary of this period is marked by extensive alterations in the settlement plan of the entire site. The two levels of architectural remains within the period correspond to two building phases of the defensive wall surrounding them (Fig. 2.18). No structures have been found outside the wall, and the cultural layer in this area is made up exclusively of ash, humus, and refuse following the steep slope of the tell.

The two upper levels (1 and 2) on the northern summit of the tell are considered the fourth and latest period of occupation at the site. No defensive wall existed during this period, and the overall settlement plan changed drastically with the total inhabited area greatly diminished. After this period, Tell Maghzaliyah was abandoned and never resettled.

Final analysis in the field revealed further insights into the above periodization. Period 3 is the most distinctive with a relatively large settlement surrounded by the defensive wall. A more detailed analysis of the lower portion of the cultural deposit (Levels 6–15) that has not yet been adequately exposed is deemed critical in reconstructing the settlement history of the site. It should also be noted that within the excavated area, Levels 1 and 2, which we have grouped together as Period 4, are in need of a thorough reworking.

North of the main tell, beyond the area eroded by rain water following the course of the road, another small mound has been located. This mound is apparently all that remains of the northern part of the settlement that has been destroyed in the intervening area. On the slopes and surface of this small tell, stone foundations similar to those exposed on the main mound can be seen. The road has cut through the entire depth of the cultural deposit down to bedrock, and thus, the stratigraphic correlation between the central area of the existing site and its northern periphery is nearly impossible.

In an open field situated in a bend of the river to the north of Tell Maghzaliyah, a considerable number of artifacts was recovered during a surface collection of the area. However, no remains of masonry structures were encountered nor were any darker patches of soil found, nor were any other indications of a cultural stratum of a significant depth. On the basis of the elevation of the cultural remains found in this area, they are probably associated with the earliest period of occupation at the main site. There are no ideal locations for test pits in the field since the cultural debris is scattered over the entire area, which is quite large, and concentrated areas of artifacts are lacking. It is quite possible that this convenient location, adjacent to the river and near the base of the main tell, was utilized by the inhabitants of Tell Maghzaliyah as a domestic activity area outside the settlement. It is interesting to note that the field has yielded a considerable amount of flint material, including debitage, and obsidian as well, though in much less quantities than on the main tell.

Excavations covering a large area of the existing mound convincingly demonstrate that the rebuilding episodes which define Levels 1–5 are not associated with any destruction or lengthy abandonment of the settlement. Due to the lack of evidence, little can be said about the lower levels. In any case, the material culture is remarkably homogeneous throughout the entire cultural deposit with only relatively insignificant changes in the material culture over time. The impressive height of the tell is due more

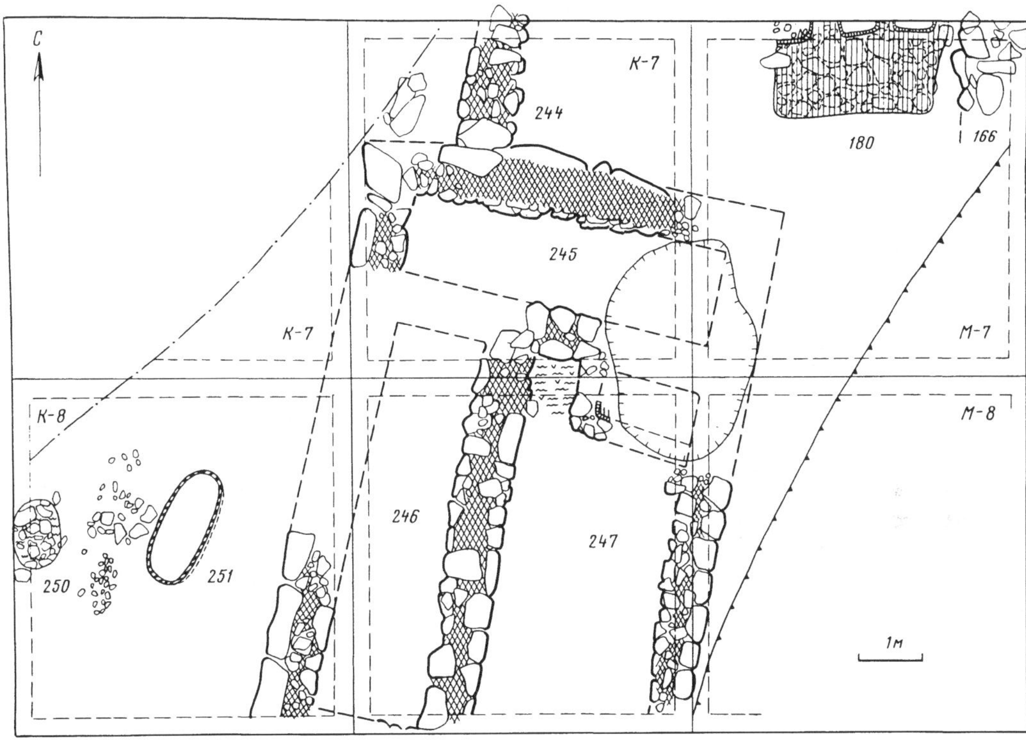

Figure 2.19. Plan of the dwelling in Level 2

to the bulk of the masonry structures found within it than an extended period of occupation. The time interval associated with each building level at the site is determined by the use-life of several contemporary or temporally overlapping structures and, therefore, should be measured not in centuries but in decades. Thus, during the relatively brief period of occupation at the site, the material culture of the rapidly developing settlement underwent few changes. Extensive alterations in the settlement plan were not accompanied by similar changes in material culture.

Architecture

The considerable amount of data recovered concerning the architectural planning of the settlement should be considered separately. We shall briefly describe the structural remains of the upper excavation area. Excavations on the western slope of the mound along the K and L lines have almost completely exposed several large buildings from Levels 2–4.

A large multiroomed dwelling has been excavated in Level 2 (Figs. 2.19 and 2.20). Its northwestern corner was destroyed by a pit originating from an upper level. The southern portion of the the dwelling continues beyond the limits of the excavation, but the overall plan of the structure can be readily obtained from the area that has been uncovered. If this structure fits the general architectural pattern of dwellings at Tell Maghzaliyah, then the area exposed is the central part of the building, and it is quite probable that the southern unexcavated portion of the dwelling contains a row of one or more adjoining rooms.

The structure is rectangular in shape with the exposed portion measuring 8.4 m × 5.8 m. Including the walls, the excavated part of the dwelling occupies an area of 50 square

Figure 2.20. Stone foundation of the Level 2 dwelling

meters. The central area of the structure is composed of two long rectangular rooms with the inner dimensions of 1.4 m × 5.3 m and 2.3 m × 5.3 m, which translates roughly into 7.5 and 12.0 square meters of habitation space respectively. To the north of these rooms is another adjoining room, with the inner dimensions of 1.4 m × 4.4 m and a living space of over 6 square meters.

The dwelling stood on a stone foundation, 40–50 cm high and approximately 80 cm wide. It was constructed in one building event, using the same technique and materials, and following a single plan. The inner and outer faces of the foundation are made of large, closely fitting chunks of limestone, and its inner core is filled with smaller stones. Atop the stone foundation, the remains of pisé walls have survived. The floor is a layer of cobbles overlain with packed clay and covered with gypsum plaster. The interior of the walls is also lined with plaster. Frequent replastering episodes were noted, and in some places the plaster reached a thickness of 2–3 cm. As is often the case, plaster fragments from the collapsed roof and walls were frequently encountered within the fill of the dwelling. The northern periphery of the house is poorly preserved. The surviving remnant of a massive wall that abuts the northern outer wall of the structure indicates that other related buildings once existed further to the north.

Near the northeastern limit of the excavation, the corner of another large dwelling was exposed, but unfortunately, the dwelling itself has been almost completely destroyed by the river. It can only be determined that the structure was also oriented along the cardinal directions. The space between this dwelling and the one described above was occupied by two ephemeral structures that probably served some domestic function. The inner dimensions of these two buildings are 1.2 m × 5.8 m and 1.8 m × 6.2 m based on the remnants of narrow and poorly preserved foundations rather haphazardly constructed of small stones.

During the course of its existence, the first dwelling was rebuilt twice, as were the related domestic structures surrounding it. In its original design the house had three rooms (245–247). The northeastern corner of room 247 was separated from the rest of the room by a low rectangular partition and was used as a storage facility. The partition was made from small stones covered with gypsum plaster. Its foundation lies considerably higher than the foundation of the dwelling. In the northwestern corner of the room next to this storage area, a large concentration of charcoal and ash and the remnants of a hearth were encountered. On the floor near the western wall of the dwelling, a large oval oven with a deep fire pit was also in use during this time.

After a segment of the western wall of room 246 had been destroyed, the dwelling plan was significantly altered. A new western wall was constructed further inside the building, and room 246 was divided by a partition wall into two nearly equal segments. The rooms were refitted with new domestic installations and their floors and walls were replastered. Near the eastern wall of room 247, three oval gypsum plaster basins, rebuilt atop each other, have been exposed. The largest of these measures 48 cm × 60 cm and is approximately 25 cm in depth. The basins replace each other on the same location and only the upper rim of each was projecting above the current floor level. Most probably the basins were first hollowed out and then lined with gypsum plaster. The basins were not portable since they were constructed in place. They were filled with a humus loam which contained several grain mortars, bone artifacts, and plaster fragments.

A storage pit or a small water reservoir was uncovered in the center of room 246. This oval hollow measured 40 cm × 60 cm and was dug from the level of the floor. It had vaulted walls carefully lined with gypsum plaster. A nearly square storage niche was encountered at floor level in the eastern wall of room 14. Its contour was easily traced since it was also lined with plaster. Opposite to this niche, along the western wall, a feature was discovered which may have served as a bench or bed platform. Additionally, several projections extending into the room along the lower part of the wall were discovered. These shelf-like features were carefully covered with gypsum plaster. Originally, only two of these projections were constructed with two additional "shelves" added, probably during the second rebuilding episode. These projections could have supported wooden planks. The floor space between these wall extensions was used as a storage area, since it was separated from

Figure 2.21. Part of the Level 3 dwelling.

the rest of the floor by plaster partitions. Near the northeastern corner of the dwelling, a pisé wall formed a narrow corridor (roughly a meter wide) with the outer wall of the dwelling. Since the former wall is poorly preserved, it is difficult to determine its purpose.

The second rebuilding episode begins the third phase of the use of the dwelling. The abovementioned wall near the structure's northeastern corner was destroyed and the partition between rooms was remodeled with the permanent wall of the second phase replaced by a thin and probably wooden partition covered with mud and gypsum plaster. Additionally, the bed platform/bench in room 13 was lengthened. The wall plaster in room 13 associated with the third and final occupation phase of the dwelling was traced to a height of approximately 50 cm above the floor.

Another structure similar in planning and design to the dwelling described above was exposed in Level 3 (Fig 2.21). Its southern wall coincides almost exactly with the edge of the excavated area, and its surviving portion has been fully exposed. However, the western part of the structure which emerges onto the modern slope of the mound has been partially eroded away. On the basis of the length of the southern wall and the remains of the wall that surrounded the Level 3 settlement, the approximate outer dimensions of the dwelling were 9.6 m x 8.8 m, and its area was approximately 85 square meters.

A Level 2 dwelling is immediately above the Level 3 dwelling described above, and both have similar floor plans. In both structures, two elongated rooms with their long axes oriented north-south have an adjoining room to the north oriented along an east-west line. In the Level 3 dwelling, the southwestern room was used as the primary habitation area, measuring 5 m × 1.8 m and occupying an area of 9 square meters. In the southern wall of this room a doorway was exposed. Within the room a plastered hearth that had been subjected to extreme heat was discovered. One of the corners was separated from the rest of the room by a thin clay partition and was probably used as a storage facility. The adjacent room (5.2 m × 0.8 m) is too small for habitation and no domestic remains were found within it. The floor and the interior of the walls were carefully covered with a thick layer of gypsum plaster, thus it was probably a storage room.

The northern part of the settlement was enclosed by a wall constructed on a massive stone foundation. The wall runs close to the dwelling described above but is not bonded to it. The northwestern corner of the defensive wall has been destroyed, though it continues further along on the western slope of the mound. It apparently surrounded the entire settlement, or at least the most accessible sector adjacent to the open field. Obviously, the wall served a function independent of the residential architecture of the settlement. This function, we believe, was defensive in nature, though to date only one U-shaped area in the northern sector of the wall can be tentatively identified as a tower foundation. Besides the residues of a large hearth, little else was found in the collapse of this hypothetical tower (Bader, Merpert, and Munchaev 1981).

The Level 4 dwelling is the largest of the excavated structures at the site, but unfortunately, it is also the least well known. Its southern part extends beyond the excavation limits, and its northwestern corner has been destroyed by erosion on the slope of the mound. Like the other dwellings that overlie it, this structure is oriented north-south along the river. It is 8 m wide, and the length of the exposed part of its eastern outer wall exceeds 9 m. Thus, the excavated area of the house is approximately 75 square meters, and its total area may approach 100 square meters. The dwelling was constructed in the usual method, with pisé walls built on a stone foundation that has not entirely survived. Gypsum plaster was used throughout the structure. The rooms of the dwelling are in the shape of elongated rectangles. In the main habitation room the remains of an oven were discovered. This room was connected to several narrow

Figure 2.22. Different examples of masonry foundations

Figure 2.23:1,2. Different examples of masonry foundations

rooms less than one meter in width, devoid of any domestic features and presumably used for storage.

The Level 4 dwelling described above is associated with the first phase of the defensive wall. Particularly impressive is the segment of the defensive wall found on the western slope of the tell where its foundation is made of huge upright blocks of limestone, up to 1.5 m in length. During the 1980 field season this foundation was traced along the entire western perimeter of the settlement (see Figures 2.22 and 2.23:1,2 for examples of masonry).

Small Finds

The material culture is remarkably uniform throughout the entire cultural deposit and can be treated as one archaeological complex. Thus, the site is considered to have been occupied by a single culture. Within the artifact assemblage, the more diagnostic forms are stemmed projectile points made of knife-shaped flint flakes, obsidian sickle blades, bone awls, polishers, needles with eyeholes, and polished stone axes. The sickle blades are made using slightly bevelled retouch and show some traces of use-wear. The working edge of the blade is often left unretouched. Basalt grain mortars are ubiquitous throughout the deposit. Complete examples are rarely more than 25 cm in length and, viewed two-dimensionally, have an oval or nearly rectangular form with rounded corners. Cobble pestles and grinding stones are less typical. A large number of unretouched obsidian flakes were also recovered from the site. Two long, straight-platformed, pencil-shaped obsidian blade cores exhibit a highly developed flint-knapping technology.

Fragments of polished marble vessels are also common. The quality of the raw material and manufacturing skill are both very high. Consciously selected bichrome marble simulates painted designs on the vessels. The most characteristic forms are deep and shallow bowls of various sizes, and goblets. These vessels usually have simple rims and flat bases.

There is a large series of crude gypsum plaster vessels and impressions of woven material on plaster and bitumen fragments (Fig. 2.24:1–6). Deep plaster bowls with broad flat bases, straight outward-flaring walls, and simple sharp-edged rims are the most common form. Basketry forms are quite diverse, including an impression of a tray made of twigs intertwined with a thin cord, and other containers made from matted or wattled reed. Several impressions of small woven bags have also been found. These forms have survived to the modern period because they were intentionally lined with gypsum or bitumen. Woven and matted material was also used as floor covers and dwelling roofs.

A number of stone ornaments were also recovered from the site during the 1980 field season (Fig. 2.24), including an unusual perforated disk-shaped object made from gray

Figure 2.24:1–6. Basketry impressions on gypsum and bitumen

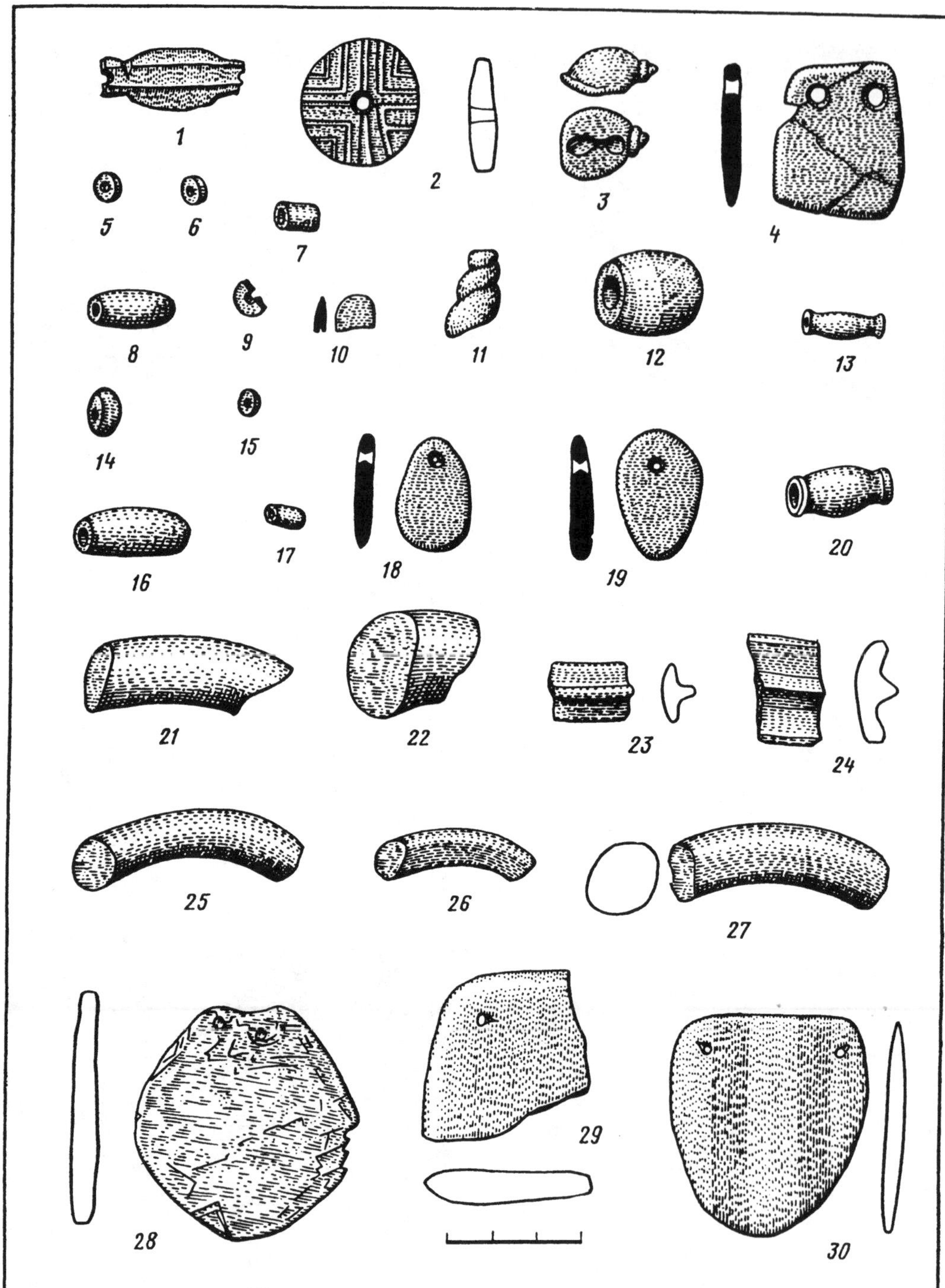

Figure 2.25:1–30. Stone and obsidian beads, pendants, and bracelets

stone with an incised design (Fig. 2.24:2). Two perpendicular lines divide the disk into four equal parts each of which is decorated with rectangular designs. Since the disk is 2.6 cm in diameter, it could be a pendant or a button and was probably also used as a stamp seal. A variety of stone beads were also found, including disk-shaped, cylindrical, and barrel-shaped examples. Larger perforated stone disks could have been used as pendants or weights. The diameter of these objects reaches 6 cm in some cases. Several of the pendants are made of obsidian, including one flat oval-shaped pendant that has been carefully polished (Fig. 2.25:30). The pendant measures 4.8 cm × 4.0 cm and has two small holes at one end. Through its slightly opaque surface the remarkable structure of the obsidian can be seen. Another obsidian fragment has been worked in the form of an oval (Fig. 2.25:28). The two small holes at its ends could be used for sewing it on a garment, though it could also have been part of a necklace. Additionally, two stone pendants were found that were worked in the form of shells. One of these is a spiral-shaped fragment with a hole running along its long axis (Fig. 2.25:11). The other pendant is in the shape of a snail shell and has a hole on its side for attachment as an earring or button (Fig. 2.25:3). Several types of polished marble bracelets were also recovered, including examples with round, oval, and flattened cross-sections (Fig 2.25:21–22, 25–27). Several of the flat bracelets have a decorative rib running along the center of their outer circumference (Fig 2.25:23–24).

Numerous fragments of anthropomorphic clay figurines have also been found at the site. In 1980, two nearly complete female figurines were discovered, which greatly aided in the reconstruction of small fragments of similar forms recovered from the site earlier. Both figurines are highly stylized. Though they differ in size and form, the manufacturing technique and artistic style are nearly identical. One of the figurines is complete except for a small portion of its head. It is 5.2 cm in height, and its torso is 2.8 cm long and 2 cm thick. The second figurine is less complete, with the massive torso surviving but the head completely missing. The figurine is 8.5 cm in length and 6.2 cm thick. Both figurines are seated and have disproportionately large torsos compared with the rest of the body, and the lower legs are not depicted. The trunk is represented by an elongated cone with a small protrusion to depict the bust. The fragments found in previous field seasons indicate that the head is usually small and only abstractly portrayed, sometimes only by pinching the top of the figurine. The legs and torso were fashioned separately and then the three parts were joined together. These three parts are often found separately in the cultural deposit.

Two pieces of copper ore or malachite and a copper awl were found in sealed stratigraphic contexts. The awl was discovered in the center of square L-8 at the depth of 2.36 mbd in a compacted and undisturbed level under the foundation of structure 19 from Level 3. This find has been carefully examined by N. V. Ryndina at the spectroscopic laboratory of the archaeology department at Moscow State University.

The awl is heavily oxidized, but its original form is well preserved. It has the shape of an elongated pyramid, square in cross-section. One end is flat, whereas the other is tapered and was probably once sharp. The angle of sharpening along the working edge is not greater than thirty degrees. The awl is 3.8 cm in length and 0.3 cm thick. The find has been subjected to metallographic and X-ray diffraction studies and has been examined under great magnification. Based on the distributions and amounts of trace elements present in the sample, one can conclude that the composition of the awl from Tell Maghzaliyah is identical to the native copper from Talmesi in central Iran. Especially distinctive are drop-like silver inclusions, measuring 0.3 microns, also characteristic of the of the native copper from Talmesi. Thus, this analysis has iden-

Figure 2.26:1,2. Burials in squares P-3 and P-4

tified the ore source of the copper awl found at Tell Maghzaliyah. Additionally, the study has established that this object was cold-forged. (See Fig. 2.26 for Tell Maghzaliyah burials.)

Environmental Reconstruction

The reconstruction of the natural environment in the Sinjar region during the sixth and seventh millennia B.C. is possible through the analysis of the paleobotanical samples recovered from the early settlements in the area. The low-lying Sinjar Plain is now located in a semiarid zone in the foothill-steppe region of northwestern Iraq. In the Sinjar Mountains, localized oak forests have survived to this day. In the Sinjar region, the summers are hot and dry, and the winters are relatively short and moderate. Annual precipitation is on the average 350 mm and occurs almost exclusively during the winter months. Based on the palynological data analyzed by E. M. Zelikson at the Institute of Geography of the USSR Academy of Sciences, the natural vegetation in this area consisted of sparse xerophytic wooded areas and savannas during the sixth and seventh millennia B.C. On the adjacent slopes of the Sinjar Mountains, bands of oak forests interspersed with steppe flora existed, and, in general, the environment of the period was more favorable for human exploitation than it is today. The faunal remains from the sites studied include common fallow deer, wild ass, wild ox-aurochs, bezoar goat, mouflon, wild boar, deer, fox, and other species that are consistent with the palynological reconstruction of the environment.

The material culture at Tell Maghzaliyah has a number of parallels at other contemporary sites in northern Syria and eastern Anatolia, including Abu Hureyra (Moore 1975), Bouqras (Akkermans, van Loon, Roodenberg, and Waterbolk 1978), and Çayönü (Cambel and Braidwood 1970), though at the latter site there are significant differences in architecture and settlement planning.

The subsistence of the inhabitants of Tell Maghzaliyah was based on a mixed strategy of plant agriculture and animal husbandry along with hunting and gathering. The continued importance of hunting is attested to by the large number of projectile points recovered from the site. As determined by D. V. Gadzhieva, 60 percent of the faunal remains belong to wild species and only 40 percent to domesticated animals. The primary game animals were mountain goat and wild ass. Less significant were aurochs and wild boar, and even less commonly exploited were bezoar goat and mouflon. Domesticated species consisted of sheep and goats, with the former somewhat more prevalent. Morphological traits indicating domestication are present only in their initial stages. Based on a limited sample, we may infer that the domestication of cattle may have taken place before the abandonment of the site. As determined by G. N. Lisitsyna, the cereals exploited by the inhabitants of Tell Maghzaliyah include wheat (*T. monococcum* and *T. dicoccum*), barley (*H. vulgare distichum*), species of wild oats, and flax. Lentils, vetch, and adonis have been also found.

The sites in northern Iraq excavated by the Soviet expedition have shed a considerable amount of light on the final stage of Neolithic development in Southwest Asia. Undoubtedly, Tell Maghzaliyah is a large, fully sedentary settlement, especially in Period 2. Its economy represents the final phase in the development of early food production, with both plant cultivation and animal husbandry present. At the same time hunting and, quite probably, gathering continued to play a significant role in the subsistence strategy. These various activities were important factors in determining the location of the settlement along the hilly flanks of the Sinjar Mountains, giving its inhabitants easy access to both highland and steppe resources.

Later settlements of the Tell Sotto-type

(Bader 1975, chapter 3 this volume; Bader 1982, chapter 4 this volume; Merpert, Munchaev, and Bader 1976; Merpert, Munchaev, and Bader 1977; Merpert, Munchaev, and Bader 1978; and Merpert, Munchaev, and Bader 1981), also excavated by the Soviet expedition, already possess a well-established mixed herding and cultivation strategy in which the role of hunting has been dramatically reduced. These later settlements are located in the fertile loess plain suitable for dry farming and reflect the first stage of developed agricultural societies. The settlement shift from the foothills to the plain can be regarded as the final victory of the agricultural subsistence mode. Its subsequent development resulted in economic specialization and intensification and a sharp increase in the number of sedentary settlements during the Sotto Phase. It is important to note that a constant and prevalent component in the Tell Sotto pollen spectrum is that derived from domesticated cereal species, and moreover, its percentage in the cultural deposit at the site increases as one moves up the stratigraphic sequence. The Tell Sotto palynological evidence points to large areas of land under cultivation and, consequently, to the growing role of agriculture in the economy of the inhabitants of the site. Thus, Tell Maghzaliyah and settlements of the Tell Sotto-type form an important sequence in the development of food-producing economies in the fertile Sinjar region. In this sequence, the settlement of Tell Maghzaliyah represents the terminal stage of the initial development of a mixed agricultural and herding economy, and settlements of the Tell Sotto-type attest to the success of this economy and the final victory of agriculture over hunting and gathering at the end of the Neolithic Revolution.

3

THE EARLY AGRICULTURAL SETTLEMENT OF TELL SOTTO

N. O. Bader

The study of the early agricultural settlements in northern Mesopotamia yields important evidence for understanding the origin of food-producing economies and has attracted the attention of scholars for several decades now. However, the accumulation of empirical data is proceeding slowly. The most important field studies in this region remain the early investigations of S. Lloyd and F. Safar at Tell Hassuna (Lloyd and Safar 1945), the regional survey studies of R. Braidwood in Iraqi Kurdistan, and the excavations at the settlement of Jarmo (Braidwood and Howe 1960). In the past few years this work has been supplemented by the excavations at Shimshara (Mortensen 1970), Telul eth-Thalathat (Fukai, Hariuchi, and Matsutani 1970), and Umm Dabaghiyah (Kirkbride 1972, 1973a, 1973b, 1974, 1975).

The beginnings of a food-producing economy in Mesopotamia can be traced at Karim Shahir and similar sites to about 9,000 B.C. Tell Hassuna yields evidence of a highly developed culture of farmers and pastoralists (about 5000 B.C.). Jarmo (7000–6000 B.C.) connects the two sites together chronologically. This sequence, however, remains fragmentary, and many links in the chain remain unknown.

On the right bank of the Abra River, at the point where it emerges onto the plain, 4–5 km to the northwest of Yarim Tepe, V. I. Gulaev discovered an open Mousterian campsite. In a valley on the western slope of the Sinjar Mountains near the small village of Kursi, a cave site and a flint quarrying site have also been investigated. The cave yielded tools typologically similar to Mesolithic assemblages. The workshop was apparently in use for a long time. Additionally, later Hassuna people living on the plain were also quarrying flint from the foothills of the Sinjar Mountains.

In 1971, V. A. Bashilov and V. I. Gulaev discovered the mound of Tell Sotto, 2 km west of Yarim Tepe. The same year, the present author conducted a test excavation at the site, which yielded distinctive remains from the pre-Hassuna period. Also in 1971, as a result of preliminary survey within a radius of 8–10 km from the expedition's base camp, ten small Hassuna sites were discovered, most of them with archaic Hassuna material on the surface. From a preliminary evaluation of this material, we note that one of these sites, Kültepe, contained remains similar to those found in the lower strata of Tell Sotto along with archaic Hassuna material. The expan-

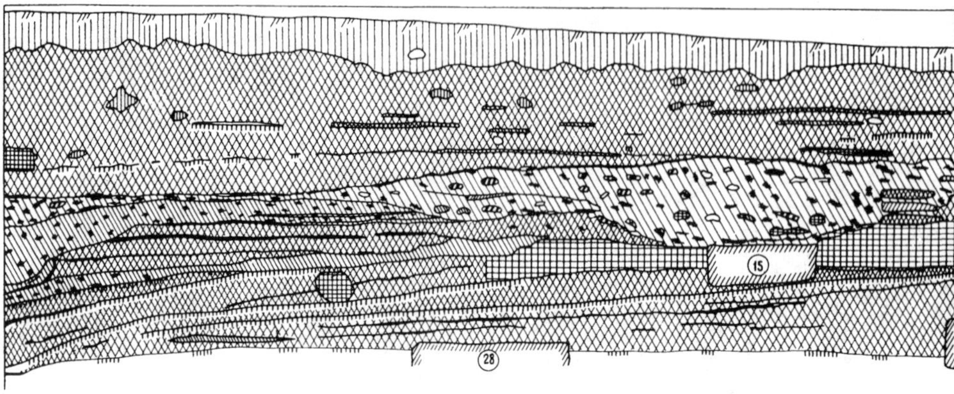

Figure 3.1. 1973 field season. Section 8 through squares 10 and 11.

sion of these preliminary excavations will make it possible to investigate an extremely interesting group of pre-Hassuna sites.

Tell Sotto rises up to 2.5 m above the modern plain and is elliptical in shape with its long axis oriented east-west, not more than 80 m in length. The mound's surface has been disturbed by plowing. The cultural deposit of the tell reaches a depth of 3.80 m, more than 1 m below the current level of the surrounding plain. Pits originating from the lowest stratum reach an even greater depth. The present author conducted excavations at Tell Sotto in 1971, 1973, and 1974. First, a 25 m × 2 m step trench was dug along the long axis of the mound just to the east of its central point. This made it possible to gain an idea of the tell's stratigraphy, architecture, and artifact assemblage while minimizing damage to the site. The lowest stratum yielded distinctive, previously unknown archaeological remains that differed fundamentally from the typical Hassuna material. In 1973 the trench was broadened in the northeastern quadrant of the mound. The total area of excavation reached 100 square meters. Seven residential strata were exposed, using the step trench as a stratigraphic guide. In the lowest stratum, the remains of a large dwelling complex were partially revealed.

In 1974 the excavation of the eastern outer quadrants of the mound continued, allowing a more detailed investigation of the lowest cultural level. The total area excavated reached 200 square meters. This entire area was exposed down to the base of the mound. As a result of three years' work, the entire eastern quarter of the mound has been excavated. However, the dwelling complex of the lowest cultural stratum remains partially unexcavated, and our knowledge of the central, transitional level is inadequate, since here the excavation encountered an open area with few architectural remains.

The excavations at Tell Sotto were carried out according to methods refined by the expedition's large-scale excavation of Yarim Tepe I.[1] A 10 m × 10 m grid was superimposed on the mound, but excavations of artifacts and features were conducted in 2.5 m squares with 20 cm balks between them. One stratum was exposed at a time, over the excavated surface. After the excavation of each stratum, the balks were removed in order to connect structures between excavated squares. Once again, the step trench served as a stratigraphic guide. Individual rooms were the basic building units of each stratum, and contemporaneous floors linked rooms together chronologically. Structure walls either abutted or were bonded to each other, and in other cases they were connected by surfaces, often plastered.

Stratigraphic control was provided by 2.5 meter-long balks. The continuous numbering

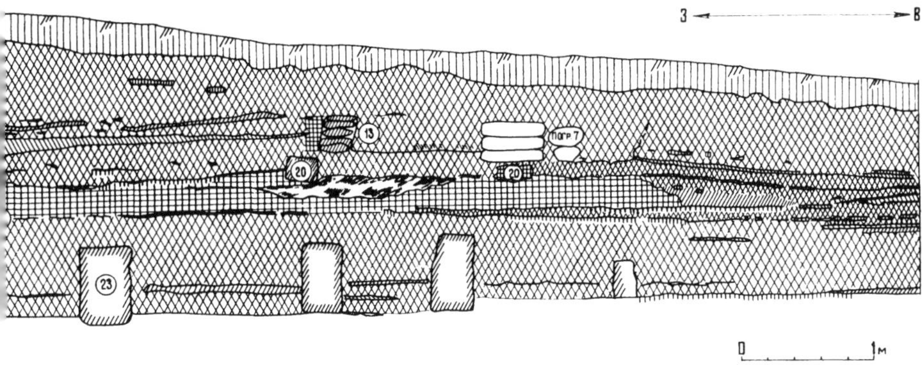

of exposed walls, in addition to precise labeling and frequent cross-sections, allowed for tight contextual control of the entire material assemblage, both features and artifacts. Examining the cultural sequence yields the conclusion that there was no occupational break in the site's settlement history (Fig. 3.1).

Several cultural strata were exposed in the excavated area. In individual cases, smaller activity areas, architectural rebuildings, and remodelings of limited areas can be distinguished within these strata. All of the strata can be grouped into four cultural phases.

A relatively thiċk lower cultural stratum (Level 7) is characterized by the remains of a large, well-preserved, multiroomed house with residential surfaces adjoining it. This structure existed for a long period and was remodeled a number of times, all of which can be grouped into a single cultural Period 4.

The structures of Period 4 are covered by an extensive series of lenses containing charcoal and ash that comprise Levels 4–6 (Period 3). It is significant to note that the charcoal layers fill in the hollow spaces around the large house in the lower stratum. Thus, their deposition began shortly after this structure was abandoned. No large structures have been found in these strata. The architectural remains found in Period 3 are mainly individual ovens and outbuildings. Over the depositional period of the charcoal layers, this sector of the site was modified a number of times, but the thinness of the layer shows that its accumulation did not take long.

The ash accumulations are covered by layers of clay soil up to the modern surface of the tell. At the base of these layers is a thick plaster surface (Level 3), which chronologically links several buildings and outbuildings. Both the surface and the structures associated with it extend into sections that have not yet been excavated. The entire complex connected with Level 3 has been labelled as Period 2.

The upper Period 1, which has been severely disturbed, is traceable in the central portion of the mound. It includes small fragments of walls, plaster floors, and layers of ash in the plow-damaged stratum and immediately beneath it. Distinctive structures from which one could define the entire phase have not yet been found. A layer that clearly divides Periods 1 and 2 is also lacking over the area that has been excavated. Only in certain small areas have we succeeded in locating fragments of floors containing standard Hassuna pottery from undisturbed contexts. We have had to assign the upper portion of the tell, composed of clay soil with a thickness of 60–70 cm and poor in finds, to a single undifferentiated stratum.

From the small area which has been excavated, it is possible to draw only preliminary

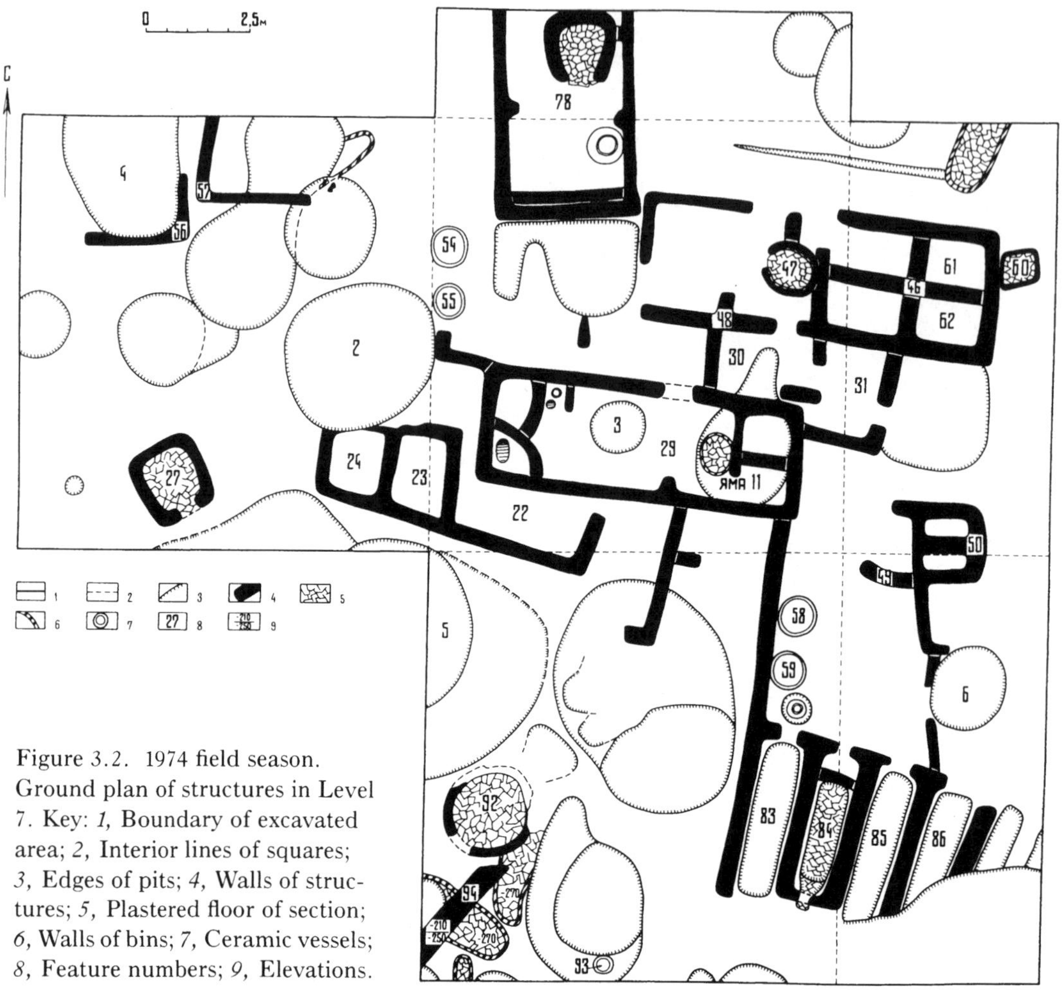

Figure 3.2. 1974 field season. Ground plan of structures in Level 7. Key: *1,* Boundary of excavated area; *2,* Interior lines of squares; *3,* Edges of pits; *4,* Walls of structures; *5,* Plastered floor of section; *6,* Walls of bins; *7,* Ceramic vessels; *8,* Feature numbers; *9,* Elevations.

conclusions as to the architectural types and general layout of the settlement. To date, about fifty structures, including dwelling rooms, ovens, storage facilities, hearths, and fragments of walls have been located. However, only the buildings of Level 7 are firmly integrated into a large residential complex (Fig. 3.2).

As we have already noted, the uppermost occupation has been almost completely destroyed by plowing. In the second period, a small rectangular room, with an adjoining thick plaster floor and Hassuna storage pots embedded in it, has been uncovered. This room is in a corner of a house, the rest of which has not yet been excavated. The thickness of the wall that has been uncovered does not exceed 40 cm. Within the same square, the remains of a half-demolished wall more than 4 m in length have also been discovered. The two structures do not form a single complex, but they are contemporaneous stratigraphically. The ruins of the long wall and the thick plaster floor extending to the limit of the excavation permit the hypothesis that large dwellings existed in this period elsewhere at the site in areas not yet excavated.

The third period also remains problematic. The accumulation of ovens, hearths, round storage pits, and large ash pits must represent the remains of a production complex. The dwellings contemporary with it may be located in areas not yet excavated. The thickness of the walls of the ovens and the storage bins does not exceed 40 cm. When the walls were taken apart carefully, their construction could in some cases be traced. They were made of large, flat mudbricks (up to 70 cm in length and 5–7 cm in height) of clay mixed with straw.

The lowest level is highly distinctive. Part of a large, rectangular, multiroomed house with a long corridor between the rooms was uncovered here. Within the rooms, accumulations of animal bones, child burials in vessels, grain storage vessels, along with grinding stones, hearths, and accumulations of pottery and tools were discovered. Around the outer walls are ceramic storage jars embedded in the floor, ovens, and devices for drying grain (Figs. 3.3 and 3.4). The study of the house has not yet been completed. As far as can be judged at present, its layout is close to that of the multiroomed rectangular houses in the lower strata of Yarim Tepe I.

The preliminary archaeological assemblage from Tell Sotto may be evaluated in the following manner. Diagnostic Halaf sherds with brightly colored patterns have been collected on the surface. However, this material, at present, cannot be connected with any stratum at the site. The first cultural period contains standard Hassuna pottery with incised ornamentation, similar to that found in the upper levels of Yarim Tepe I (Merpert and Munchaev 1971b). The second period is characterized by large, crudely fashioned Hassuna pots. This phase is comparable to Levels 6–8 of Yarim Tepe I.

The third period of Tell Sotto yields round-rimmed vessels and burnished pottery colored with ochre. This combination is not found at Tell Hassuna but has been noted in the lower levels of Yarim Tepe I. The lowest level of Tell Sotto is earlier than Yarim Tepe I and shows a number of close parallels to Umm Dabaghiyah (Kirkbride 1972, 1973a, 1973b, 1974, 1975). Among the characteristic finds of the lowest level we may note the following: large-scale architecture; large quantities of crude, ribbed cooking ware, including some with simple and zoomorphic molded designs on the rims; other crudely made vessels, including chalices, oval tubs, and rectangular vessels; thin-walled burnished pottery painted with ochre (Figs. 3.5–3.9); a large number of stone vessels (Fig. 3.10), bracelets, and beads, including pendants with a sculptured representation of a ram; and two clay female and zoomorphic figurines (Fig. 3.11). Among the tools are wedge-shaped stone axes, awls made from fragments of long bones, needles, polishers, clay and stone spindles, and clay pellets for slings (Fig. 3.12).

There are as yet no absolute dates for Tell Sotto and Umm Dabaghiyah. The radiocarbon date for Telul eth-Thalathat 15 is 5,570 ± 120 B.C. (Fukai, Hariuchi, and Matsutani 1970). Braidwood obtained a radiocarbon date of c. 6750 B.C. for the ceramic stratum of Jarmo. According to Waterbolk's data, this stratum dates to 6000 B.C. (Waterbolk 1971).

Tell Sotto's imported artifacts are also of great interest. In the third phase of the tell, we found a number of small sherds which differ sharply from the remaining ceramic assemblage in color, ware, and firing technique. These sherds are undecorated, have a rich, dark, cherry color on the surface and in section, and are extremely dense. They are tempered with shell and have a burnished surface. Two similar sherds were also found at Yarim Tepe I in Level 8. N. Ya. Merpert explains their presence in this lower stratum of Yarim Tepe I by identifying them as imports from the Sotto culture (Munchaev and Merpert 1973). However, cherry-colored ceramics are exotic to Tell Sotto as well.

Hassuna pottery, as a rule, is organic tempered. All pottery from Tell Sotto is also tempered with organic material. Black and red burnished thin-walled pottery, either sand or shell tempered, is widespread in the early agricultural sites of Anatolia. Close analogs to the cherry-colored sherd material from Tell Sotto and Yarim Tepe I can be found at Mer-

Figure 3.3. Dwelling 29 with storage area for grain and cooking pot

sin (Stratum 26: Garstang 1953), Çatal Hüyük (Stratum 2: Mellaart 1967), and Sakçe Gözü (Strata 1 and 2: Taylor, Williams, and Waechter 1950). The cherry-colored sherd material has no parallel in the rest of the Tell Sotto ceramic assemblage and is obviously imported. It is interesting to note that metal has been found in the same strata as this cherry-colored ware at both Tell Sotto and Yarim Tepe I. These imports can be used to chronologically link the Tell Sotto and Yarim Tepe I stratigraphic sequences.

Until recently, Tell Hassuna was considered the earliest agricultural site in the plains of northern Iraq. However, earlier sites are now being investigated in the same region at Tell Sotto and Umm Dabaghiyah. Kirkbride interprets evidence from Umm Dabaghiyah as representing a new culture which arrived from the west, and is thus foreign to the local Hassuna culture. N. Ya. Merpert concurs with Kirkbride on this issue (Munchaev and Merpert 1973).

Kirkbride posits that at the beginning of the sixth millennium B.C., significant ethnic movements occurred in upper Mesopotamia, and that one of these may have proceeded from the Khabur region eastward. To Kirkbride, the settlement at Umm Dabaghiyah is an example of this migration. Furthermore, the distinctive and sophisticated pottery of Umm Dabaghiyah, in Kirkbride's view, may represent an earlier stage of the Halaf culture, of which little is yet known. In fact, the lowest stratum of Tell Sotto differs from the archaic Hassuna culture of Tell Hassuna, Yarim Tepe I, and Tell Sotto in several essential characteristics. The thin-walled, burnished pottery covered with ochre, the zoomorphic and anthropomorphic molded

Figure 3.4. Dwelling 78 with an oven, large jar, and a bench near the wall; excavations of 1975.

figures on the rims of crude vessels, the stone vessels, the stone bracelets, the beads, the small, wedge-shaped, polished axes, and the female and zoomorphic clay figurines all distinguish the earliest occupation of Tell Sotto from that of the archaic Hassuna. However, after the work at Yarim Tepe, the conception of the Hassuna culture itself has changed significantly, and the third phase of Tell Sotto has yielded a number of transitional elements in the culture.

The large, multiroomed houses so typical of Kültepe, Umm Dabaghiyah, and the lower strata of Yarim Tepe I have not been discovered at Tell Sotto (Ed. note: according to a letter from N. O. Bader, 8 March 1992). They must also be present in the lowest stratum of Tell Hassuna. Female figurines, apparently of rather archaic forms, are known from the Tell Hassuna assemblage. The ceramic assemblage of the Hassuna culture and of the lowest stratum of Tell Sotto are very similar in ware and manufacturing technique. We have already noted that the cherry-colored sherd material from Tell Sotto and Yarim Tepe I differ sharply from the rest of the assemblage. This would seem to indicate that certain elements of the assemblage were derived from a western source, rather than a western origin of the Tell Sotto culture in its entirety. The crude cooking ware and the thin-walled vessels of the second, third, and fourth phases of Tell Sotto include a number of transitional forms, such as ribbed vessels and oval tubs with corrugated bottoms ("husking trays").

The previous discussion leads to the conclusion that there are not sufficient grounds for considering the lowest level of Tell Sotto as an independent culture. A significant amount of continuity can be traced between

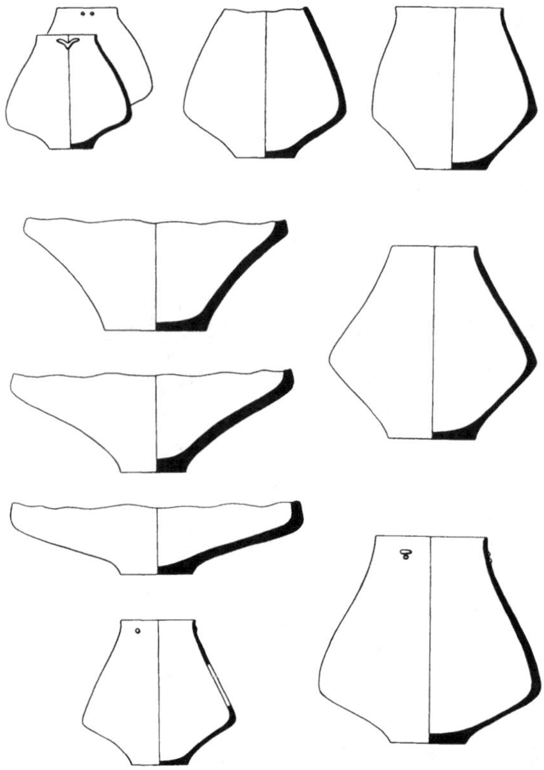

Figure 3.5. Crude vessels from the lower levels

archaic Hassuna sites and the earliest phase at Tell Sotto. Tell Sotto represents an unknown early stage of Hassuna, perhaps a Proto-Hassuna. Tell Hassuna, Tell Sotto, and Umm Dabaghiyah constitute early agricultural settlements within a single cultural sphere.

The parallels between Tell Sotto and Jarmo are also of great interest. The chronological gap between the Tell Hassuna and Jarmo sequences makes their comparison difficult. The excavation of Tell Sotto partially bridges this gap. Let us briefly compare the collections from Jarmo, Tell Hassuna, and Tell Sotto in the Iraqi National Museum in Baghdad. Jarmo yields a large and varied collection of stone vessels, including carinated vessels, carinated and noncarinated cups with straight or flaring rims, and cups with a simple vertical rim (Braidwood and Howe 1960). In many vessels the edge of the rim is tapered, with a horizontal groove running around the upper circumference of the vessel. The bases of the vessels are round and flat. The flattened bases and flaring rims of the stone vessels are probably modeled after clay prototypes. Stone vessels, on the whole, are not characteristic of the Tell Hassuna and Yarim Tepe I assemblages. However, there are many of these vessels at Tell Sotto, primarily fragments of small bowls and platters. The ceramic corpus from Jarmo and Tell Sotto, although differing significantly, also have comparable elements. At Jarmo, one of the most ubiquitous forms of the ceramic assemblage are vessels with carinated sides, both of the crude cookware variety and small, delicate vessels and platters made from well-levigated clay. At Tell Sotto carinated forms are also common. The form and proportions of the vessels may differ, but the carinated side itself is commonplace at both sites. Among the crude vessels in Jarmo we find large oval troughs, similar to forms found at Tell Sotto and Tell Hassuna but with a smooth base rather than the pinched- or pitted-base variety found at Tell Sotto. In vessels from this type found at Yarim Tepe I and Tell Hassuna, the base is more often corrugated. Jarmo and the archaic Hassuna are both characterized by small, poorly fired, smooth cups. At Jarmo, both stone and ceramic vessels of oval or rectangular shape are frequent. At Tell Sotto a whole series of similar vessels has been found.

At Jarmo, in addition to poorly manufactured pottery, one also finds thin burnished sherds of a finer quality. Some of these are covered entirely with ochre on their exterior and then burnished. The ochre was apparently applied after firing. Precisely the same method of manufacture is observed for polished sherd material from the lowest level of Tell Sotto. One fragment of a large vessel, with a burnished surface covered with parallel diagonal stripes of ochre thickened in places, was found at Jarmo. At Tell Sotto and Umm Dabaghiyah this decorative style is not

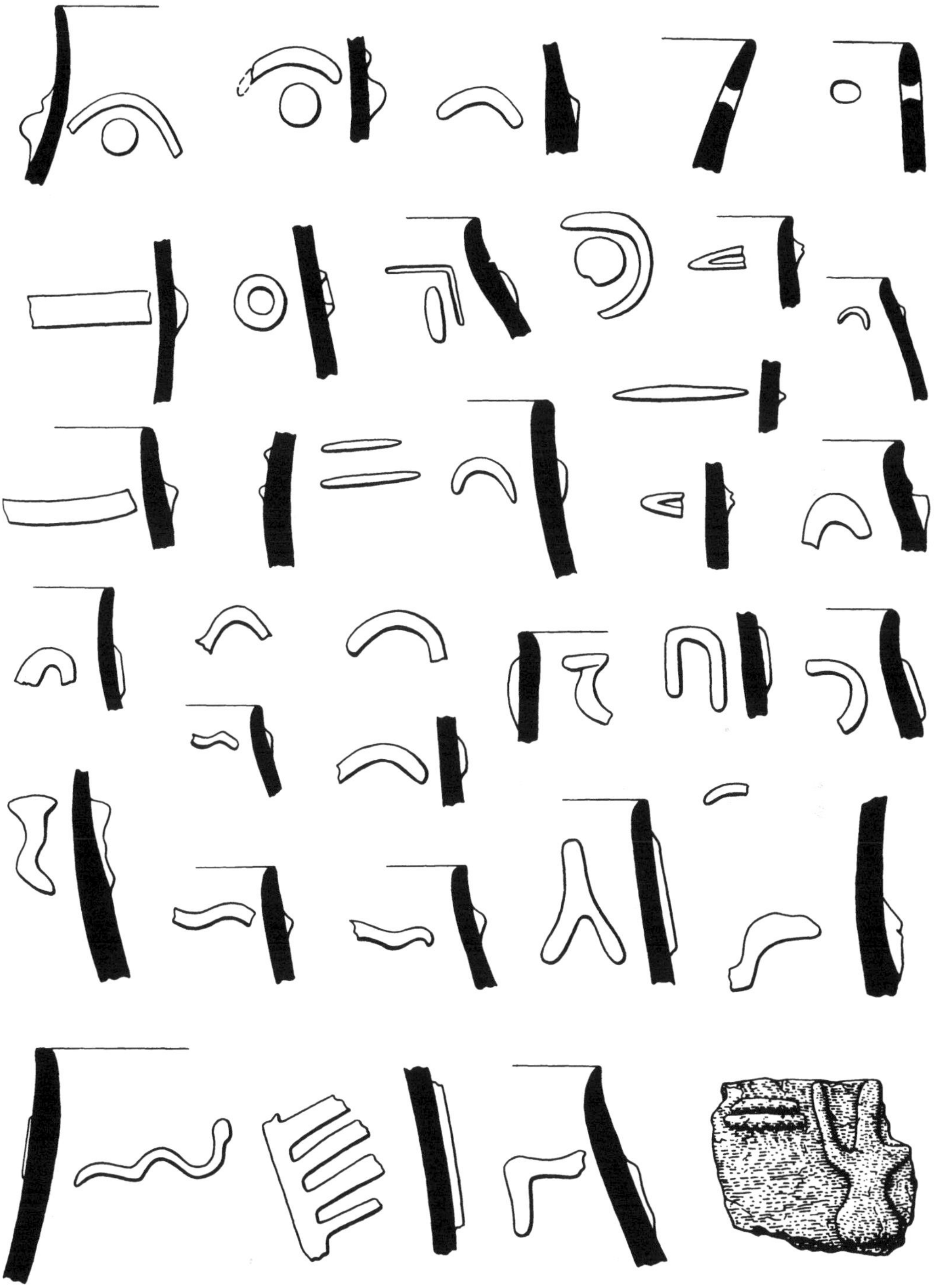

Figure 3.6. Various appliqué designs on crude vessels from the lower levels

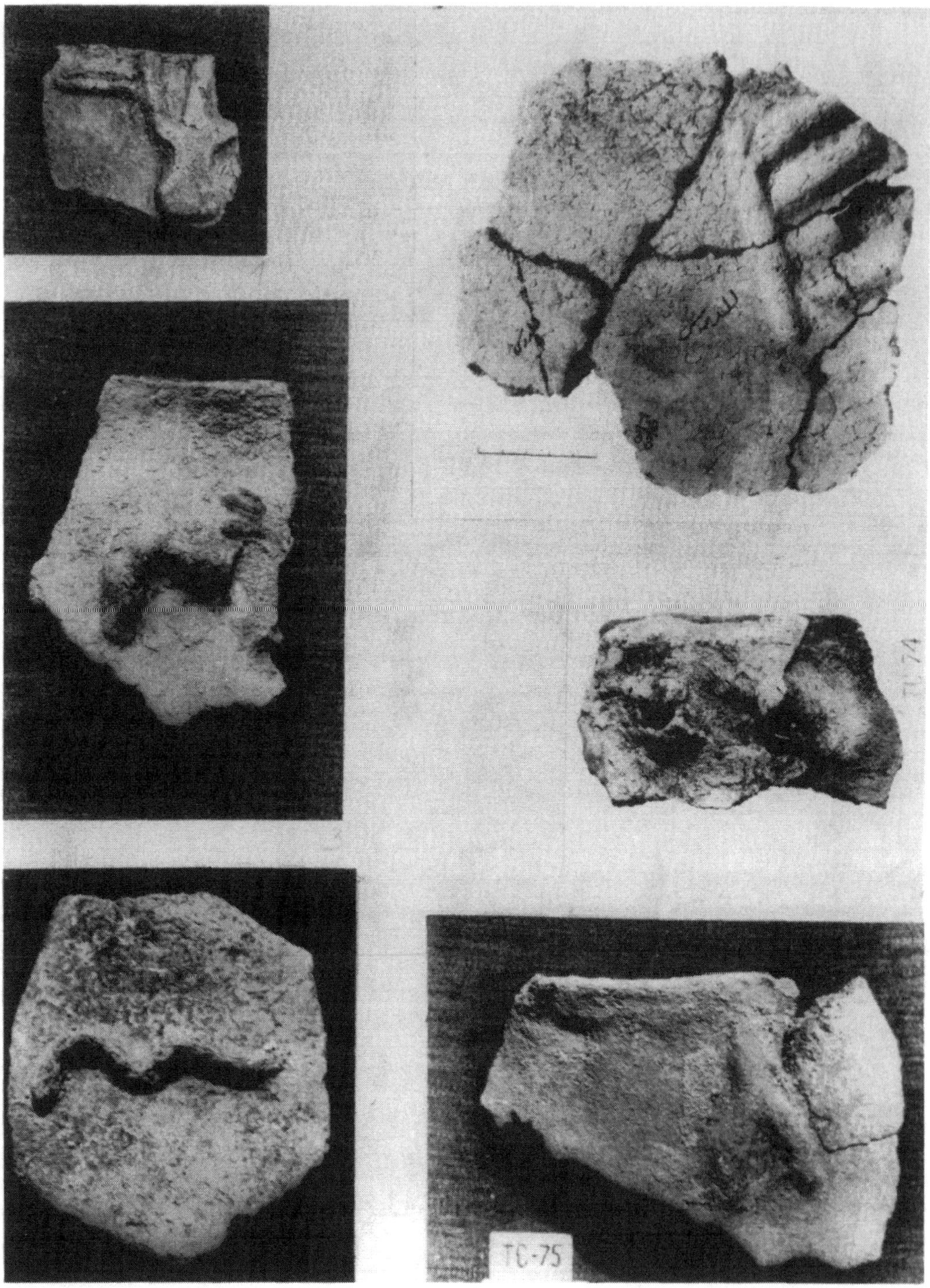

Figure 3.7:1–6. Photographs of appliqué designs from the lower levels

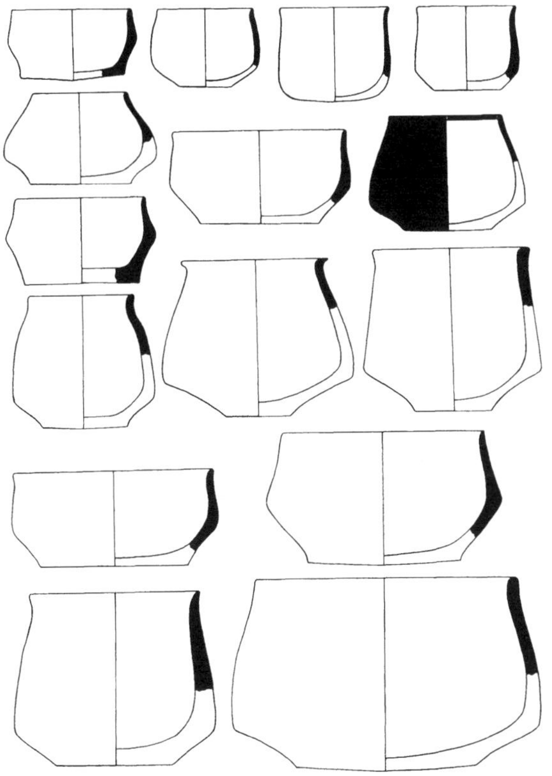

Figure 3.8. Typical pottery forms

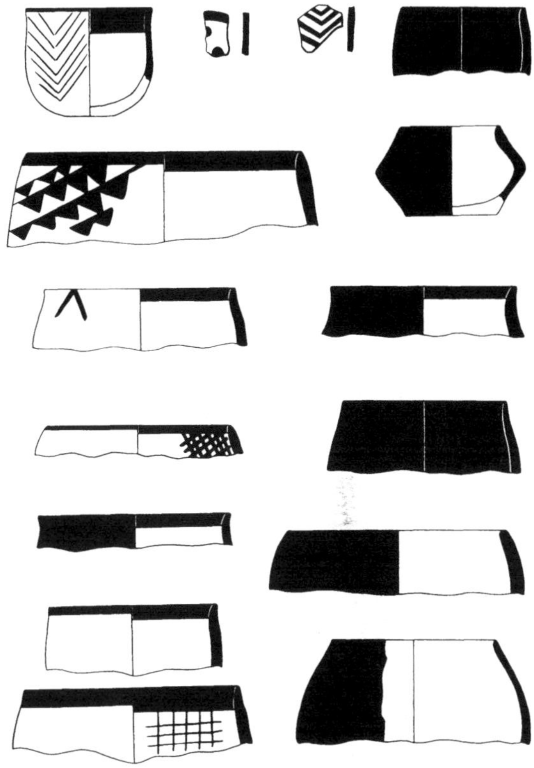

Figure 3.9. Painted fine wares

found. However, it is known from Shimshara (Mortensen 1970, Fig. 96), Tepe Guran (Meldgaard, Mortensen and Thrane 1963, 34), and other sites in Iranian Kurdistan (see also Oates 1973, Pl. 17). At Jarmo, the stone and ceramic vessels with horizontally everted rims stand out in the assemblage. The same type of rim is also encountered in the lower levels of Tell Sotto.

Finally, the molded decorations on the vessels from Jarmo and Tell Sotto are beyond question an important connection between the two sites (Figs. 3.6 and 3.7). Thin, perforated decorations on the rim, vertically molded and stretching upward, are commonplace in the Jarmo assemblage. This style is found in both ceramic and stone vessels. At Jarmo, there are also simple molded decorations in the form of round knobs. In some cases, these knobs are produced by pressing from the inside of the vessel. The molded decorations on the vessels from Tell Sotto are varied, including zoomorphic and anthropomorphic examples. Knob projections are fundamental elements of the molded decorations found on these vessels. A small stone drinking vessel with a single perforated vertical "molded decoration" was found at Tell Sotto and is closely related to the abovementioned molded decorations known from Jarmo.

The stone beads from Jarmo exhibited in the Baghdad Museum are particularly interesting. These beads are flattened, oval in cross-section, with a small, drilled opening through the long axis. One of the beads has slightly offset grooved ends and another approximates a rhombus in cross-section. Parallels to these forms can be found at Tell Sotto and Tell Hassuna. Cylindrical stone beads are also found at Jarmo. A straight-lined

Figure 3.10:1–5. Stone bowls and clay objects

stone seal from Jarmo is no less interesting. This artifact may serve as the prototype for clay and stone Hassuna seals from Yarim Tepe I. Stone bracelets, undecorated or with ornamentation on the outer surface, and either round, oval, or rectangular in cross-section, are quite numerous at Jarmo. Some of the bracelets from Tell Sotto are quite similar to these. Finally, small, wedge-shaped stone axes are quite common both at Jarmo and Tell Sotto (Fig. 3.12). This form is practically indistinguishable at both sites. The axes are usually entirely polished, and have broad blades and wide, short butts. Sometimes the blade has additional sharpening in the form of a wedge. Small adzes have also been found at Jarmo but to date not at Tell Sotto.

The above parallels indicate a definite cultural similarity between Jarmo and the lowest level of Tell Sotto (Period 4). As we have already noted, the latest radiocarbon dates permit us to place these sites close to each other chronologically. Typological analysis presently leads us to the same conclusion. It would be tempting to see a direct cultural connection in this similarity. However, further work in the form of broad-scale comparative analyses is required before this can be verified. Let us remember that ochre diagonal line patterns thickened in places, observed in the Jarmo assemblage, are widespread in Iranian Kurdistan but absent in northern Mesopotamia.

Miniature clay sculpture is one of the major sources of comparison among the Tell Hassuna, Yarim Tepe I, Tell Sotto, and Jarmo assemblages. At Yarim Tepe I, the majority of figurines were found in Level 5, and none was found below Level 6. The anthropomorphic examples are generally standardized with a stylized form, undivided flattened base, the breast barely defined, and a high headdress tilted backward. N. Ya. Merpert attributes the appearance of the figurines in Level 5 of Yarim Tepe I to Samarran influence (Munchaev and Merpert 1973). Certainly, the absence of prototype forms in the lower strata of this rich and extensively excavated site demands explanation. The figurines of Tell es-Sawwan, however, also appear rather suddenly.

It is useful to compare the Hassuna figurines from Yarim Tepe I with those from Tell Hassuna itself. The female figurines from Tell Hassuna are more varied in form. The most distinctive group consists of stylized figures of the Yarim Tepe type. In addition to this rather standardized series, the collection from the Baghdad Museum also includes a fragment of the torso from a standing figurine with legs divided by a straight line. Finally, one finds realistically depicted seated figurines in the exhibit as well. Both of these forms have their hands and the breasts barely defined; in one case the legs are indicated, and in the other, the lower body remains undivided. Unfortunately, the heads of the statuettes have been broken off. We have not succeeded in separating out this material stratigraphically. Typologically, this material is divided

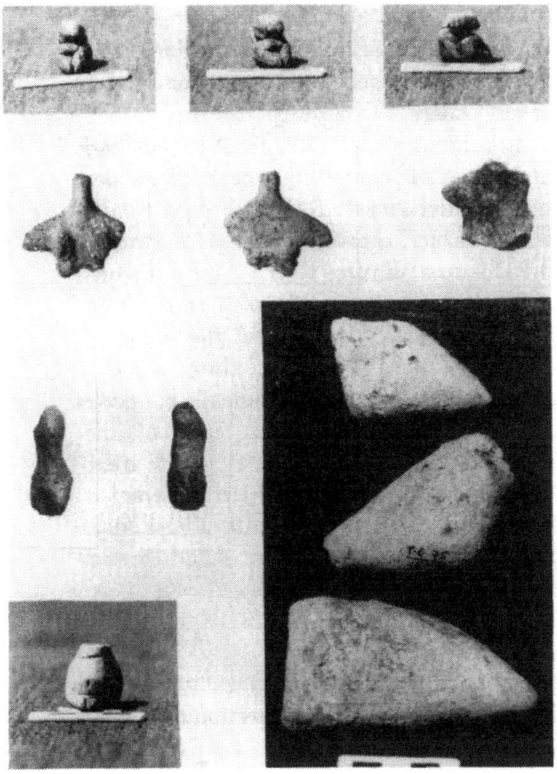

Figure 3.11:1–5. Clay female figurines from the lower levels (5 from the surface)

into three groups: realistic seated figurines, stylized figurines with divided legs, and stylized figurines of the Yarim Tepe type. Logically, we may assume that the realistic statuettes are the original form. The museum collection also includes a zoomorphic figurine from Tell Hassuna; the head and the short legs are broken off, with only the long torso depicting a large pectoral zone surviving.

At Tell Sotto (Period 4), two female human and two zoomorphic figurines were found (Fig. 3.11). The realistically depicted female figurine has only the bust preserved. It was found in the lowest stratum and is rather distinctive typologically. It has broad, sloping shoulders, a well-formed neck, and breasts in relief. The arms have been broken off and the head, also missing, was apparently adorned with a headdress tilted slightly backward. The second female figurine is of the stylized type. The flattened torso, the scarcely defined bust, the lack of arms, and the elongated head are on the whole reminiscent of the figurines of the Yarim Tepe type. One of the zoomorphic figurines is similar to the previously described figurine from Tell Hassuna. The other is represented only by a blank or poorly preserved head fragment. Thus, at Tell Sotto, one encounters statuettes of an earlier form than those at Yarim Tepe I. Certain stylistic features suggest that the Tell Sotto Period 4 figurines are the prototypes for the Yarim Tepe group.

The most varied group of anthropomorphic figurines comes from Jarmo (Braidwood and Howe 1960). One form possesses a realistically depicted head with a high headdress, eyebrows, nose, and eyes executed in the

Figure 3.12:1–9. Stone implements from the lower levels

appliqué method. There is also an intact seated statuette with a face executed in the same technique. The hair is plainly visible on the back of the neck. The breasts are indicated by a pinching technique. Delicate hands are sketched out along the abdomen. The legs are depicted by a vertical line and the torso is large and flattened.

In later forms, the torso of the seated statuette becomes massive. The head has often not survived. The hands and the breasts are barely indicated, with the former sketched out as in the previous case. On the whole, the figure is usually depicted in a considerably more stylized fashion. On one figurine with an undivided, flattened base, the eyes and the mouth are rendered by simple slits. Schematic, cross-shaped or cylindrical figures, with bases in the form of shoes, are distinguished from the realistic or other form of stylized statuettes. The female statuettes of this type from Jarmo are certainly distinctive in many respects. However, here, as at Tell Hassuna, we can note the evolution from realistic to stylized seated statuettes with massive, flattened torsos.

The zoomorphic figurines are more numerous at Jarmo. They have, as a rule, short legs, large pectoral zones, and withers. The hair standing on their backs is indicated by a pinching technique. It is difficult to say with confidence what animal(s) are represented. However, the figurines with ribbed manes on the back of the neck are reminiscent of dogs or wild boars.

In summary, the Tell Sotto Period 4 assemblage does not constitute a new culture derived from the west, as Kirkbride postulates, but an earlier, previously unknown stage of the Hassuna culture. N. Ya. Merpert compares the lower levels of Yarim Tepe I with Tell Hassuna 1a. In our view, the former are earlier, comparable with the third period of Tell Sotto. Both sites are culturally connected and coexisted for a significant length of time. However, the lowest level of Tell Sotto is earlier than the lowest level of Yarim Tepe I. This earliest period is the Sotto Phase, directly comparable in a cultural and chronological sense with Umm Dabaghiyah. The high degree of similarity between the material from Tell Sotto and Jarmo permits us to place these sites close together culturally and chronologically. However, the question of a direct cultural connection remains unanswered.

Note

1. For more detail, see Bashilov and Kuza (1972). In this work, the same methods of analysis and generalization from the stratigraphic data are successfully used.

4

RESULTS OF THE EXCAVATIONS AT THE EARLY AGRICULTURAL SITE OF KÜLTEPE IN NORTHERN IRAQ

N. O. Bader

The Soviet archaeological expedition of the Academy of Sciences, USSR, has developed over the past ten years of excavation a general chronology and cultural sequence for the early agricultural settlements of northern Mesopotamia. One of the basic tasks of the expedition was to search for the earliest agricultural sites in order to study the initial development of food-producing economies. Toward this end, from 1971 to 1975 excavations were carried out at the small site of Tell Sotto, situated on the northern Mesopotamian plain at the foot of the Jebel Sinjar range in the survey region of the expedition (Bader 1975; chapter 3 of this volume). A previously unknown early Hassuna stratum, genetically connected with the archaic Hassuna stratum which covered it, was exposed. Excavations at Kültepe were conducted to verify our findings at Tell Sotto.

The author began work at Kültepe in 1975, with another field season conducted in 1976. A. V. Kuza also participated in the work at the site. Kültepe is situated approximately 6 km due west of the Yarim Tepe group. The site consists of two tells located on the right bank of the Wadi Sharai, to the north of the village of Sharai and south of the village of Sharai Shemali. Ahmed Fottotepe, which contains Halaf remains, rises to the northwest of Kültepe. The locality of Kültepe or "ash mound" is named after a small tell, partially eroded away by the adjacent river. The ashy cultural deposit is plainly visible in the area that has been cleared. An unnamed tell 120 m to the southeast of Kültepe was chosen as the site of the excavations. This latter site was named Kültepe I or simply Kültepe whereas the tell eroded by the river was named Kültepe II.

Kültepe II is a about 2.5 m high and has a diameter of about 60 m. The mound is roughly circular with long, gently sloping sides. Judging by what can be seen from the section cut by the river, the site belongs entirely to the archaic Hassuna culture.

Kültepe I is located approximately 100 m from the river and has been preserved in its entirety. The mound is slightly ovoid with the long axis extending west to southeast, covering an area of 60 m × 80 m. The tell is apparently perched atop a small natural rise over the surrounding plain. The height of the mound is c. 2–2.5 m and the depth of the cultural deposit is 2.25 m. The surface of the mound has been frequently plowed, and archaic Hassuna pottery can be collected on the surface. Coarse ware pottery was especially

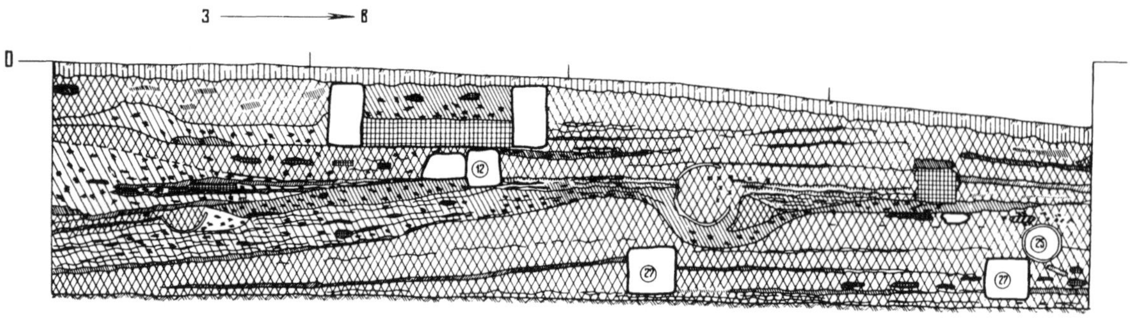

Figure 4.1. Cross-section 2 along northern balk of square 55. Key: *1,* Zero-grade on summit of mound; *2,* Top soil; *3,* Material level; *4,* Contour of wall; *5,* Clay; *6,* Number of wall; *7,* Loam; *8,* Dense clay; *9,* Black humus; *10,* Greenish humus; *11,* Plaster floor; *12,* Charcoal-ashy level.

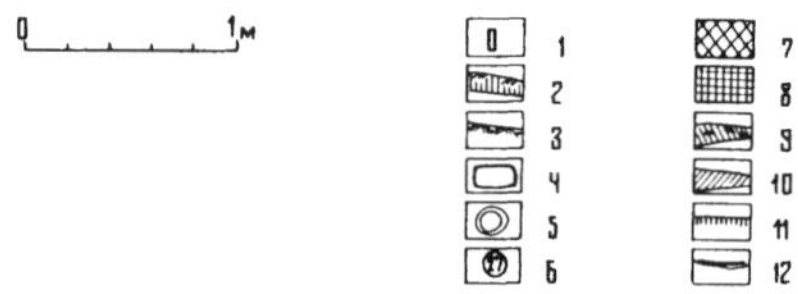

plentiful on the steeper eastern slope. The mound's settlement probably expanded westward toward the river with time. If this were the case, the center of the tell's earliest settlement would have been located in its eastern portion. Subsequent excavations have proved this assumption correct.

The 1976 excavations exposed an area of 125 square meters. A stratigraphic trench measuring 20 m × 2.5 m was dug from the summit of the mound all the way to its eastern edge. The trench formed the orienting line for the site's grid system aligned along the cardinal points. Later, this trench was expanded into a 100 square meter excavation, rectangular in shape, with its northwestern corner at the summit of the mound. On the western slope of the mound, along the line of the trench, a 2.5 m square test pit was excavated. The excavation of Kültepe followed he strategy established at Tell Sotto. The grid was laid out at intervals of 2.5 m and 5 m. The standard excavation unit was a 2.5 m square.

As previously mentioned, the depth of the cultural deposit of the tell does not exceed 2.25 m. At least four building levels were exposed in the trench. The thick lowest level (1) yielded material of the Sotto type, whereas the three levels above it belonged to various phases of the archaic Hassuna culture. There are no obvious breaks in the the site's settlement history (Fig. 4.1).

The earliest level (1) was exposed in the eastern half of the mound. The entire excavated area consists of the remains of a single residence. This stratum gradually diminishes to the west and does not show up at all in the test pit on the western slope of the mound.

Level 2 consists of several circular ovens on the western slope of the original settlement and a thick layer of charcoal and ash associated with their use. This stratum is easily distinguished by color and consistency from Level 1 beneath it.

Above Level 2, Level 3 consists of several small rectangular structures with well-defined plaster floors. The fill within these structures is primarily aeolian-deposited loess. The top level (4) is comprised of a thick layer (up to 1 m) containing fragmentary remains of small rectangular structures, rebuilt several times.

The architecture and layout of the lowest level deserve special attention (Fig. 4.2). They are distinguished from those above them by their excellent state of preservation and durable construction. All of the structures which have been excavated constituted a single complex consisting of a residence and its associated storage facilities. The structures have a regular rectilinear form and are symmetrically arranged. The thickness of the walls is highly uniform (40 cm). In the central part of the settlement the walls are preserved to a height of 70–80 cm. The technique employed in the construction of the walls can be determined only with great difficulty, but in a number of places we nevertheless succeeded in establishing that they were constructed in the usual manner (pisé). The floor and walls of the rooms, primarily of the residence, are mud-plastered. There is no evidence of gypsum or lime plaster in this level. The structure was built directly on sterile soil and has a single foundation. The external walls have only a single perpendicular abutment. However, there is no doubt that the entire structure was erected at one time according to a single plan.

The residence itself consists of a rectangular structure with the internal dimensions of 2.5 m × 5.5 m. The entry is 1 m in width (70 cm after a rebuilding episode) and is cut into one of the end walls. The structure is divided into three relatively equal parts by internal partition walls arranged symmetrically in order to buttress the external walls. The internal walls are clearly later additions as indicated by their thick interior mud-plaster.

The first third of the structure is divided to the right of the entrance by a partition running parallel to the long axis of the residence. This space functioned as a storage area, probably for grain. The cross-wall abuts with the other walls of the residence, and its foundation rests, not on the ground, but on the clay floor of the house in contrast to the external walls. This compartment is carefully mud-plastered on the interior, and its floor shows traces of burning and the impressions of carbonized grain.

In the central part of the house, a coarse carinated vessel with a diameter of 80 cm and a dozen simple molded decorations on the rim is embedded into the floor adjacent to the wall. In the corner, between the wall and the vessel, more than a dozen prepared flint cores are also embedded in the floor.

The easternmost outbuilding associated with the house has the internal dimensions of 1.25 m × 1.4 m and 1.4 m × 1.7 m, thus it is a slightly irregular quadrilateral. Its floors and walls are also mud-plastered. The fill is primarily clay soil, conspicuously devoid of

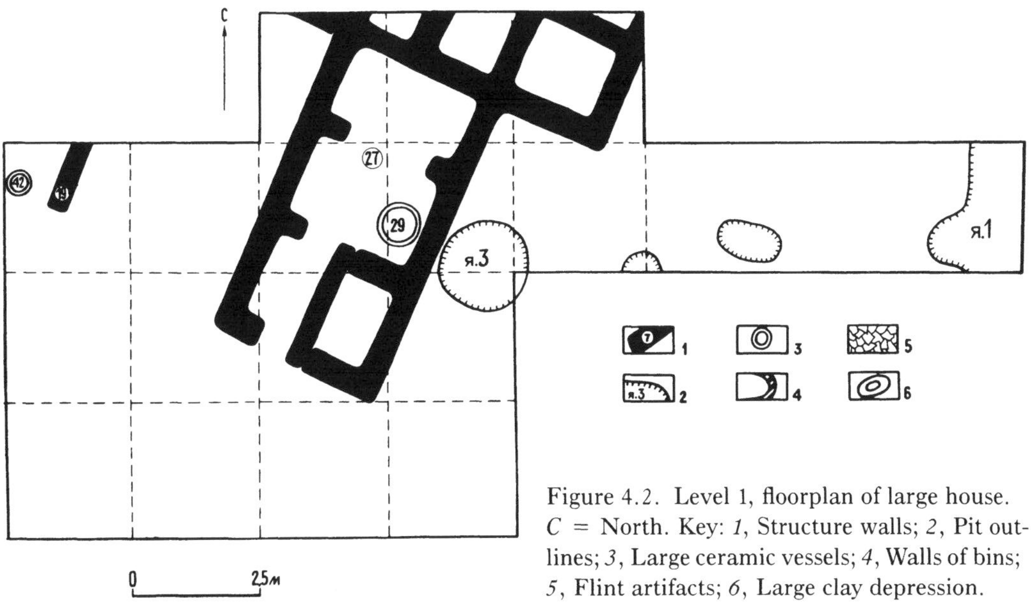

Figure 4.2. Level 1, floorplan of large house. *C* = North. Key: *1*, Structure walls; *2*, Pit outlines; *3*, Large ceramic vessels; *4*, Walls of bins; *5*, Flint artifacts; *6*, Large clay depression.

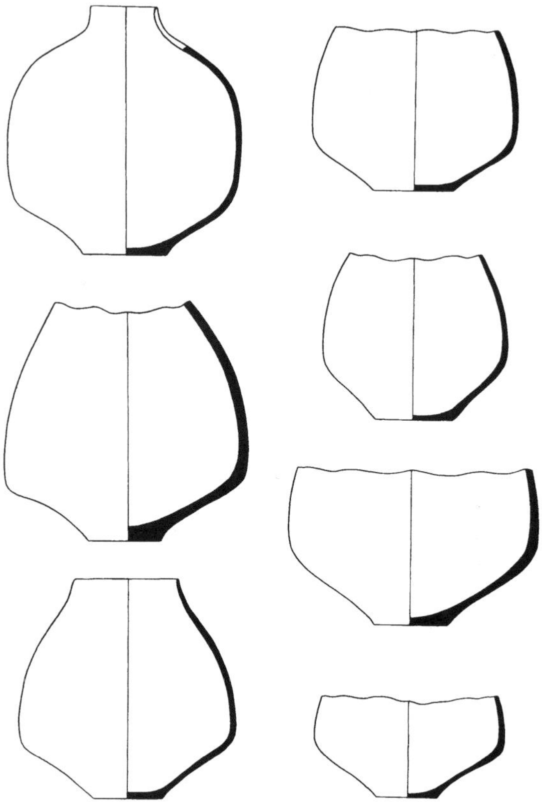

Figure 4.3. Coarse ware cooking vessels from lower levels

cultural debris. On the floor in the middle of this area, two intact stone vessels were found, apparently abandoned by their owners.

Despite the small area excavated, the groundplan of this architectural complex is of great interest. The residence is reminiscent, in its dimensions and the details of its household assemblage, of the dwellings in the lowest level of Tell Sotto (Merpert, Munchaev, and Bader 1976). The assemblage as a whole can be compared to the assemblage from the lowest stratum at Umm Dabaghiyah. The architectural construction techniques are also very similar.

A distinctive ceramic assemblage has been collected from the excavation, making it possible to draw conclusions about the uninterrupted development of the settlement of the site over the entire course of its existence. The lowest level (1) yielded an overwhelming preponderance of undecorated, coarse pottery with barely discernible rims and molded decorations (Fig. 4.3). Thin-walled, burnished, and often painted pottery made from well-levigated clay with organic temper is rare but present (Fig. 4.4). The firing temperature of this pottery is rather low, and the clay in cross-section has a reddish or gray color and is easily broken. The painting consists of a broad red band around the rim, or the vessels are covered entirely with ochre. The entire ceramic assemblage in the lowest level of Kültepe is closely related to that found in the earliest layer of Tell Sotto.

Levels 2–4 at Kültepe give an idea of the development of these traditions, which represent the archaic Hassuna culture. The devel-

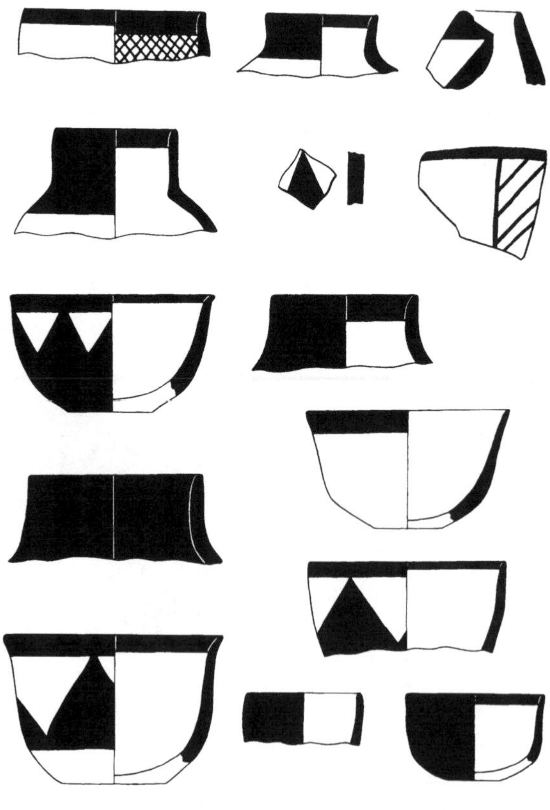

Figure 4.4. Painted fine ware pottery

opment of carinated pots into spherical vessels with high, narrow necks and vertical rims is clearly visible from the three upper strata. The quantity of thin-walled pottery increases and the motifs become more complex. A complete picture of the development of the Kültepe ceramic assemblage, from one stratum to another, and also a comparison of the assemblages between Kültepe and Tell Sotto will be possible after the completion of the processing of recovered sherd material and subsequent statistical analysis.

Other artifacts recovered from the lowest stratum are also of great interest. Here we found several biconical ceramic spindle whorls of the usual shapes, simple awls made from the epiphyses of long bones, and a bone needle. Many artifacts were made of stone. Among these are a discoidal bead, a fragment of a stone bracelet, a fragment of an oval marble palette bearing traces of ochre, several fragments of marble vessels, and one fragment of a thick-walled alabaster vessel. Finds of pestles and grinding stones made from coarse black stone are common.

A small slab of highly polished black stone, in parallelepipedal shape, with cross-hatched engraving and two grooves, has also been recovered (Fig. 4.5). It was found adjacent to the eastern wall of the house in Level 1 on one of the lower floors at a depth of 1.93 mbd. The tool may have been used for straightening bone or wooden shafts, or as a polishing stone.

Especially interesting are the three intact marble vessels recovered from the lowest stratum of Kültepe. As has been previously stated, all three were found together in a well-defined stratigraphic context on the floor in the easternmost subdivision of the large house. This group consists of: (1) a deep bowl made from a yellowish marble with a well-defined flat base, slightly flaring walls and a simple rim. The diameter is 7.8 cm at the bottom and 18 cm at the rim, and the bowl is 7.2 cm high (Fig. 4.6); (2) a spherical vessel of pink marble with a round bottom, convex walls, and a narrow, flaring rim. The diameter of the vessel is 13 cm at the rim, and approximately 11 cm at the base with a height of 10 cm. The vessel's interior has been carefully carved out, leaving thin walls (Fig. 4.7:1); and (3) the most distinctive vessel is a flat-based oval bowl of white marble. Its base is oval-shaped with the dimensions of 9.5 cm × 7.5 cm, and its rim is an oval of 12 cm × 8.5 cm and its height is 11 cm. The walls are almost vertical and the rim is defined by a small groove around the edge of the vessel. The vessel is asymmetrical and massive, with thick walls and base (Fig. 4.7:2).

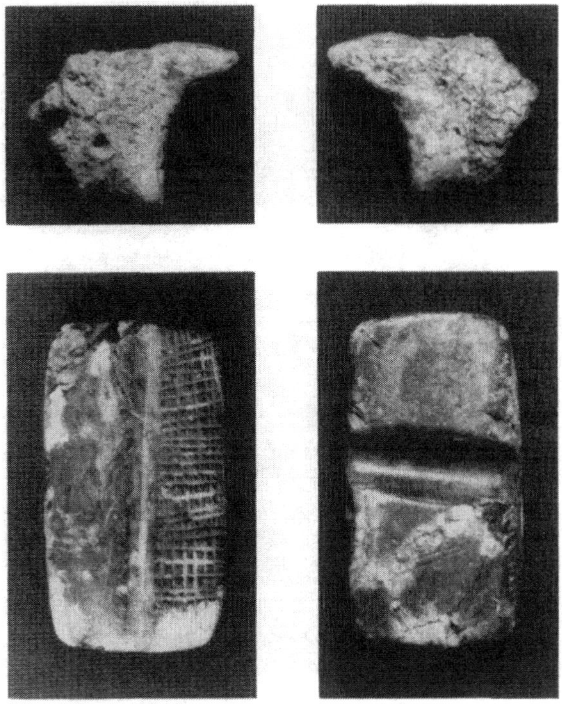

Figure 4.5. Level 1, ceramic bull head and black stone slab with cross-hatch incisions

The stone vessels of the lowest level of Kültepe described above have close analogs at Tell Sotto, Umm Dabaghiyah, and Tell es-Sawwan. The yellow marble bowl from Kültepe is analogous to examples from Tell Sotto and Umm Dabaghiyah. Certain stone vessels of red marble or of pink marble with red veins, identical to vessels found at Kültepe, are well known from Umm Dabaghiyah and

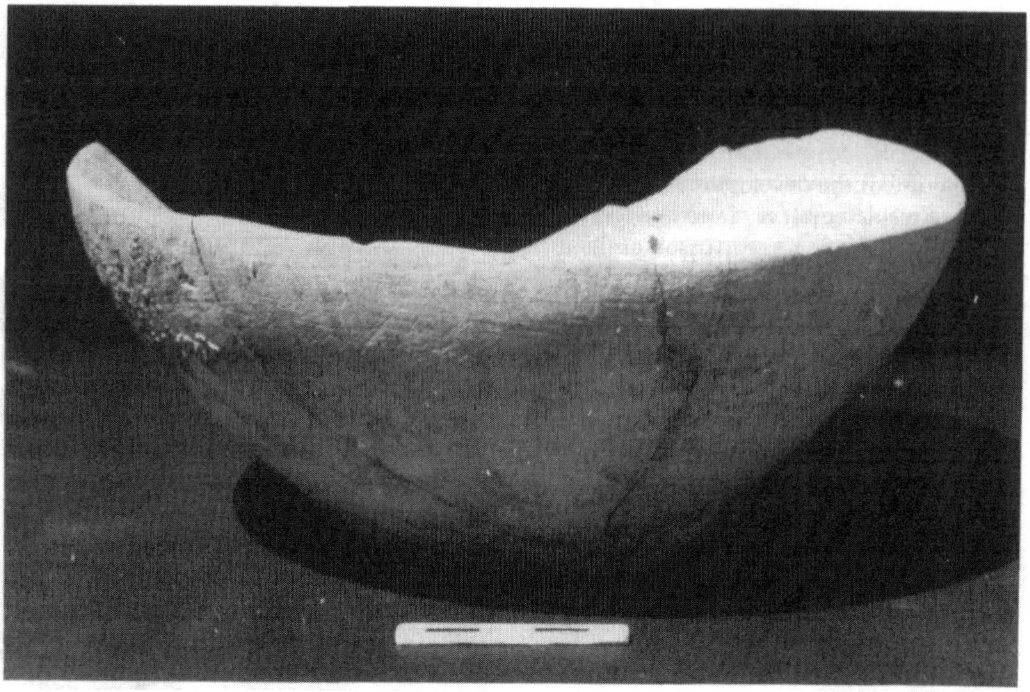

Figure 4.6. Level 1, yellow marble bowl

Figure 4.7:1,2. Level 1, marble vessels

Tell Sotto. All three sites have practically identical forms of vessels of this type, but it is not encountered at Tell es-Sawwan. However, the oval white marble vessel from Kültepe is reminiscent of oval vessels frequently found at Tell es-Sawwan. At Umm Dabaghiyah, one whole vessel of a similar shape was found. Many oval ceramic vessels are known from Tell Sotto, but stone vessels of this shape were not found there. It is interesting to note that the lowest stratum of Tell es-Sawwan yields a large quantity of coarse, ribbed cooking ware which is on the whole similar to the pottery of the Hassuna phase. Detailed comparisons of these assemblages have not been undertaken. Although the early Hassuna sites in the Jezira yield one type of stone vessel, the lowest stratum of Tell es-Sawwan apparently yields an entirely different assemblage. However, both assemblages are seen at Kültepe, suggesting that they came from a common tradition.

The excavations at Kültepe have further confirmed the view that the lowest level Tell Sotto is an early phase of the Hassuna culture. At this time, a similar interpretation can be given to the material remains of Umm Dabaghiyah and Telul eth-Thalathat.

It is quite probable that sites of the Tell Sotto type reflect the settlement of the northern Mesopotamian plain by early agriculturalists. Parallels between the lowest strata of Kültepe and Tell es-Sawwan show that this process proceeded along similar trajectories over a considerable territory within the Zagros foothills. In order to answer the question concerning the origin of the Hassuna and Samarra cultures, sites earlier than Kültepe, including aceramic settlements, must be examined. The aceramic site of Tell Maghzaliyah, on the slopes of the Jebel Sinjar, is an example of one of these initial settlements that may provide such an answer (Bader 1979; Munchaev and Bader 1979).

5

SUMMARY OF *THE EARLIEST AGRICULTURALISTS OF NORTHERN MESOPOTAMIA* (1989)

N. O. Bader

Prior to the work of the Soviet expedition, Jarmo was the only aceramic Neolithic site known in northern Iraq, and the Hassuna culture was known only from the excavations at the site of Tell Hassuna. The new materials uncovered by the Soviet expedition, together with the results of excavations by the British expedition at Umm Dabaghiyah and by the Japanese at Telul eth-Thalathat, have dramatically increased the amount of information pertaining to these periods and have thus altered our conceptions of the development of agriculture and animal husbandry in northern Mesopotamia.

History and Research Techniques

The basic goal of the expedition was to study the development of food production in northern Mesopotamia. Chosen as research area was the plain adjacent to the Jebel Sinjar, which, although little studied, was known to lie at the crossroads of both the north-south and east-west routes of communication in West Asia. Before this work, the earliest village site known in the Assyrian steppe was Tell Hassuna and, therefore, the Hassuna settlement Yarim Tepe I was chosen as the main research object of the expedition.

As a result of the archaeological surveys conducted over a territory of 100–150 square kilometers around the expedition's base at Yarim Tepe more than fifty Ubaid, Halaf, Hassuna, and pre-Hassuna sites were registered. About a dozen small tells, hardly visible above the plain, but with archaic pottery on the surface, were explored. In 1971 and 1973–75 the expedition excavated one of these, called Sotto. Under its archaic Hassuna level, a small, early agricultural settlement with mud brick architecture and extremely primitive ceramics was discovered. The 1976 excavations at Kültepe revealed, on the whole, material similar to that of Tell Sotto and confirmed the connection between the Sotto complex and the subsequent archaic Hassuna.

Studies of the Sotto-type sites have pushed the emergence of settled villages in northern Mesopotamia back in time but still did not reach its origins. Our careful surveys of the plain failed to reveal any older settlements there. It was logical, therefore, to expect that the Sotto-type settlements were the earliest in the plain and that prior sedentary activities should be located in the hilly spurs of the surrounding mountains. This supposition was confirmed by the discovery in 1977 of the fortified aceramic settlement Tell Maghzaliyah in the upper reaches of the river Abra. Tell Maghzaliyah was extensively investigated from 1977 to 1980.

This series of early agricultural settlements enables us to study the economy and social relations during the Neolithic. Of special interest are the plans of structures and settlements, the succession of building and reconstruction levels to which we paid particular attention. The sites with mud brick architecture were excavated in squares of 2.5 m × 2.5 m, level by level, with all the sections recorded at intervals of 2.5 m. Tell Maghzaliyah, with its large stone structures, was excavated in squares of 5 m × 5 m with sections at intervals of 5 m.

Since the work on the larger sites continued for several years, we were able to devote special attention to studying the sections in laboratory conditions. The results thus obtained were summarized in stratigraphic diagrams that reflected the succession of buildings and the relation among the main structures over time.

Each site's cultural deposit was subdivided into occupation surfaces, cultural and natural levels, and cultural periods of each site's settlement history. A cultural level is determined by a complex of contemporaneous structures, with changes in levels defined by significant reconstructions, whereas periods are marked by drastic rearrangements in the site plan as a whole.

All archaeological material was collected by structures and the stratigraphic levels connected with them. Then, during analysis, the data were classed into phases and periods. This thorough procedure has made it possible to follow in detail the development of culture, the economy, and social life.

Tell Maghzaliyah: An Early Agricultural Settlement of the Seventh Millennium B.C.

The settlement is located 7.5 km northeast of Yarim Tepe on the right bank of the Abra River, approximately one kilometer away from the point where it comes to the plain and 50–70 m above it. The tell stretches along the river which runs in a deep canyon at this point. After the settlement was abandoned, its eastern slope was eroded by the river and its northern slope was damaged by a ravine and a road crossing it. The surviving part of the settlement occupies some 4500 square meters. The cultural stratum is 8.2 m deep in the center of the settlement.

The three areas of the tell we investigated were at its northern part, on the eastern slope, and at the center. We joined all operations into a single excavation of 625 square meters. Altogether, the excavations include about one-tenth of the tell's surface area. Three hundred twenty-five square meters were excavated at the top of the tell. This was the latest settlement, which extends beyond the defensive wall, occupying some 1500 square meters. Thus, about one-fourth of the total area of this occupation was studied.

The tell is formed entirely of mud brick ruins on stone foundations. As a result of the work, the depth of cultural deposits was divided into sixteen building levels, each being some 50 cm thick on the average, with the surface material comprising a seventeenth level. The record of accumulation of a layer is actually determined by the life of the structures that form it. It is unlikely that the large residences of Tell Maghzaliyah stood for more than fifty years each, whereas smaller ones were remodeled more frequently. According to these estimates, the duration of the settlement can be calculated to be about seven hundred years. No cultural discontinuities have been detected in the stratigraphic sequence.

At present three periods of the settlement can be distinguished. The earliest settlement

(Levels 1–5) was formed by a chain of dwellings along the river bank. Structures in Levels 6–12, by contrast, run mostly, not along, but at an angle to the river, and on the strength of this attribute they form the second period of the settlement's existence. The plan of the settlement of that time, however, remains unclear.

The third period includes Levels 13–16 at the top of the tell. At this time the settlement consisted of rectangular multiroomed structures that were oriented along the river. Significantly, the settlement was surrounded by a defensive wall (Levels 13–14). The wall did not continue into Levels 15 and 16, during which the site was largely depopulated.

Typical of the lower part of the cultural stratum are small rectangular single-room houses of 10–12 square meters. Their wall socles were made of stone, floors were faced with fine stones and compacted clay or, sometimes, gypsum plaster. In house 120 we recovered a large oven with a gravel foundation and a flue which led to the outside through the outer wall of the house. Next to dwellings large and deep oval hearths filled with ashes and stones were frequently found. Even in the first period, along with the smaller dwellings, some larger structures (205, 216) were found. In Levels 5 and 6 of square M-8 supporting walls, presumably of structures, were traced along the edge of the settlement (185, 191); in Level 12 remnants of the first wall around the settlement (174) were found.

The settlement at the top of the tell consisted of large, rectangular, multiroomed houses of up to 100 square meters each, and the entire site was surrounded by an impressive wall. All the structures had stone socles and the walls were usually 50–60 cm high and up to 80 cm thick. The most commonly used building technique consisted of laying three rows of stones, with the outer and inner rows made of carefully faced, large limestone slabs, and then filling the space between with smaller stones. Floor foundations were usually filled with rubble and small stones and were mud-plastered. The floor and wall surfaces were plastered with several layers of gypsum. Frequently found among the debris of the structures were pieces of gypsum with imprints of reed mats on them, presumably the remains of roofs. In smaller structures only the outer surface of socles was sometimes plastered, and all the inner space for the floor was filled with small stones flush with the masonry to form a socle projecting above the ground. We have recorded only one case in which the socle masonry was sunk below the ground, and room 253/258 is made of upright slabs. The walls resting on stone socles were made in the pisé technique.

The economic data indicate that rooms were used either for living, food preparation, or storage. Surviving in room 239 (house 2) is a gypsum-plastered partition for grain and a small open hearth faced with stones and mud-plastered. In room 249 (house 3), by tracing the remnants of gypsum plastering, we were able to define a bench and several storage niches in the wall. The habitation rooms were usually 10–12 square meters in area, and often found next to them were long rooms, not more than 1–1.5 m wide with clean plaster floors but with no hearths. These seem to have been used for storage. An almost square room (232) in house 1 occupies an area of almost 25 square meters. Its roof was supported by posts whose symmetrically arranged postholes have survived on the floor. A room of similar dimensions (242) probably existed in house 2. In Level 15 at the northern edge of the settlement a rectangular platform with a related outbuilding has been found. Its floor was lined with limestone slabs and gypsum plaster, in which postholes can also be traced.

The settlement of Levels 13 and 14 was surrounded by a wall which has been exposed over a length of 60 m. Outside the excavation, in the southwestern part of the settlement, the large stones of its foundation are traceable on the tell's surface. In the northern sector of the settlement, a horseshoe projection in the wall (structures 223 and 236) forms a possible tower, and in the western sector there was a gate into the settlement (structure 271). Having been built according to a single plan, this impressive wall surrounded the houses, but was structurally apart from them. During the reconstruction of the dwellings, the wall was reconstructed as well. Different parts of the wall were built in different techniques and from stones of different sizes. These differences, however, do not disrupt the overall unity of its construction, since the different sections of the wall were, as a rule, bonded together. The "tower" and the part of the wall adjoining it from the west are actually socles made of flat slabs a meter or more across. The part of the wall east of the "tower" and its western part were connected with walls made from small stones. However, it is here that the wall has been traced to the height of 1.8 m. The northern end of the western part of the wall (structure 264) is made of upright blocks up to a meter and a half across. They seem to have been sunk slightly below the surface level. Apart from other uses, the wall performed here the function of supporting other masonry structures, and the territory behind it was leveled for further construction. Remnants of mud brick walls have survived in parts of the masonry foundation. The unity of its plan, its independence from the dwelling units, and its size lead to the conclusion that the encircling wall was built for defensive purposes.

The settlement behind the wall occupied about 1500 square meters. That territory included eight to ten multiroom houses. Each of these would have accommodated from 10–15 people. The population of the settlement probably did not exceed 100–150 persons, including children.

In the third level of squares P-3 and P-4 eight burials were unearthed. All of them were found below the level of masonry structures 117, 122, 123, and 124, and all of them are within the settlement. Topographically, the graves form two groups, northern and southern. The internments were not contemporaneous since the newly dead were interred next to older, but not yet forgotten, burials. The grave pits, which are not more than 40–50 cm deep, were found under the floors of the structures. The pit walls were usually faced with flat limestone slabs. In burials 4 and 7 the heads of the deceased rested on flat stones. In some cases the fill of the graves also contained several stone slabs. Only in burial 5 was a grave offering found, a marble bead located under the upper humerus of the right arm. None of the other burials had any objects at all. Out of the eight interments there were five adults, an adolescent, one child three to five years of age, and one infant. All the bodies were buried in a flexed position. The child's body rested on its right side with the head to the east; of the remainder, four had their heads oriented to the south (two on the right and two on the left side), and two others to the north (one on the right and one on the left side). The infant's orientation was not determined.

The material culture at Tell Maghzaliyah is characterized by a highly developed technique of flint knapping and bone working. All the levels are rich in imported obsidian. The proportion between flint and obsidian gradually changed over time. In the first period of the settlement's existence, obsidian tools and production waste accounted for nearly three-quarters of the total amount of chipped stone. In the middle period (Levels 6–12) the amount of obsidian diminished to two-thirds. At the time of the walled settlement (Level 13) the amounts of flint and obsidian were equal, and beginning in Level 14, flint predominated. It is a very tempting idea to connect the construction of the defensive wall around the settlement with difficulties that arose in the import of raw materials. Perhaps the abandonment of the site was due to the disruption of trade routes that had previously brought obsidian to Maghzaliyah.

Small obsidian flakes were unretouched and used, probably, for composite tools.

However, the settlement has yielded only two geometrical flint inserts. A diagnostic series of artifacts consists of flint projectile points, some with tangs and some leaf-shaped. The edges and hafts of these implements are retouched whereas their lateral edges are not. Grain reaping knives are of two types. The first group includes large unretouched blades that show sickle sheen whereas the second is made of obsidian on which sheen is also visible. End scrapers are more numerous than side scrapers.

Of the groundstone assemblage, mortars and pestles are the most numerous types, some showing traces of ochre. There are also marble vessels, bracelets, pendants, beads, and stone wedge-shaped axes, adzes, maces, and pendants.

Bone working had reached a high technological level. Many awls made of long bone fragments, spatulas of ox ribs, thin bone needles with eyeholes, and long bones with their ends removed (parts of necklaces?) have been found. Some of these bone implements were decorated, and one with a carved head of a gazelle naturally depicted on the finial was found.

Imprints of reed mats and of woven baskets on pieces of plaster and bitumen were frequently encountered. Plastered baskets were apparently the predecessors of ceramic vessels, the latter being rare finds at Maghzaliyah. The settlement has yielded only one sherd and one large storage vessel of unbaked clay.

Represented in the aesthetic category of artifacts are marble bracelets, barrel-shaped marble beads, various stone and obsidian pendants, and perforated stone disks. A carved limestone button-seal with a hole in its center was also found. Two stone pendants worked in the shape of a snail shell were also recovered.

Female and zoomorphic figurines of unbaked or poorly fired clay are also quite numerous. Several of the zoomorphic figurines are horned, and there are many fragments of sitting female figurines. Two intact figurines help to reconstruct the modeling technique. The stylized body has the form of a high cone, and the breasts are represented by oval-shaped protrusions. The top of the figurine is pinched to indicate its head. The bodies of these figurines have parallels at Telul eth-Thalathat and Jarmo, and the legs and torsos are identical with those found at Tell Sotto.

Finally, a cold-hammered copper awl and a few pieces of malachite have been found within the walled settlement. According to spectrographic and trace-element analysis, the copper came from the Talmesi region in Luristan, Iran.

The collection of cereals and fauna from Tell Maghzaliyah indicates a diverse economy. Not only did the inhabitants of Maghzaliyah collect wild barley and pulses, they also practiced agriculture, as the existence of domesticated wheat and barley shows. The osteological assemblage includes ten species of wild and three species of domesticated animals. Of the recovered faunal assemblage, 58.4 percent was comprised of domesticated species. Among the hunted animals were steppe species (e.g., onager), forest-steppe species (aurochs, deer, common gazelle) and mountainous species (goat, mouflon), with onagers the most numerous. The large number of flint projectile points indicates the important role hunting played in the subsistence of the settlement. The sheep and goat bones show only a few morphological traits suggestive of domestication. Thus, animal husbandry was in an early stage of development.

The Proto-Hassuna Settlements of Tell Sotto and Kültepe

As a result of surveys conducted within the radius of 10 km around Yarim Tepe, ten small Hassuna settlements with very archaic ceramics on the surface were discovered. Two of them, Tell Sotto and Kültepe, were excavated.

Tell Sotto

This tell is 3 km west of Yarim Tepe, only 2.5 m above the plain, and its cultural stratum 2.8 m thick. The tell occupies some 3500 square meters, although its central part is not more than 2000 square meters. Through time the settlement became considerably smaller, and the Hassuna settlement at the top of the mound was not more than 500 square meters in area. The tell's earliest settlement was situated under the eastern slope and together with its sparsely developed periphery occupied not more than 500–600 square meters of which 475 square meters have been excavated.

The cultural stratum has been divided into seven building levels, with the eighth consisting of materials collected from the top soil and surface. Levels 1–6 yielded a type of very archaic carinated cooking ware with organic temper, which developed successively from one level to the next, until in the seventh level were found some archaic Hassuna vessels.

The first (lowest) layer is comprised of over a dozen pits of various sizes. Several of these preceded the first structures of the second level, the earliest in the settlement. We have failed to determine the relation of the pits to the structures, but some were probably dwellings. Thus, for instance, at the bottom of pit 11 a vessel was found with several fragments of grain grinders. Many sherds were found in the charcoal and fill layers at the bottom of pit 19. Pit 10 was filled with ashes and ceramic slag and could have been used for baking ceramics.

In Level 2, three contemporary dwelling units and a complex of "reservoirs" connected with them were found, as well as a number of auxiliary structures. The rectangular one-room houses ranged from 12–16 square meters in area and typically had hearths, ovens, compartments for storing grain, and storage vessels dug into the floors. The mud brick walls were *tauf* constructions. The "reservoirs" were shallow plaster-coated pits between parallel mud brick walls. Their purpose can be deduced from the plaster-coated drain found at the butt-end of "reservoir" 84 and the wooden platform over them that was presumably used for drying grain. Similar lattice constructions, but without pits between them, have been discovered in other Hassuna settlements.

Approximately three-quarters of the Level 2 settlement has been exposed. Only in the southwestern part of the tell one or two houses more might have been found; there were not more than four or five houses in the entire settlement. If we assume that each house could accommodate five to seven persons, the total population of the settlement would amount to twenty-five to thirty-five people, including children.

The overlying levels (3–7) do not present a clear picture of the site. Two houses with storage vessels have been recovered in Level 4. Level 6 contained several storage vessels and ovens that indicate domestic economic activities. It is possible that some dwellings were situated in the unexcavated western portion of the tell, but it is also possible that there was no permanent settlement at this time and that the storage facilities and ovens were built by people who only periodically visited the site. It should be pointed out, however, that there are no hiatuses in the stratigraphic sequence of the site.

Nine burials were found in Tell Sotto under the floors of the houses or next to them, many of them in vessels. Level 1 contained one burial; Level 2, six; Level 4, one; and Level 7, one. Eight of the burials were infants or children aged one to three years; six of the bodies were interred in vessels and two in shallow pits. Two of the bodies lay with their heads to the east (on the left and right sides), and one was oriented to the west (the body lay on its right side), whereas the bodies interred in vessels were not oriented along any of the cardinal points. In three cases the inhumations were secondary, and the remainder were primary and tightly contracted. In one of the pits the body was carelessly thrown in head first. Accompanying one body was a crude clay cup, whereas another burial had a small basin with animal bones in it (remnants of food) and a rich necklace made from stone beads, including marble and lapis lazuli, and a copper plate bent into a small tube. None of the other burials had any associated grave goods. Only one adolescent burial was encountered, a youth aged ten to fourteen years. The dismembered body was interred in a pit with no mortuary offerings.

The pottery from Levels 1–6 forms a unified ceramic assemblage. The crude cooking ware was made from straw-tempered clay and the finer wares were made of clay tempered with other organic materials. All the sherd material is poorly fired. A few slipped cherry-colored and gray sherds are mineral tempered. In Levels 7–8 several fragments of dense and highly fired Hassuna ceramics were recovered.

The ceramic assemblage from Tell Sotto has been divided into twelve basic types according to its forms: carinated vessels, spherical vessels with a straight vertical rim, flat, oval troughs for winnowing, troughs with corrugated bottoms, oval troughs with smooth bottoms, flat platters, carinated bowls, rounded bowls, straight-walled basins, round basins, carinated cups, and round cups.

The first and second strata have produced an extensive and diagnostic collection of crude carinated pottery with diverse anthropomorphic and zoomorphic figures molded onto them. Typologically, these clearly precede the archaic pottery of Hassuna, and have been grouped by us into the "Tell Sotto-Umm Dabaghiyah" complex. In Levels 3–6 one can trace the evolution from carinated to spherical vessel forms. Level 7 has yielded several large, crude, spherical vessels that are characteristic of the archaic Hassuna complex.

In the lithic assemblage microlithic production was on the decline. Stone tools were primarily made from local material, and obsidian was very rare. Only one flint projectile point has been found, and scrapers are few. Awls were made from fragments of long bones. There are also several small polished axes and adzes, marble vessels and bracelets, and beads of stone and clay. Clay was widely used for various artifacts, including sling balls, spindle whorls, and female figurines.

In terms of the subsistence economy barley is common and sheep and goat are now fully domesticated.

Kültepe

The site, actually consisting of two tells, is found in the plain on the left bank of the Wadi Sharaya, 6 km due west of Yarim Tepe. The locality was named for the tell immediately adjacent to the wadi, but we excavated the second site 100 m to the southeast. Kültepe II is 60 m × 80 m, 2 m in height above the plain, and is situated on a slope that inclines slightly toward the river. The excavations began with a stratigraphic trench that was later expanded into a total area of 120 square meters. The entire cultural stratum of 2.25 m in depth has been divided into four building levels. No hiatuses have been detected in the stratigraphic record.

Unearthed in Level 1 was a portion of a dwelling house, 14 square meters in area with a grain storage compartment and a storage vessel dug into the floor. Adjoining the house were several small rectangular storage structures. Under the floor of one of these, three beautiful marble vessels were uncovered. Level 2 includes several hearths and large ovens with large storage vessels around them. The size of the ovens and the extent of ash around them seems substantial. Since the adjoining territory of the tell is too small to accommodate any large houses, it is possible that the tell was not inhabited permanently but was occupied during the agricultural season. A similar scenario has been envisioned for Level 6 of Tell Sotto. Among the poorly preserved structures in Levels 3–4 were small one-room houses with storage vessels, compartments for storing grain, and querns.

The Tell Sotto and Kültepe ceramic assemblages belong unquestionably to one and the same culture, but Kültepe was occupied somewhat later than Tell Sotto. Levels 1–4 of Kültepe correspond to Levels 3–6 of Tell Sotto and characterize the final stages in the development of the Tell Sotto complex. The inventory of tools at Tell Sotto and Kültepe are very close typologically. The stone vessels found at Kültepe have direct analogues at Tell Sotto and Umm Dabaghiyah.

The Economy of Northern Mesopotamia in the Seventh to Sixth Millennia B.C.

This section provides an overview of the early agricultural settlements of northern Mesopotamia (including Qalat Jarmo, Gerd Ali Aga, Tell Hassuna, Tell Shimshara, Telul eth-Thalathat, Çayönü Tepesi, Bouqras) which are compared to Tell Maghzaliyah, Tell Sotto and Kültepe. An attempt is made to determine the historical-cultural continuities and to reconstruct the cultural ties of the population of north Mesopotamia in the seventh to sixth millennia B.C.

Tell Maghzaliyah and Jarmo are physically close and display a certain cultural similarity. We can even suppose some contacts between the populations of these sites, although differences between them are significant. Thus, for example, no projectile points typical of Maghzaliyah are known in Jarmo. Maghzaliyah belongs, rather, to the westerly group of sites that includes Çayönü Tepesi, Jafar Hüyük, Gritille, and Shimshara. These sites are all in the foothill zone of the Zagros and Taurus mountains and on the routes through which raw materials passed. The obsidian blades, marble bracelets, bone needles, and leaf-shaped projectile points that are distributed throughout this area provide evidence about the constant ties among people and probably constitute a historical and cultural entity.

The research at Tell Sotto and Kültepe together, with the materials from Umm Dabaghiyah and Telul eth-Thalathat, have brought to light new information about these early agricultural sites of the sixth millennium B.C. that preceded the archaic Hassuna of Tell Hassuna. Tell Sotto and Kültepe have made it possible to trace the evolution of this culture and its connection to the culture of the archaic Hassuna. In fact, Tell Sotto- and Umm Dabaghiyah-type sites represent the earliest stage of the Hassuna culture or the proto-Hassuna.

The spread of Tell Sotto-type sites over the plain was connected with the adoption of an agricultural way of life and the spread of early farmers across the northern Mesopotamia plain.

Between the Tell Sotto-type sites and Tell Maghzaliyah there are some distinct traits of similarity, but there is also a certain chronological gap. Pointing to the local origin of the Hassuna culture is the presence of crude organic-tempered ceramics that is typical to sites of the Assyrian steppe. The western origin of sites of the Tell Sotto and the Umm Dabaghiyah type is based on imported ceramics from the west that contain mineral temper, and that are absolutely atypical of the area.

On the Euphrates the Tell Sotto complex has parallels in the materials of the ceramic levels from Buqras, which have also yielded several crude vessels with legs. In the Zagros and along the Tigris similar material is found in ceramic Jarmo and Ali Aga, and in the lower levels of Tell es-Sawwan, Shimshara, and in the Mandali area. Present throughout this wide area are crude, carinated, organic-tempered pottery. However, for all the great similarities, the cultures of the Assyrian steppes and of the foothills of the Zagros were not homogeneous. Thus, for instance, a typical Jarmo pot with holes is represented in Tell Sotto by only a single sample on a stone vessel. Clay figurines of these sites are also different.

Sites of the Tell Sotto and the Umm Dabaghiyah type found on the right bank of the Tigris and those of the ceramic Jarmo type on the left may represent variations of the early agrarian culture of Mesopotamia of the early sixth millennium B.C. The Tell Sotto-type sites gave rise to the Hassuna culture, whereas from sites on the left bank of the Tigris the Samarra culture was formed.

Origins of Food Production in the Near East

This section [of Bader's book] is devoted to the physicogeographical conditions, historical and socioeconomic prerequisites for domestication and the initial stages of food production. The author explains the emergence of agriculture and animal husbandry as an historical phenomenon and discusses the socioeconomic regularities of this development.

The emergence of a food-producing economy is not treated as a gradual and seamless process. One distinguishes, rather, the formation of preconditions, a time of transition, and the early consequences of the reliance on food production in the Neolithic. The accumulation of prerequisites first proceeded at a slow rate within the framework of a hunting and gathering economy; the transition to food production was characterized by relatively rapid qualitative changes; and the progressive reliance on agriculture and animal husbandry was accompanied by a radical reformulation of the entire economy and society.

Although the groups practicing a complex economy of hunter/gatherer subsistence survived to a certain degree through the Neolithic revolution, a sedentary way of life is distinctly traceable in the archaeological record. The process of sedentism is marked by the intensification of agricultural activities and by the expansion into the fertile plain by these early farmers. This process also included the development of regional cultures and the networks of exchange between them that make them part of a larger interregional system.

Conclusion

The work of the Soviet archaeological expedition at the sites of Tell Maghzaliyah, Tell Sotto, and Kültepe has made it possible to follow the closing stages of the Neolithic Revolution, the initial phase of the settlement of the plain by agriculturalists, and the formation of a new type of economy and society.

The settlements of Maghzaliyah, Çayönü, Jafar Hüyük, Gritille, and Shimshara found in the foothill zone of western Zagros and south Taurus, were linked by the exchange of raw materials and the development of cultural ties.

The sites of the Tell Sotto and the Umm Dabaghiyah type characterize the earliest, previously unknown, phase of the Hassuna culture. In the sixth to fourth millennia B.C. northern Mesopotamia was embarking on a complicated historical path, one that resulted in the transformation of the Hassuna, Halaf, and Ubaid cultures into the emergent Mesopotamian states.

6

YARIM TEPE I

N. Ya. Merpert and R. M. Munchaev

In comparison with other regions of Mesopotamia the Sinjar Plain has not been adequately explored. Archaeological investigation was first undertaken here in the late thirties of this century, under the direction of Professor Seton Lloyd, who identified more than two hundred sites and conducted limited excavations on single sites of relatively late date. On the basis of excavated material he distinguished twelve chronological groups among the pottery found on survey. Of these, Group I, the earliest, he compared with Samarra and Nineveh 2 wares dating to the latter part of the Hassuna period, which had not yet been identified (Lloyd 1938, 128). Group 2 he associated with the Halaf culture, noting that sites of this phase are extremely rare in the region by contrast with their frequency east of the Tigris, where Ubaid sites were found to be numerous (Lloyd 1938, 124–26). In the past twenty years investigation of the area has been continued by David and Joan Oates. The number of known sites has more than doubled, and material has been found that might be related to the Jarmo culture of the seventh millennium B.C., one of the earliest stages in the development of agriculture.[1] But excavation has again been confined to relatively late sites, notably the second millennium town of Tell al Rimah, and the third millennium town of Tell Taya, excavated by Dr. Julian Reade (D. Oates 1965; Reade 1968, 1971). However, no early agricultural settlements have been investigated, and the Soviet archaeological expedition set out to remedy this omission.

As the prime objective of our work we selected a group of tells situated 10 km southwest of Tell Afar and locally known as Yarim Tepe. This group was first mentioned by Lloyd as number 34 in his list of sites, although he refers only to a "low mound with distinctive pottery," among which he identified Samarra, Halaf, and a few northern Ubaid sherds (Lloyd 1938, 138). There are in all six mounds forming a compact group less than 1 km in extent. Three of these are of considerable size and have produced abundant material indicating that they represent settlements of three of the most important early agricultural cultures of northern Mesopotamia, Hassuna, Halaf and Ubaid. During the first three seasons, from 1969 to 1971, excavation has concentrated on two of these, Yarim Tepe I and II, dating to the Hassuna and Halaf respectively.

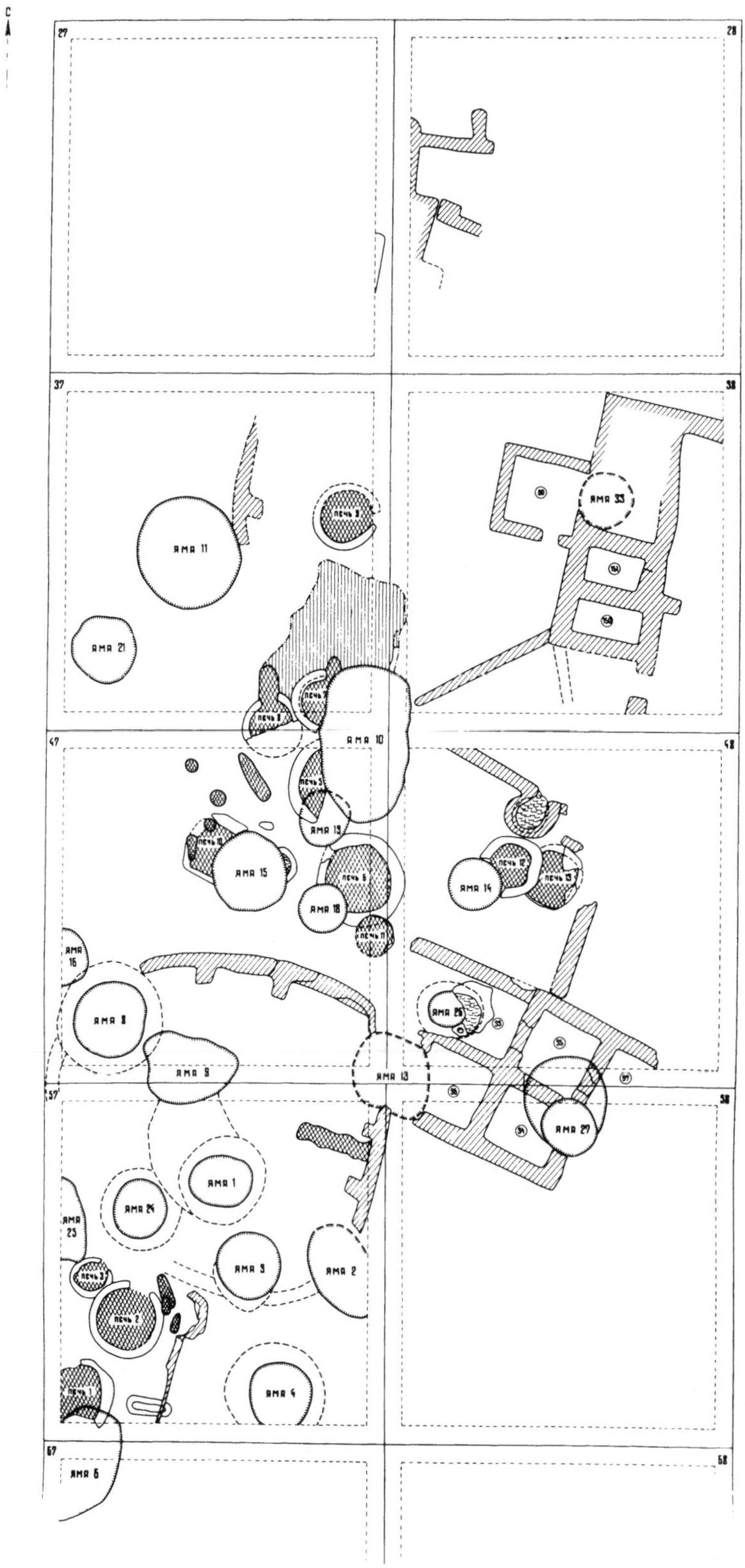

Figure 6.1. Level 4 in the central part of the settlement with a 10 m grid. Key: печь = oven; яма = pit.

a. Yarim Tepe I: The Upper Hassuna Levels

Yarim Tepe I, the extreme eastern mound of the group, is located at a distance of 250 m from the west bank of the brook, called Joubara Diariasi. It has a maximum diameter of 100 m and rises 4.5 m above the present field level, and the surface pottery is wholly of Hassuna type, with the exception of a few sherds. Investigation of the mound and its surroundings, and in particular the cutting of a trench down the slope for stratigraphic purposes, revealed that the occupation levels extend beyond the apparent limits of the mound at a depth of 1.5 m or more below the present surface of the surrounding fields, and that the total depth of deposit exceeds 6 m. Clearly, there has been a rise in the level of the surrounding land associated with a rise in the water table. At the period of its maximum extension the area of the settlement reached two hectares, and its dimensions are thus closely comparable with Tell Hassuna, the type site. The elevation of Tell Hassuna (Lloyd and Safar 1945, 271–75, Fig. 35) above the bedrock of the neighboring stream is 7 m, but the actual depth of Hassuna occupation is 5.2 m, overlaid by Halaf, Ubaid, and Late Assyrian deposits, and it covers an area some 200 m × 150 m. Of the fifteen building levels identified at Hassuna, five belong wholly to the Hassuna period, and Halaf pottery makes its first appearance in the sixth. At Yarim Tepe I the whole deposit is of the Hassuna culture, and in the stratigraphic trench eight major building levels, each incorporating several subphases, have been distinguished; although there are a number of graves and storage pits of later periods, no trace of later occupation levels was found. The number of building levels can only be determined after the excavation of considerable areas of the settlement and will no doubt be greater in the center of the mound than in the stratigraphic trench. Level 8, reached there in 1971, is not the earliest. The levels are numbered from top to bottom.

Excavation began in the eastern sector of the mound, and by the end of the third season 1440 square meters had been exposed; work proceeded on a grid of 10 m squares (numbered from 1 to 100, increasing from west to east and north to south) with intermediate balks at intervals of 2.5 m to ensure a complete stratigraphic record of the relationship between all major building complexes and the minor alterations and additions that were made to them. In defining successive "levels" of occupation we have relied on architectural relationships, observed both in the foundations, walls, and associated floors of individual buildings and the gypsum surfaces of the open spaces between them; these are discussed in more detail below. Each of the levels represents a considerable lapse of time, and is marked by consistency in the overall layout of large architectural units despite modifications of detail that are often extremely complex. The transition from one level to another is, conversely, characterized by a contemporary replanning of large units in different parts of the settlement, although it may be bridged by other units. These transitional units were often those founded in the later phases of one level, and survive into the next with, at most, some internal remodeling. Other lesser evidence of architectural continuity are also found—the reuse of individual rooms or even single walls within a unit, or the leveling of a structure that is then packed with debris to provide a solid foundation for its successor. It must therefore be emphasized that the precise definition of building levels may vary from unit to unit and that the detailed sequences sometimes have only a local validity.

Level 1: 0.0–0.7 mbd (meters below datum). The uppermost building level is represented only by isolated wall remnants immediately below the surface on the crest of the mound; on the flanks it has been entirely destroyed by erosion and ploughing. Only in squares 37 and 47 can fragments of a coherent plan be discerned, representing a multiroom complex not less than 15 m in overall length, aligned from north to south with an adjoining open courtyard, of which only a small area on the west side is preserved. Within the complex can be traced a rectangular room 3.5 m × 2.5 m, apparently living quarters, two subsidiary chambers 2.5 m square and parts of two other rectangular rooms; in one of the rooms were the remains of a round kiln.

Levels 2 and 3: 0.7–1.0 mbd and 1.0–1.4 mbd respectively. Within the excavated area the levels seem to reflect earlier versions of the Level 1 plan, although some reconstructions were observed. Structural remains of these levels occurred in most squares but their state of preservation was again poor; particularly in the central sector they were badly damaged, not only by erosion, but by later burials and domestic rubbish pits of varying date. Square 47 was a rectangular room of the same dimensions, 3.5 m × 2.5 m, and approximately in the same position as the Level 1 room already referred to, with an oven in its southwest corner. A door in the north wall gave access to an internal courtyard, and on this side there were also subsidiary structures, including very small storerooms, 1.5 m × 1.2 m and 1.0 m × 0.7 m, and a blind alley 50 cm wide and at least 3.0 m long. The living room was connected by another doorway to a second room of the same length but only 1.5 m in width, from which an external door opened southwards onto a passage over 4.0 m wide. The passage separated this unit from its neighbor, of which a pair of rooms, each 3.5 m × 3.0 m, was uncovered in square 57. At this time the central part of the settlement was intensively occupied. In square 46 was found a complex of structures including small rectangular store rooms, 1.5 m × 1.0 m, as well as long corridors, 4.0 m × 1.5 m. One of these corridors led to a round oven of tannur type, and in the other a large jar for the storage of grain or water, 70 cm in height and diameter, had been set into the floor and packed around with stones. Ovens were found both in the rooms and in the small internal courtyards. One observation of special interest may be made here, although it applies to all the upper levels. The passages and open spaces that separated the buildings had been repeatedly surfaced with gypsum plaster to form a kind of road surface. In square 57 between Levels 3 and 4 there were five such plastered surfaces, each 2–3 cm thick, at intervals of 4–8 cm, and similar sequences have been found in Level 5. Equating these surfaces with the associated structures has enabled us to establish a "microstratigraphy," giving a framework of relatively short subphases for the dating of both buildings and finds. The discovery also reflects a high degree of organization in the measures taken by the community to cope with the rainy climate of this part of Mesopotamia in the winter and spring months.

Level 4 was again seriously disturbed by later pits and graves, but structural remains have been found in eight squares (Fig. 6.1), and shed light on specific features of the development of the eastern half of the settlement at this time. In two areas, the southwest corner of square 57 and the middle of squares 37, 47, and 48 were found at least thirteen large kilns, originally domed, some of which were more than 2.0 m in diameter. They were situated outside the houses, although they may have been sheltered by some flimsy structure, and clearly served some industrial purpose, probably the firing of pottery (since kiln wasters were discovered nearby). Near the kilns were several round or oval pits, up to 50 cm across and 1.0 m long, filled with ashes and charcoal, and beside the kiln in square 57 was a well-preserved trough-like construction of clay, oval in shape (1.5 m × 0.5 m). Not all of the kilns were in use at the same time, but their arrangement in groups, sometimes marked off by boundary walls, suggests that this was an established industrial quarter, and indeed, Level 5 kilns were found in the same area.

The central group of kilns was flanked on the north, east and south by large houses, each with several rooms. The most character-

Figure 6.2:1. Yarim Tepe I. Levels 4–5

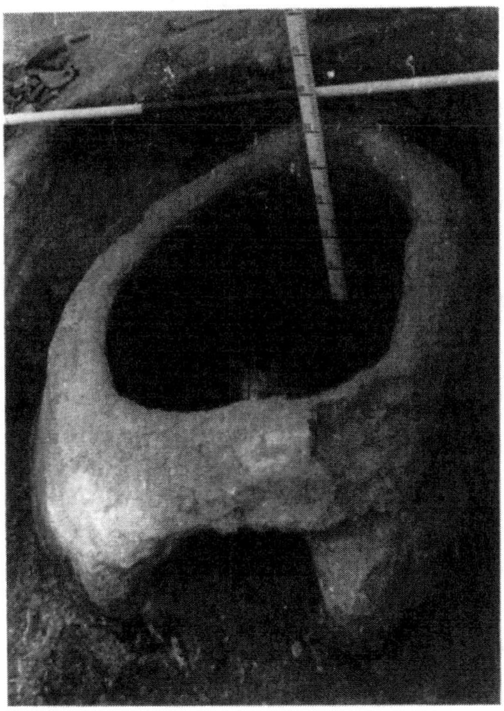

Figure 6:2:2. Round oven, Level 4

Figure 6:2:3. Yarim Tepe I. Two-stage pottery kiln, level 7.

istic is complex 7, in squares 48 and 58. It was oriented from southeast to northwest and consisted of two rows of small rooms of approximately the same size, 2.2 m × 1.75 m, connected by doors. In one of the rooms there was a round oven of tannur type (Fig. 6.2:2), in front of which a stone mortar for pounding grain had been plastered into a clay base. This house originally contained not less than six rooms, and measured at least 8.0 m × 5.0 m overall. Farther to the north, in squares 38 and 28, the houses of this level slope downhill, indicating that the site was already a mound, and in two other areas we found evidence of the architectural continuity between levels as mentioned above. In the northeastern sector, squares 27 and 37, the main features of a layout already elaborated in Level 5 were preserved in Level 4, while in the central area, square 47, the buildings of Level 5 survived into the early phase of Level 4, but were then superseded by a quite different, almost circular, structure. This building is 10 m in diameter, with the stubs of internal partitions surviving against the inner face of the perimeter wall (Fig. 6.2:1); it directly adjoins the southern group of kilns and presumably served some special function, though there are no relevant parallels on known Hassuna sites; it affords an instance of the variety and complexity of Hassuna architecture. (Round buildings are known from Hassuna Stratum 1c, but they are very different from ours in

Figure 6.3. Yarim Tepe I. Building plan of Level 5. Roman numerals indicate building complexes.

size and, apparently, in purpose. See Lloyd and Safar 1945, 272 and Fig. 28).

Level 5: None of the later graves and few of the pits penetrated to the fifth building level, which has consequently yielded a relatively complete picture of the layout of the settlement (Fig. 6.3). In this respect it may be compared with Level 5 at Hassuna itself, which was far better preserved than the other levels and also yielded basic information about the planning and construction of the settlement (Lloyd and Safar 1945, 274–75). In Yarim Tepe I the house walls were preserved to a height of more than 1 m, and from the structural evidence we were able to plan units in detail and to identify the stages of their development, from the initial nucleus through the later additions. This yielded not only the history of individual complexes but also a picture of the development of whole areas. In the 1400 square meters so far exposed (the buildings of Level 5 have been discovered throughout this area with the exception of the stratigraphic trench), no less than twelve large buildings have come to light, all of them containing numbers of rooms; certain common elements of planning can be discerned. The initial nuclei were oriented with respect to one another, often with their long axes parallel and separated by open spaces in the form of passages or small courts, plastered with gypsum. But the growth of the population and the subdivision of families into separate units brought about a parallel elaboration of the house plans (see Nissen 1968, 107 for a similar extension of houses in modern Iraqi villages), and to the original nuclei were added new areas both for functional and living purposes. Space for these was gained by encroachment on the passages, courts, and blind alleys, some of which were roofed to serve as rooms. Some separate units merged, density of occupation constantly increased, and the consequent obliteration of the original layout was one of the factors that imposed the need for radical replanning and a new building level. This complex picture of the evolution of a whole area has been built up by the observation of architectural details such as unbonded joints between walls, differences in the level of their foundations and floors, and the relation between structural additions and the sequence of replastering in the courts and passages. The history of doorways has proved particularly revealing, since they usually lead from the living rooms into passages and courtyards, and the provision of a new doorway marks the final stage in the creation of a dwelling unit designed to house a new family.

Most of the structures seem to have been intended for domestic purposes. As a rule one of the rooms of each house contained an oven; the room was usually the largest, with doorways leading to the open air, rarely into other rooms. This may be confidently identified as the principal living room, with the other, smaller chambers serving subsidiary functions, principally the storage of food. These had no ovens, and sometimes lacked doorways; in such cases they may have been bins with access over the top of low partitions or from the flat roof of the buildings, in the manner known from other early agricultural settlements in the Near East (Mellaart 1965, 81ff.; Lloyd and Safar 1945, 273–74) and also modern villages in northern Iraq. Usually in the room containing the oven, or in a chamber directly accessible from it, there was a massive stone mortar for pounding grain, set into the floor and wedged in position with large sherds and stones which were then plastered over. Even when the ovens were located in small open courtyards, the mortars and grinding stones were placed under cover.

Some buildings, however, yielded no ovens or mortars, nor any of the usual debris of domestic occupation, and these must be excluded from the category of living quarters. The best example is a structure in the northeast corner of the settlement (complex 1), 6 m wide and not less than 16 m long overall, oriented from east to west. It contained at least fourteen rooms arranged in two rows, of which twelve are of consistent size, 2.4 m × 2.0 m, whereas the two rooms at the east end are somewhat smaller, 2.0 m × 1.5 m. There was only one doorway, opening into the southwest corner room, and the whole building looks like a storehouse with, to judge by its size, some communal function. It was originally separated from the main residential area to the south by a passage more than 2 m wide, but later additions to the dwelling houses narrowed it to 1 m or even 0.7 m. It should be noted that complex 1 itself resulted from the reconstruction of the northwest sector in the later phase of Level 5; in the earlier phase of the same level its site had been occupied by a house of the normal type. South of the passage lay three other multiroom complexes, numbers 2, 3, and 4. The first two were separated by a north-south passage which was later partitioned into rooms; the original entrances to both complexes were blocked. Complex 4, which existed from a late phase of Level 5 into early Level 4, affords an unusually complete picture of the process of development. It was an irregular quadrilateral, 7 m × 6 m, with two rows of three rooms on the north side terminating in a single long room at the east end. The rooms in the north range varied from 1.25–1.5 m × 1.1–1.2 m and were interconnected. Those in the southern row were slightly larger, but only the southeast room had a doorway, leading into the courtyard that occupied the southern half of the complex. The long east room, 3.5 m × 1.0 m, also opened off the courtyard, and contained an oven and mortar. The courtyard originally measured 6.0 m × 2.5 m, but was later subdivided by a partition wall, and a storeroom about 1.2 m square was built in the southeast corner. To the west of the courtyard was an open space, plastered with gypsum, south of which lay five more complexes. Four of these, numbers 5–9, comprising not less than fifteen rooms, were bonded together by later additions to form a virtually separate unit divided only by internal courtyards. On the east this unit was separated by a passage from the continuation of the residential area, in which a typical complex is number 10, found in squares 48 and 58.

The nucleus of this complex, some 10 m × 6 m and aligned from northwest to southeast, was laid out in the transitional phase between Levels 5 and 6 and consisted of ten or twelve rooms with average dimensions of 1.6 m × 1.5 m. The northwest room, however, was double this size and obviously contained the oven, although it was badly damaged by later construction. There were originally open spaces on the east and south, but the addition of two rooms and boundary walls surrounding courtyards, one with an oven, brought the eastern space within the confines of the unit. The southern space was similarly enclosed, with a number of new rooms around it, one again containing an oven. This addition was sufficiently important to require a change in the plan of the original nucleus, one of whose walls was displaced to make way for it, and both afford excellent examples of the elaboration of an original plan that resulted from the subdivision of family units. A similar process can be observed in complex number 11, immediately to the north. The nucleus, 6 m × 2.5 m and again oriented from northwest to southeast, also dates from the Level 6 to Level 5 transition. It contained four rooms, but the entrance is unusual in Hassuna buildings, consisting of a passage 4.3 m long and 0.7 m wide with an outer doorway at the north end and an inner doorway leading into the north side of one of the rooms. Halfway through Level 5, annexes were added to the east and south sides of the nucleus, which underwent a general reconstruction before the end of this building phase; some walls and partitions were moved, doorways were blocked and a large rectangular oven was built on the site of one of the southern annexes. At the beginning of Level 4 further drastic changes were made. Partition walls were demolished and part of the rooms themselves were filled up to provide a solid foundation for new construction. The outer walls of the house were also rebuilt, but their outline was not substantially altered; the general plan of the nucleus thus survived for the beginning of Level 5 to the end of Level 4. Evidence of Level 5 build-

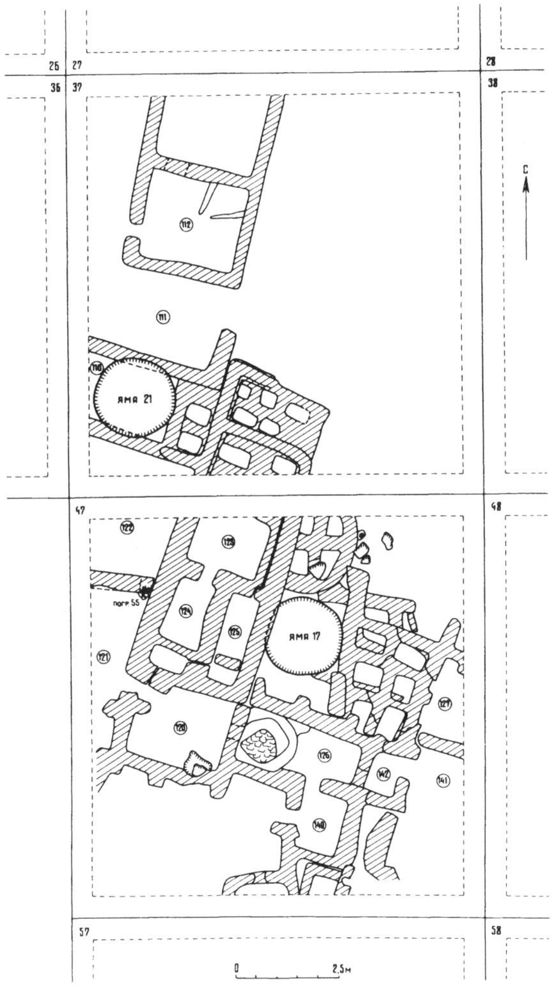

Figure 6.4. Building plan of Level 6 in the central part of the settlement. (яма = pit)

ing activity has also been observed in the northern sector of the settlement, west of complex 1, where the remains of multiroom houses, with elaborate alterations and additions, have come to light; and we have noted above that the alleys and open spaces of Level 5 were frequently replastered with gypsum.

Level 6 presents a no less complicated architectural history in the 400 square meters so far investigated (Fig. 6.4 and Fig. 6.5:1, squares 17, 27, 37, 47). One complex of great interest is number 15 in square 47. It is 10 m long and up to 4 m wide, with a pair of living rooms each 2.5 m × 2.0 m at the south end. In the eastern room was found a vaulted oven, with a firing shaft opening through the partition wall into the next chamber. Domestic quarters were added on the north side of this nucleus that contained accumulations of grinding stones, pestles, sherds of thick-walled storage jars and animal bones. At the north end of this unit was an interesting cellular structure of eight or nine rounded compartments, hollowed out of a stamped clay layer and designed to hold pottery or gypsum containers for foodstuffs. Similar structures, one of semicircular plan and about 2 m in diameter, adjoined the east side of the unit, whereas on the west were large open courtyards (considered by Lloyd and Safar [1945] to be cattle enclosures). The second large complex of Level 6, number 16, was explored in squares 37 and 27. The nucleus was a rectangle 8 m × 6 m, divided into the rooms with interconnecting doorways. The suite was entered by a door in the west wall of the southernmost room, which had an open hearth shielded by clay screens in the northeast corner. In the middle room an unusual installation adjoined the west wall, consisting of a rectangular plastered surface, 48 cm × 35 cm, slightly below floor level and surrounded by a raised clay lip, with a second similar platform a little above it to one side. These may represent the emplacement for a large grinding stone and the working space in front of it. The third, northern room was certainly a kitchen. In the northeast corner was a well-preserved oven, 65 cm × 60 cm at the base, with a dome 20 cm thick and 33 cm high. There was an opening in the top, as is normal in tandir type ovens, and also an ash vent at the base. In the west wall of the kitchen there was a second external doorway. All these structures were associated with the earliest house floor of Level 6, but the floor was then repeatedly raised to keep pace with the overall accumulation of occupation debris on the site, and corresponding changes took place in the internal arrangements. The oven, which

Figure 6.5:1. Level 6 from the south

had been founded at 3.25 mbd, was partly dismantled and filled in, and another constructed on a new floor 30 cm above and obstructing the west door of the room, which was not sealed. Later, but still within Level 6, the floor was raised by a further 15 cm and the oven again removed, this time to the northwest corner. Finally the whole suite was packed with the debris of its own walls, thus preserving their lower parts, and used as the foundation for Level 5 construction. In general, the buildings of Level 6 exhibit some distinctive features, notably the cellular structures which held storage vessels, but the principles and techniques of construction have much in common with Level 5, notably the use of rectangular multiroomed pisé buildings as the nuclear units.

The buildings of Levels 7 and 8, however, show marked differences even in the limited area so far exposed (square 47), and are clearly associated with the earliest development of the settlement and of the Hassuna culture (Fig. 6.5:2). The rectangular buildings are smaller and simpler in plan than in the later levels, with thinner walls and no trace of plaster. Most important, they are found together with round structures; for instance, in Level 8 a two-roomed rectangular house, 10 m × 2.5 m, was almost surrounded by a number of round buildings up to 2.5 m in diameter with, originally, conical roofs. In one of these were found three crouched human skeletons, but the others yielded no trace of burials, and there is no reason to suppose that they were funerary structures. The question of their purpose and the functional difference between them and the rectangular houses is still unsolved, but their relative frequency in the early levels is of great significance, for they represent the continuation of a widespread and ancient Near Eastern tradition that originated when the brushwood shelters of nomadic hunters were transformed into the first permanent dwellings. Round houses, either semisubterranean or constructed of clay and stone above ground are known in Palestine from the Mesolithic proto-agricultural settlements of the Natufian culture at Eynan, Hagoshrim and Beidha, and in prepottery Neolithic contexts at Beidha, Jericho Ia, and Wadi Falla, and in the same period at Mureybit and Ramad in Syria, Zawi Chemi Shanidar in northern Iraq, and Ganj Dareh in southwest Iran. The later appearance of rectangular houses is in many of these cases stratigraphically attested, at Mureybit (van Loon and Skinner 1968, 267–75), Beidha (Kirkbride 1968, 264–65), Ramad (de Contensen 1936, 19), Jericho Ib (Kenyon 1957, 53–54, 71) and Ganj Dareh (Smith 1968, 159 and 1970, 179). On the other hand in certain areas, notably southern Turkey (Braidwood, Çambel, Redman, and Stewart 1974, 1236–40), the rectangular house tradition has been shown by recent investigations to be very ancient, and it is not surprising to find the two types coexisting on several of these sites at certain periods, e.g., Beidha Level 4 and Ganj Dareh I; we postulate a similar explanation for their simultaneous use in the lower levels of Yarim Tepe I. In the later Hassuna culture the rectangular house type was dominant, whereas in other cultures the round plan continued to develop and remained the principal house form, most characteristic of the Halaf culture; many Halaf examples were found in the excavation of our second mound, Yarim Tepe II, discussed in chapter 8. One other unusual architectural feature of the lowest levels of Yarim Tepe I should be mentioned here, a clay platform about 1 m high that was found in Level 7, square 47, near the middle of the settlement. It had been pounded flat, and the base was faced with cakes of clay. It served as the foundation for some domestic structure, probably a granary.

Figure 6.5:2. Levels 7–9 from the south

Architecture and Furnishings — General Observations

The buildings of the six upper levels of Yarim Tepe I, taken as a whole, illustrate the elaborate overall planning and organization of the settlement as well as relatively sophisticated structural techniques that reflect a developed tradition of domestic architecture. All these levels are characterized by large rectangular houses of several rooms. Such houses may be considered characteristic of the more advanced Hassuna and Samarra settlements. At Hassuna they have been found from Level 2 onwards, and houses of Level 5 had eight, nine, or even more rooms (Lloyd and Safar 1945, 273, Fig. 32 and Pl. VIII:1). The houses so far found at Matarrah had up to seven rooms (Braidwood, Smith and Leslie 1952, 6). The large houses in the lower level at Tell es-Sawwan had more than fourteen rooms (el-Wailly and Abu es-Soof 1965, 20, Pls. XIX–XX). Well-planned houses with up to twelve rooms have been discovered at Choga Mami (J. Oates 1969, 115, Pl. XXIV). The walls of the houses and the vaults of the large ovens were built of mud bricks, and in some individual buildings we have been able to detect a regular pattern in the laying of these bricks as well as approximate uniformity of size. This suggests that the development of standard units leading at a late stage to the use of sun-dried bricks took place within the context of Hassuna building techniques. The outer walls of the houses were massive, usually 45–50 cm thick, and were often reinforced by buttresses at the angles and opposite the junctions of internal partitions. Such buttresses are typical of Samarra architecture, and later became a characteristic feature of Sumerian buildings. The authors of the Hassuna report assert that they served not only to strengthen the walls but to support the beams that carried the flat or low-pitched roofs (Lloyd and Safar 1945, 275). The method of roofing, with poles laid across the beams and covered with reed matting and finally a thick layer of clay, is universal in village houses in the area to the present day. The floors of the houses are of hard-packed earth, often speckled with ash and charcoal; floors as well as walls were sometimes coated with clay plaster, and instances of a more elaborate technique have been found, e.g., within house 89, Levels 4–5, in which the plaster was laid on matting, which rested in turn on a bed of small stones and sherds. A standard feature of doorways, known also from Hassuna, is the provision of a high sill of clay, often surfaced with large sherds. In another house, number 90, clay benches were found against the wall of one room (Lloyd and Safar 1945, 274).

Ovens and Kilns

Among the best-preserved and most characteristic elements of domestic equipment are the ovens, used for baking bread and also for heating in the cold north Mesopotamian winters. A number of different types have been found throughout the excavated levels, but they fall broadly into two categories, round or almost rectangular. Among the latter are open hearths with high walls, and vaulted ovens with an aperture taking up almost the whole of one end wall. A well-preserved example of the vaulted oven was found in house 57, Level 5; it measured over 1 m long, 80 cm wide and the vault was 30 cm high. The round ovens include open ring-shaped hearths, tandirs with an upper aperture and probably also an ash vent near the base, as well as the cupola-shaped ovens with side apertures; large pottery kilns of the cupola type are up to 2 m in diameter and consequently cannot have had an upper aperture for access. A well-preserved tandir was found in building number 86 (Level 5). It was 1.45 m in diameter and 50 cm high, with walls 18 cm thick; the floor was well plastered with clay, on which were strewn small stones to improve heat retention, and outside the aperture was a small pit to facilitate removal of the ashes. Another oven of the same type was found in complex 15 in Level 6. In Hassuna the earliest ovens in the shape of "a simple clay barrel tipped towards the ground at an angle of 30 degrees" were found in Stratum Ib (Lloyd and Safar 1945, 267). In the later levels vaulted ovens similar to those of Yarim Tepe I are widespread. The authors of the report call them "bread ovens" and remark that "they are only slightly different from the modern Arab tannur (Lloyd and Safar 1945, 262, 272, 275, etc.) Similar ovens are known from settlements of the Samarra culture. At Matarrah an open hearth was found (Braidwood, Braidwood, Smith, and Leslie 1952, 7). At Tell es-Sawwan vaulted ovens have been discovered, including large kilns for firing pottery. The most notable discoveries, however, were the two-stage pottery kilns found in 1971, in Level 5, square 48, and another in Level 7, square 47. The latter has an internal diameter of 1.2 m and is relatively well preserved (Fig. 6.2:3). The combustion chamber was sunk into the ground and roofed with a massive plate of clay up to 10 cm thick, pierced by some fifty apertures, 15 cm across, which conducted hot air into the domed upper stage containing the pots to be fired. Hitherto, the first occurrence of such sophisticated two-stage kilns had been ascribed to the Uruk period, and even to its latest Jemdet Nasr phase, which the Yarim Tepe example antedates by at least two thousand years.

Burials

In the area so far excavated six burials of infants have been discovered within the houses. Four of them belong to Level 5, one in a small pit beneath a doorsill, two beneath the floor and the fourth in a recess cut in a wall to receive it. In the last case the skeleton lay on a sherd from a large jar, and the recess was blocked by another sherd. One burial was found in Level 4; two newborn children, probably twins, were interred in a small pit dug for the purpose beneath the floor of a room. The sixth grave, also beneath a room floor, was found in Level 7. In all cases the skeletons were contracted, lying on the right side with the head to the east or south. The practice of infant burial beneath house floors was widespread among early agricultural populations, including the earliest Mesopotamian cultures.

At Hassuna, from Stratum 1b onwards, twelve infant burials in jars were discovered, one a pair of twins (Lloyd and Safar 1945, 263, 267–68). At Tell es-Sawwan many infants were buried in shallow gypsum-lined pits in the corners of rooms (el-Wailly and Abu es-Soof 1965, 24). At the same settlement, in buildings 1 and 2 of Level 1, more than 130 contracted skeletons, almost all children, were buried in narrow oval graves beneath the floors, accompanied by a rich inventory (el-Wailly and Abu es-Soof 1965, 23 and al-Adami 1968, 57). Isolated adult burials were also found at Hassuna (Lloyd and Safar 1945, 267, 272), but no cemetery of the Hassuna culture has yet been discovered.

The Pottery

The habitation levels of Yarim Tepe I yielded a large quantity of material, principally pottery, which adds substantially to our knowledge of this most important aspect of Hassuna material culture. Considerable light has been thrown on the earliest Mesopotamian wares by the seventh millennium settlement of Jarmo in the mountains of Iraqi Kurdistan. In the five upper levels (Jarmo II) primitive but recognizable wares made their appearance, superseding a prepottery assemblage (Braidwood and Howe 1960, 43–44). A similar situation can be observed at Tepe Guran in Luristan (Meldgaard, Mortensen, and Thrane 1963, 113ff) and at Ali Kosh in the Deh Luran Plain of southwestern Iran (Hole, Flannery, and Neely 1969, 352). Pottery of Jarmo type has now been discovered over a wide area of the Zagros mountain region, e.g., Tepe Sarab (Braidwood 1960, 695), as well as peripheral sites on the plain, notably Tamerkhan (J. Oates 1968, 3–4). It marks the

Figure 6.6:1–6. Yarim Tepe I ceramic vessels Levels 3–6, scale 2:5

first known stage in the development of pottery in regions bordering on Mesopotamia; the second stage is represented by the Hassuna culture in northern, and Samarra in central Mesopotamia. This evidence points to highly developed and widespread ceramic traditions in the lowlands, linked to some extent with those of the Zagros (Masson 1964, 51). Pottery of Hassuna type was first found at the bottom of the deep sounding at Kuyunjik, Nineveh (Campbell Thompson and Mallowan 1933, 127), in Levels 1 and 2a, but its systematic study was possible only after the excavation of Hassuna itself had produced a much richer assemblage (Lloyd and Safar 1945, 261–62, 276–83). The excavators here distinguished four principal groups, which they labeled "Coarse Neolithic" (Level 1), "archaic" (1b–2), "standard" (3–6) and "Samarra" (Level 3 and above). The same classification has more recently been retained and elaborated by Dr. Takey Dabbagh (1965, 93), and it is in general confirmed by the evidence from Yarim Tepe I. The great majority of pottery discovered here in Levels 1–6 belongs to the Hassuna standard group, with an admixture of the imported Samarra style, but in Level 6 sherds with the gloss and burnished red paint of the archaic group began to appear, and in the lowest level of the stratigraphic trench we found specimens of the Coarse Neolithic wares.

Among the Hassuna standard wares the most characteristic Yarim Tepe I shapes are: (1) open bowls with rounded or flat bases and a curved or sharply carinated transition to straight sides, terminating in straight or everted rims (Dabbagh 1965, Pl. I, types 2–4; Pl. II, types 20–22, 24–35); (2) jars with flat or rounded bases, globular bodies and cylindrical or, more rarely, flaring necks (Dabbagh 1965, Pl. IV, types 53–64); (3) jars with flat or rounded bases and sides with a straight or ogival profile; some very large examples were clearly used for the storage of grain (see Dabbagh 1965, Pl. I, type 15; Pl. II, type 36; Pl. III, types 43–45); and (4) high, flat-bottomed and straight-sided beakers (Dabbagh 1965, Pl. III, type 42).

Also worthy of special mention are the large oval trays of coarse ware, with thick sides and carinated bases. The excavators of Hassuna suggest that these were used in winnowing to remove the husks of wheat and barley (Lloyd and Safar 1945, 262). Sherds of these trays have been found at all levels in association, as at Hassuna, with both archaic and standard wares.

The decoration of the pottery from Yarim Tepe I is in the main characteristic of Hassuna standard ware (Fig. 6.6). By contrast with the archaic material, decorated only by painting and burnishing, it includes three types of ornament, painted, incised, and a combination of painting and incision (Lloyd and Safar 1945, 279). The incised ornament is found on the bowls and, more commonly, the jars. It is executed with a thin stone or bone blade, usually occupies much of the surface of the vessel and employs geometric motifs, lines, chevrons, "fir-tree" designs and triangles with oblique cross-hatching. Painted ornamentation occurs on vessels of almost all shapes and is applied in a range of colors from deep red through pink and brown to black over a light gray, cream, or pink slip. The most common painted motifs are oblique cross-hatching, sometimes in opposed panels, bands, chevrons, solid triangles, and "caterpillars". Painted and incised ornament has been met with on bowls and jars. The usual combinations are incised and painted triangles with incised and painted bands, painted oblique cross-hatching with the incised "fir-tree" pattern and painted solid triangles with incised hatched triangles. Incision on the outside of a vessel is sometimes combined with painting on the inside.

From Level 6 onward examples of the imported Samarran style are found (Figs. 6.7: 1,2 and 6.8:1–4). Samarran pottery has been the subject of much discussion since its first discovery by Herzfeld (1930). It has been related to the Hassuna complex (Braidwood,

Figure 6.7:1. The pottery from Level 5. Samarra vessel.

Figure 6.7:2. The pottery from Level 5. Samarra vessel.

Figure 6.7:3. The pottery from Level 5. Depiction of a scorpion on a Standard Hassuna vessel.

Braidwood, Tulane, and Perkins 1944, 65) and suggested by Braidwood in his prefatory remarks to the Hassuna report to be the product of travelling craftsmen. Lloyd and Safar (1945, 282) regard it as either a derivative of Hassuna or a direct import from outside, and the last suggestion may now be considered proven. The excavations of Tell es-Sawwan and Choga Mami have proved the existence in central Mesopotamia of the Samarran culture, akin to that of Hassuna and at times in close contact with it, but undoubtedly independent. The shapes differ little from those of the Hassuna range, but the style is distinctive; typical are the complex combinations of many different geometric motifs, and stylized scorpions (Fig. 6.7:3), in chocolate or black over an almost white slip. Some sherds decorated in two colors, red and chocolate brown, should be mentioned as a rare exception to the usual monochrome (Lloyd and Safar 1945, 281, only one similar sherd). One painted vessel is of very unusual shape, resembling a bivalve shell with a narrow neck in the center of the carination and an orifice at one corner; a similar pot was found at Hassuna (Iraq Museum no. IM 50267).

In the lowest levels excavated to date, 7 and 8, the pottery shows some features that distinguish it from the later material. Although the repertoire of shapes and the techniques of decoration show an overall consistency that confirms the essential unity of the sequence represented by this settlement, incised ornament is less common, paint colors are more vivid and the combination of painting and incision, common in the late levels, is completely absent from Levels 7 and 8. At the same time we find vessels with a glossy red surface, a wide red band beneath the rim, rows of pecked incision or appliqué horseshoe motifs that disappear entirely from Level 6 onwards (Lloyd and Safar 1945, Fig. 6; at Hassuna these ornamental elements are characteristic in lower levels). The absence of Samarran pottery in the lowest levels is a most significant fact; it is not found until Level 6 or even, perhaps, the transition from 6 to 5. The active impact of Samarra culture from central Mesopotamia was clearly an important landmark in the development of Hassuna settlements, apparent not only in the spread of its characteristic pottery but also in the changes we have observed in the layout of the settlement, in the form of houses and in building techniques. Another phenomenon that coincided with this development is the first appearance in the Level 6–5 transitional phase of extremely interesting anthropomorphic figurines, discussed in more detail below.

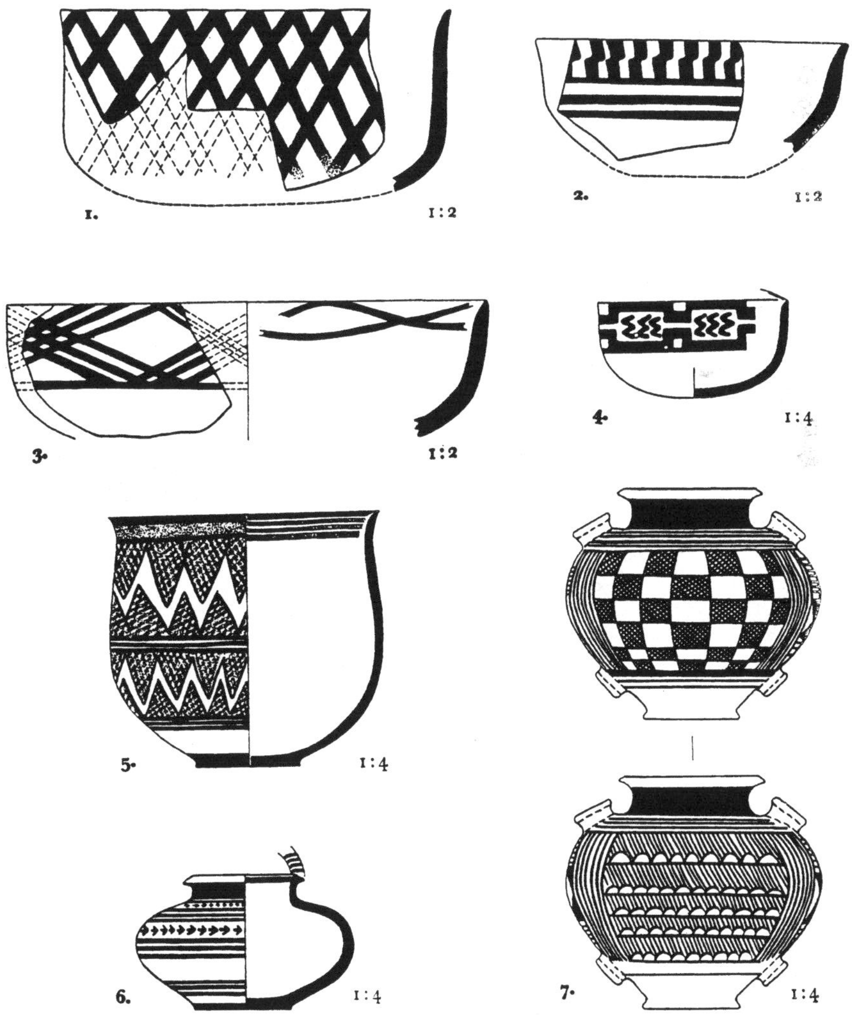

Figure 6.8:1–7. Yarim Tepe I and II Pottery. Yarim Tepe I: 1–4: Samarra pottery; 7: Halaf vessel from grave. Yarim Tepe II: 5: Halaf vessel.

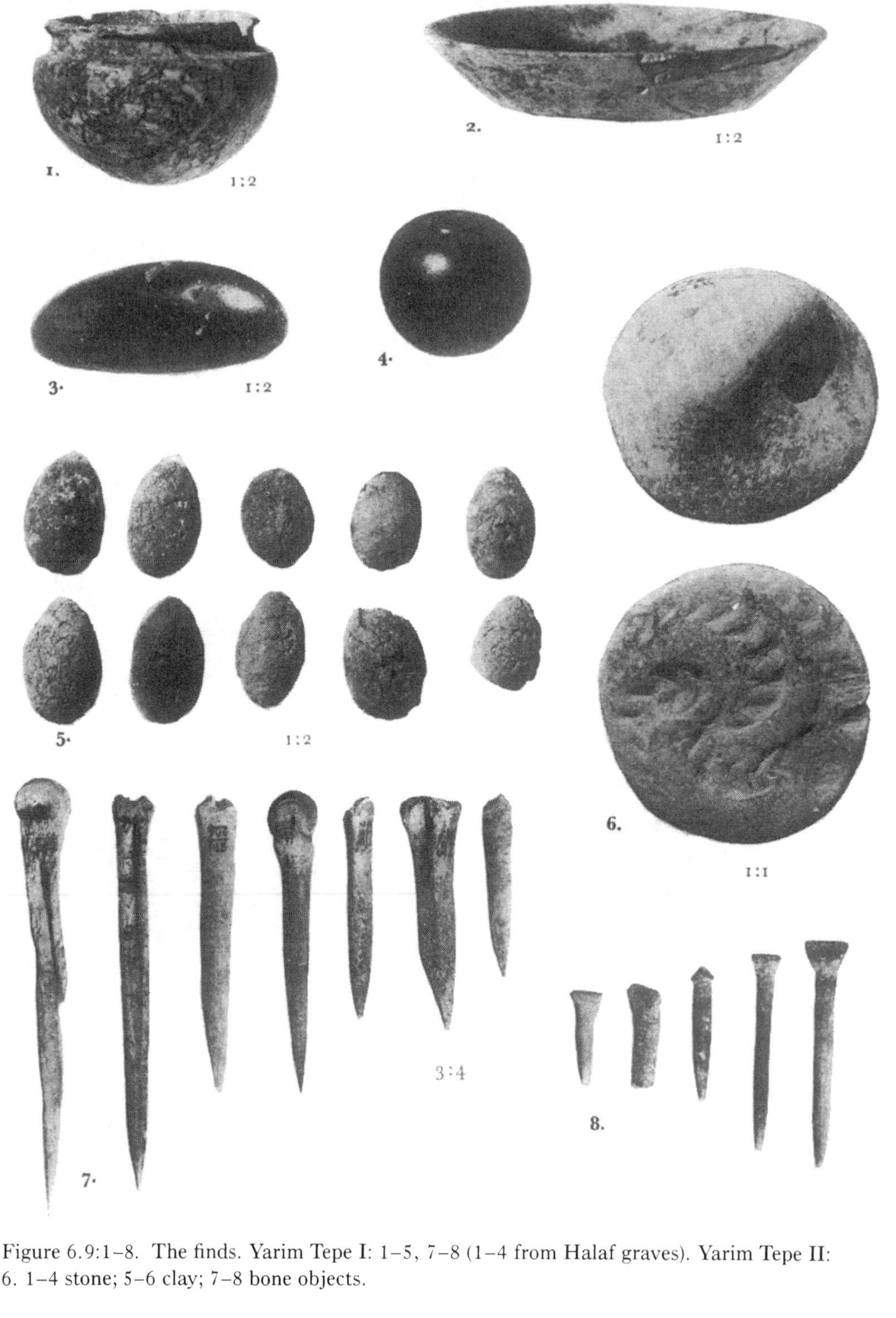

Figure 6.9:1–8. The finds. Yarim Tepe I: 1–5, 7–8 (1–4 from Halaf graves). Yarim Tepe II: 6. 1–4 stone; 5–6 clay; 7–8 bone objects.

Tools and Ornaments

Tools include knives, other cutting implements, and sickle blades of flint or, more often, obsidian, as well as mortars, pestles, querns, and hammers of crystalline rock; it should be noted that none of the large hoes characteristic of Hassuna has yet been found. Among other implements are bone awls, needles (Fig. 6.9:7,8), seven polishers and handles, sometimes decorated, as well as clay spindle whorls and "sling bullets" (Fig. 6.9:5 [Lloyd and Safar 1945, 267; Safar suggests these numerous "sling pellets" might be heated in ashes and used as pot boilers]). Small articles worthy of mention are the stone toilet sticks, some superbly fashioned large beads of marble and chalcedony (Fig. 6.9:4 and Lloyd and Safar 1945, Pl. XI, 1 and 2), flat, oval and spindle-shaped, and clay and stone stamp seals (Fig. 6.9:6). Of outstanding importance is the discovery in Level 7, square 47, of a bead rolled from thin sheet copper. There is no possible doubt of its stratigraphic position, for it lay beside the two-stage kiln described above, in a deposit sealed by several house floors of Levels 5 and 6, and must consequently be assigned to the early phase both of the Yarim Tepe I settlement, and of the Hassuna culture.

Figurines

Of special interest is the group of clay female figurines, of which thirty-eight pieces have so far been found (Fig. 6.10:1–10,12). Most of them came from Level 5 and represent a well-developed and clearly defined type, a standing figure with a high conical coiffure or headdress, oblong eyes set obliquely, a narrow waist, hips of exaggerated breadth and vestigial legs; symmetrical breasts are sometimes schematically modeled, and the figurines are decorated with horizontal incised bands. A painted figurine, a small seated figure, and a tiny head with almond-shaped eyes in relief were found in the upper levels. This is the largest collection of figurines strictly assignable to the Hassuna culture that has yet come to light. From Hassuna itself there are a few figurines, and many fine examples have been found at contemporary Samarran settlements (el-Wailly and Abu es-Soof 1965, Pls. XXVI, XXVII, XXXI; J. Oates 1969, Pls. XXV–XXX; Lloyd and Safar 1945, 270 and Pl. XVIII; see also Goff 1963, Fig. 53). One figure has a grain of wheat beside one of her eyes, demonstrating the association of the whole group with the cult of fertility and the image of the Mother Goddess. Another find of particular interest from a Level 5 room is a unique terra-cotta object in the shape of a cigar, terminating at one end in a sharp cone painted with ochre, at the other in a ram's head (Fig. 6.10:11); the body is pierced with regular holes, suggesting that it was a musical instrument or whistle. At the same time its general shape implies a phallic symbolism and, again, some relationship with a fertility cult, with which the ram's head might also be associated (see el-Wailly and Abu es-Soof 1965, 22 and Pl. XXVI illustrating the phallic symbols found in the graves at Tell es-Sawwan).

Evidence for the Economy

The economic basis of life in Yarim Tepe I was clearly an already advanced combination of agriculture and animal husbandry, both of which are well documented. For agriculture we have the direct evidence of numerous grains of two varieties of wheat, *Triticum aestivum* G. and *Triticum spelta,* and of multi-row barley, *Hordeum vulgare* G. (determined by S. W. Ganushevich and L. W. Prishchepenko), as well as the indirect testimony of sickle blades, mortars and hand mills, and the specialized storage complexes within the houses; it is in any case implied by the permanent character of the settlements. The evidence for animal husbandry is equally definite. Together with the bones of wild animals — gazelle, wild ox, wild boar, hare, and

Figure 6.10:1–12. Yarim Tepe I: 1–11; and II: 12. Clay anthropomorphic figurines: 1–10, 12; and one zoomorphic object: 11.

other species—we have numerous and unquestionable examples of domesticated goat, sheep, pig and, most important, cattle (examined in the field by S. Bökönyi and in the lab by the late distinguished paleobotanist Prof. B. J. Tsalkine). The stratigraphical position of the cattle bones is well established, for they were found in a room of Level 5, sealed by an overlying Level 4 floor.

b. Yarim Tepe I: The Lower Hassuna Levels

In 1973 the journal *Iraq* published the results of our first three seasons of excavations at the Hassuna settlements of Yarim Tepe I and the Halaf levels of Yarim Tepe II, situated in the Sinjar Plain of northern Iraq (Merpert and Munchaev 1973, chapters 6a and 8a in this volume). Following that publication the Soviet archaeological expedition, under the direction of R. M. Munchaev, continued its extensive studies in the area up to and including 1980. Prior to 1976 its major efforts had been centered on the separate tells of Yarim Tepe I and II, which were excavated over large areas down to virgin soil. This section presents the general results of our studies in the earliest levels of Yarim Tepe I.

General Data and Cultural Levels

Excavations were conducted on the eastern part of the tell. The upper levels have been completely excavated over 1660 square meters, and the lower levels, down to virgin soil, over an area of 400 square meters. In addition, on the southern and northern slopes of the tell a number of stratigraphic trenches have been excavated, totalling 60 square meters. Within the Hassuna occupation twelve building levels have been identified and numbered consecutively downwards. Each level corresponds to a comparatively long period of time and contains a number of internal remodelings without, however, any radical change in the layout of the settlement as a whole. In distinguishing these levels, the evidence from the stratified building remains was combined with that from all categories of finds.

The upper levels are greatly disturbed by graves and pits. The remains of the first (uppermost) level are represented solely by unconnected wall segments, whereas the second level yielded some identifiable rooms, and the third, some odd building complexes whose ruins lay directly over several massive structures, i.e., over the houses and pottery kilns of the fourth level. The main structures in all building levels were directly overlain by the debris of subsequent ones. The settlement's general layout has survived most fully in the fifth level which yielded eleven complexes of over 150 rooms. The structures below these are equally densely situated. In all the levels mentioned so far the center of the settlement remained invariably unoccupied, the complexes being grouped around an open central square. The Level 7 structures are less frequent, and Levels 8–12 have yielded only individual and widely scattered buildings that seem in the beginning to be quite simple but gradually show signs of greater complexity. In the six upper levels the "standard" group of Hassuna ceramics predominates; the seventh level is transitional and here Hassuna pottery is found mixed with an "archaic" painted variety which is plentiful from the twelfth level upwards. In this earliest level it is found together with "coarse" ceramics of Hassuna 1a–1c type. The five earliest levels correspond to Hassuna I a–c (i.e., to stages A–B, after Mortensen 1970). The materials recovered from these levels are discussed below.

Domestic Structures

The earliest signs of human activity are large pits up to 4.5 m long and 1.0 m deep; these are sunk below the ancient ground surface (6.0–6.2 mbd). These represent areas where

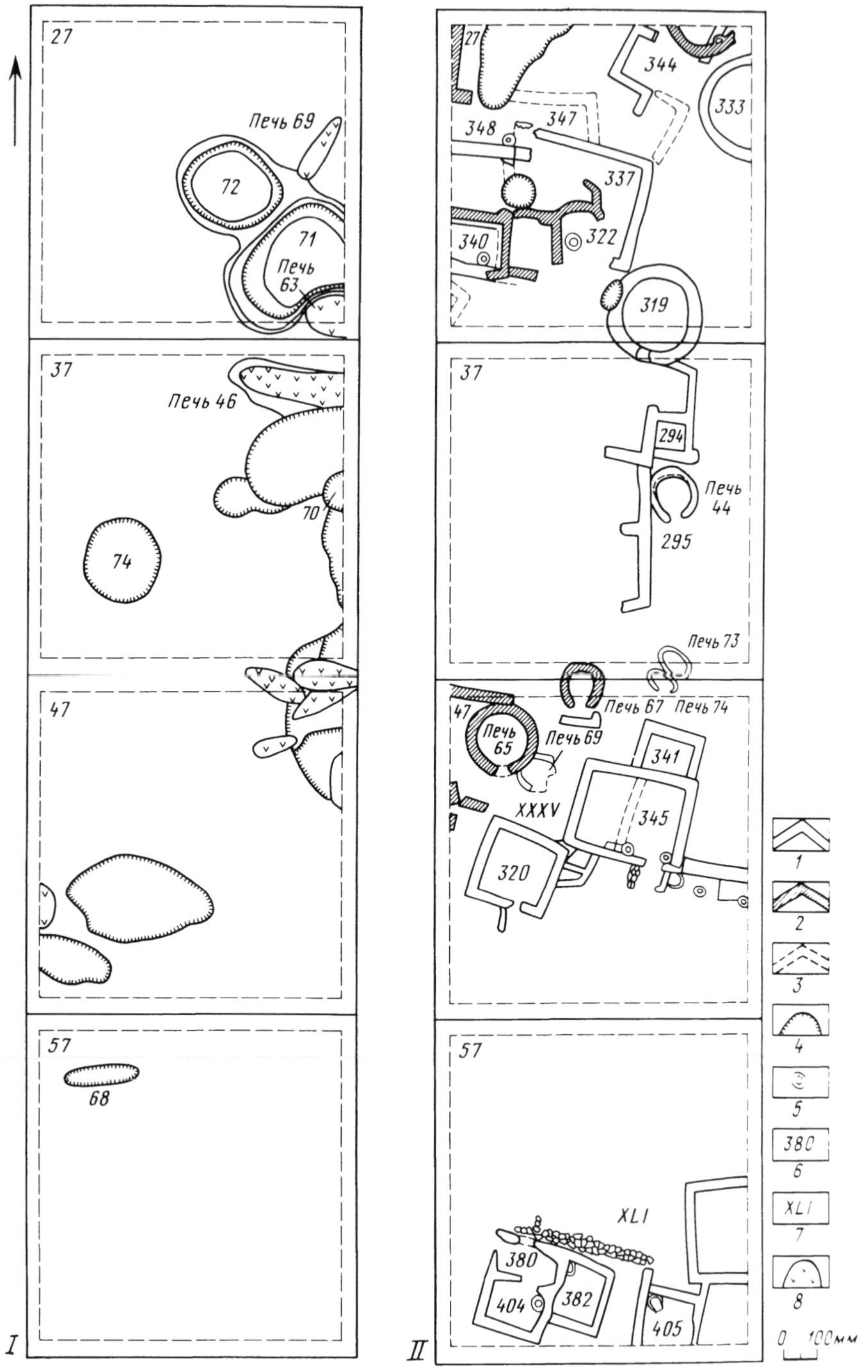

Figure 6.11. Plans of building levels. I. Pits and ovens. II. Level 12. Key: *1*, Structures of the main level; *2*, Structures of the upper level; *3*, Slight traces of wall; *4*, Holes and pits; *5*, Vessels set in floor; *6*, Room numbers; *7*, Numbers of building complexes; *8*, Subterranean ovens; ovens are indicated by circles.

clay was quarried for the construction of the earliest settlement at the site. Some of them, together with "trough-shaped" ovens that were sunk into the ground, appear to have been associated with a temporary camp of builders (Fig. 6.11:I). All of the seven oval and round pits discovered were filled with irregular layers of debris that contained great accumulations of ash together with pieces of tempered clay from the temporary ovens and hearths that had probably been destroyed when the area was prepared for subsequent construction. The pottery from this and the overlying Level 12 is identical; together with some coarse storage vessels with the characteristic small, flat bases and applied ornament, were found fragments of jars, pots and cups—some painted (with elements reminiscent of Tell Sotto and Umm Dabaghiyah)—as well as some fragments of marble vessels. Similar hollows or pits have been found to underlie the earliest structures at other early settlements such as Umm Dabaghiyah, Ganj Dareh, Tell Ramad, Beidha, and their excavators are quite justified in connecting them everywhere with the camps of the first builders at these sites (Mellaart 1975).

From the earliest levels of the Hassuna culture its house-building tradition appears to have been quite developed. Level 12 (5.7–6.2 mbd) is represented by building complexes 41 (square 57), 35 (square 47) and 31 (square 37) (Fig. 6.11:II). Its structures are substantial with walls up to 35 cm thick. Both single- and multiroomed houses occur: complex 41 consisted of four rooms, complex 35 of three, whereas complex 31 included a minimum of one round and five rectangular-shaped rooms (Fig. 6.15). Individual rooms were rectangular and rather small (measuring an average of 2 m × 2 m). With these were found structures apparently intended for some ritual purpose (containing burials and some dismembered skeletal remains), grain-drying platforms, open hearths and domed ovens. The various structures stood well apart from each other over the site.

All of this, we repeat, points indisputably to the existence of a developed house-building tradition from the earliest level of the Hassuna settlement. From the very beginning we encounter multiroomed structures with substantial plastered walls. These force us to reconsider the premise that the Hassuna house-building tradition started from very primitive single-roomed structures (Hassuna Ib, with no building activity attested in Hassuna Ia, Lloyd and Safar 1945). We can now assert not only that the building tradition was already rather developed at the time, having been in existence since the pre-Hassuna culture of Tell Sotto, but also that the structures were themselves diverse. The residential area (squares 37, 47, 57) is characterized by multiroomed houses with rectangular rooms. By contrast, no such complexes have been discovered in the northern area excavated (square 27); the structures there are small and, together with the rectangular ones, some round buildings of tholos type have been found. The latter are known not to be typical of the Hassuna culture; only one other such building has been found, at Tell Hassuna itself (Stratum 1c), where it was described as a dwelling. By contrast, the round structures from the lowest level of Yarim Tepe I cannot be classed as dwellings either by their size, form, or contents. It is premature as yet to fully evaluate their function or significance, but they are undoubtedly connected with burial practices. In tholoi 319 and 333 burials were found both inside the structures and under their floors. The finds from the tholoi are also very interesting (skeletal remains of sacrificial animals, necklaces deliberately deposited, stone vessels and the like). The other burials of Level 12 are characteristically concentrated in the same northern section.

Level 11 (5.3–5.8 mbd) includes complexes 40 (square 57), 34 (square 47), 30 (square 37) (Fig. 6.12:II). The structures are complex with many rooms (up to seven) measuring up to 12 square meters, the massive walls coated with plaster. The building complexes were well planned. Here, too, some structures, both round and rectangular, were found to

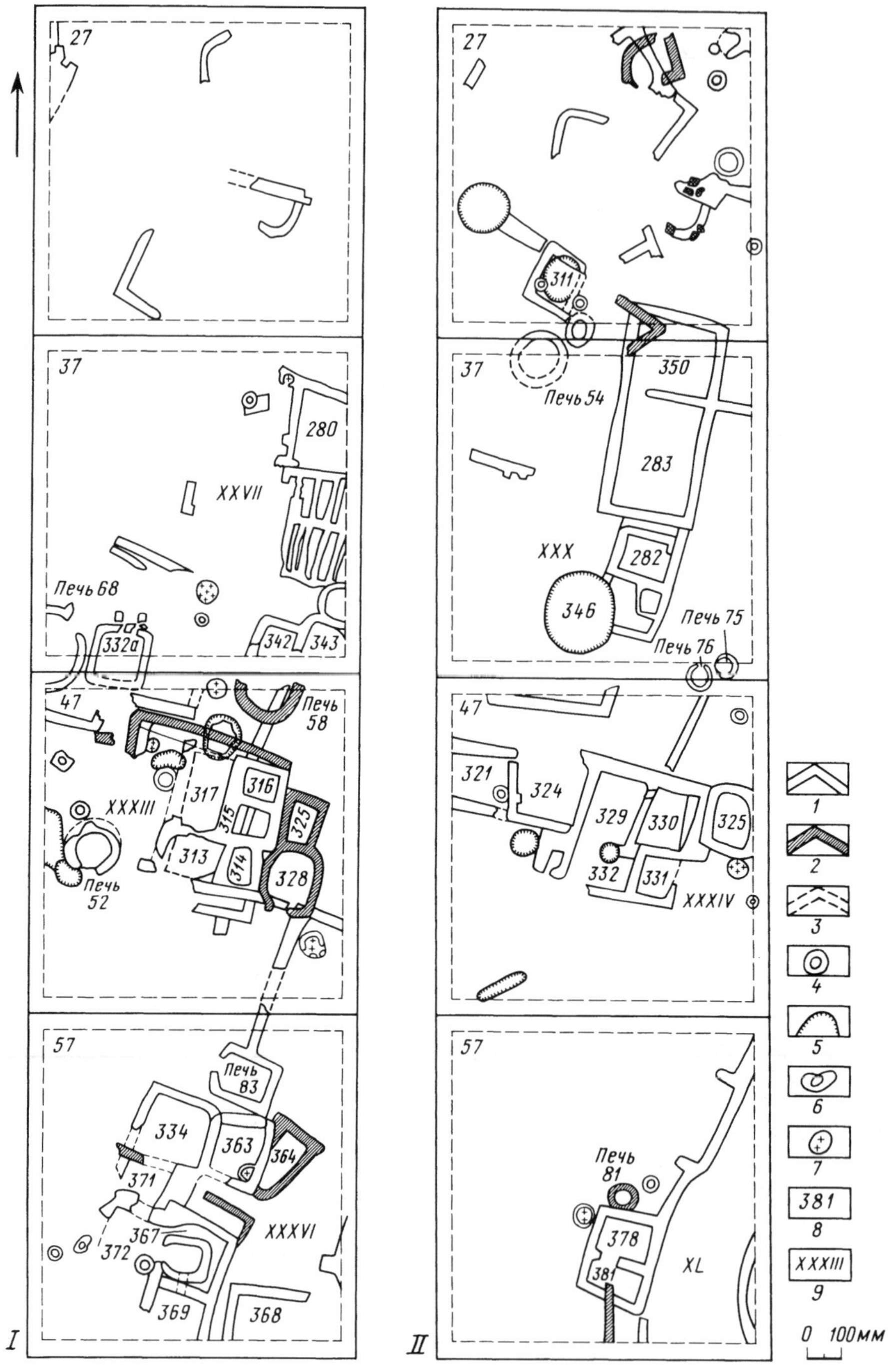

Figure 6.12. Plans of building levels. I. Level 10. II. Level 11. Key: *1*, Structures of the main level; *2*, Structures of the upper level; *3*, Slight traces of walls; *4*, Vessels set in floor; *5*, Holes and pits; *6*, Mortars; *7*, Hearths; *8*, Room numbers; *9*, Numbers of building complexes; ovens are indicated by circles.

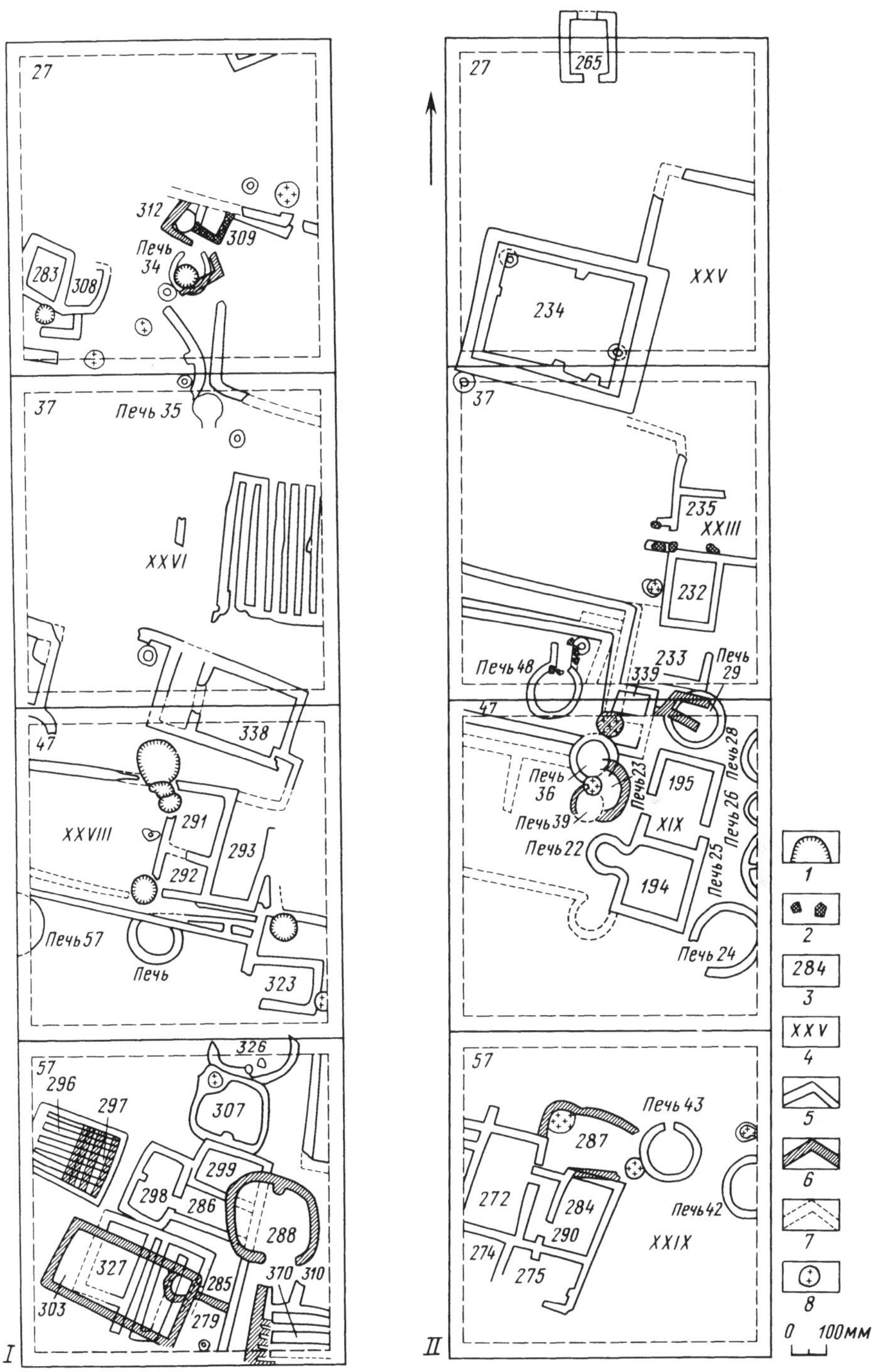

Figure 6.13. Plans of building levels. I. Level 9. II. Level 8; Key: *1*, Pits; *2*, Stones; *3*, Room numbers; *4*, Building numbers; *5*, Structures of the main level; *6*, Structures of the upper level; *7*, Wall traces; *8*, Hearths and ovens.

Figure 6.14:1. Excavation of the lower levels from the south

contain dismembered skeletons, one of which lay on a gypsum-plastered platform (room 282). No substantial dwellings were discovered on the northern periphery of the excavated area (square 27).

Numerous reconstructions and completely new buildings constitute Level 10 (5.00–5.30 mbd) (Fig. 6.12:I and Fig. 6:16.2). The greatest concentration of structures was recorded in the southern part of the site. Now the buildings become more complicated in design. Complex 36 (square 57) consisted of seven rooms and two big rectangular ovens, complex 33 (square 47) of eight rooms and several ovens, whereas complex 27 included a grain-drying platform with a system of parallel walls and a number of ancillary structures. The northern periphery remained unoccupied.

Structures in Level 9 (4.70–5.10 mbd) are even more densely distributed than in underlying levels and its complexes are more complicated in design (Figs. 6.13:I and 6.16:1). The southern part of the area (squares 47 and 57) is entirely occupied by building complex 32, which comprises nine rooms, three grain-drying platforms and an open yard with at least three ovens; complex 28 (square 47) is enclosed with long walls and consists of four rooms and an open yard with two ovens; in square 37 is a domestic complex (26) with a grain-drying platform similar to and overlying that in the previous level (10). Level 8 is distinguished by considerable reconstruction and a new general layout (Fig. 6.13:II). Densely packed buildings now occupy the entire southern periphery of the site. Square 57 contains the remains of multiroomed complex 29 (with no less than ten rooms); square 47 is occupied by the remains of complex 19 (two rooms) and a number of big ovens and round structures, and square 37, by the remains of complex 23. The first monumental structure was excavated in this level (complex 25, square 27) in an area which had previously yielded a concentration of burials and ritual structures. These were now replaced by a massive single-roomed structure (234) which served as the "nucleus" for complex 25 with its several open yards. The structure measured 4.14 m × 3.32 m, and its double walls were strengthened with buttresses. Inside the structure were flint tools and cores, marble palettes, mortars, some accumulations of ochre, and, under the floor, an infant's burial, horns of oxen and goats deliberately buried, and a painted jar.

Level 7 (depth 3.6–4.4 mbd) is the latest of the lower levels with sparsely set houses in marked contrast to the extremely dense patterning of buildings in the sixth and later levels (Merpert and Munchaev 1973, 97ff., Pls. XXXVII–XXXVIII, see chapter 6a in this volume). Present are compact building complexes well separated from each other. Such complexes occur in both the northern and southern peripheries of the excavated area. The southern complex comprised of houses 13 and 14 is separated from residential complex 17 by a wide square with sparsely set ovens and round structures. Complexes 17 and 22 were separated by a passage over 2 m wide, and complexes 22 and 24 by wide open yards. Complex 17 is of particular interest. In its complete form it was a house with eight rooms, 11 m long and 6 m wide at the maximum. The house was not regular in form; six of its rooms ran west to east in two rows, forming a rectangle, whereas the two others were situated at right angles to the main structure, one eastwards and the other to the south. The structure underwent several reconstructions: originally it had five rooms with an oven in the yard; later the oven was destroyed and overlain with a large platform that served to reinforce the structure from the south and east, the enclosing walls were doubled, their thickness now reaching up to 50 cm, the central room were divided in two,

Figure 6.14:2. Excavation of the lower levels from the north

Figure 6.15:1. Building 31 of Level 12

Figure 6.15:2. Building 35 of Level 12

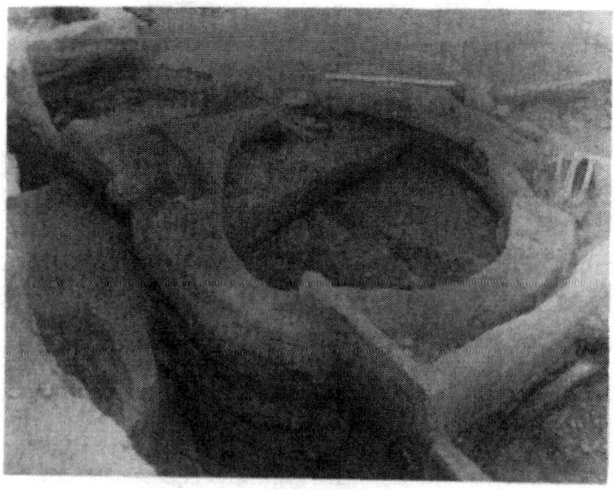

Figure 6.15:3. Tholos 319 of Level 12

and two new rooms were added, one of which was later filled up with earth and made into a kind of buttress that reinforced the structure from the east. Some jars were found buried under the walls and floors of the house, and in its central room were the remains of two dismembered human skeletons, a fact which suggests that the building had some special purpose.

Character and Building Techniques of Hassuna Structures

From the very beginning the building remains found in the Hassuna strata of Yarim Tepe I indicate that the tradition of multi-roomed structures (Fig. 6.14) was well developed at the time and that its roots should be sought in earlier cultures (in the case of the Sinjar area, in the cultures of Tell Maghzaliyah and Tell Sotto). In the earliest level (12) all the complexes (31, 35, 41) consisted of several rooms and provided evidence of well-developed building techniques. It has been established that prior to building, the occupants of the site leveled an appropriate stretch of land, filling all its cavities and hollows with earth or fill derived from earlier building remains. Mats and layers of reed were used for the wall beddings, or foundations were constructed of large pottery fragments (the latter technique was common in the construction of substructures for floors, thresholds, and especially, ovens). On their exteriors, facing the slope of the tell, structures were reinforced with massive clay platforms, often with enclosing walls and rubble filling. Similar plat-

Figure 6.16:1. Buildings of the lower levels: Level 9

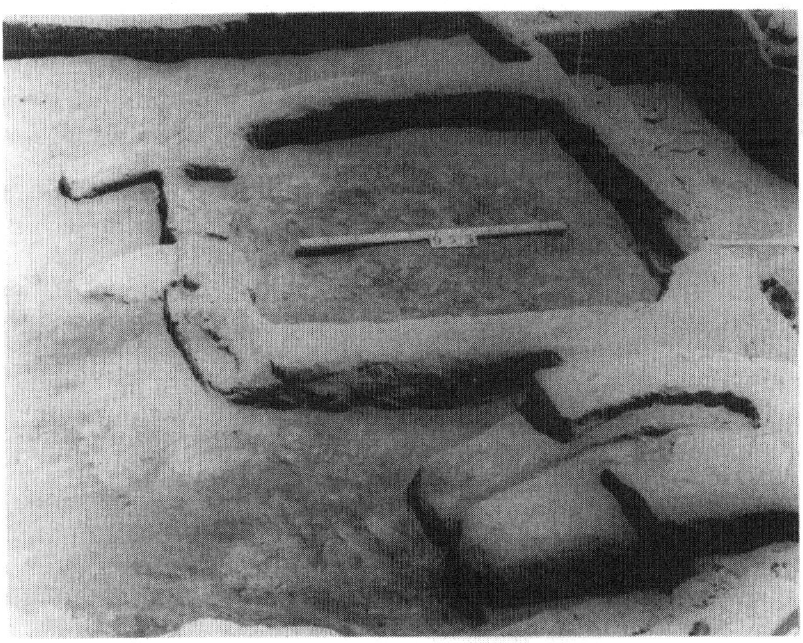

Figure 6.16:2. Buildings of the lower levels: Level 10

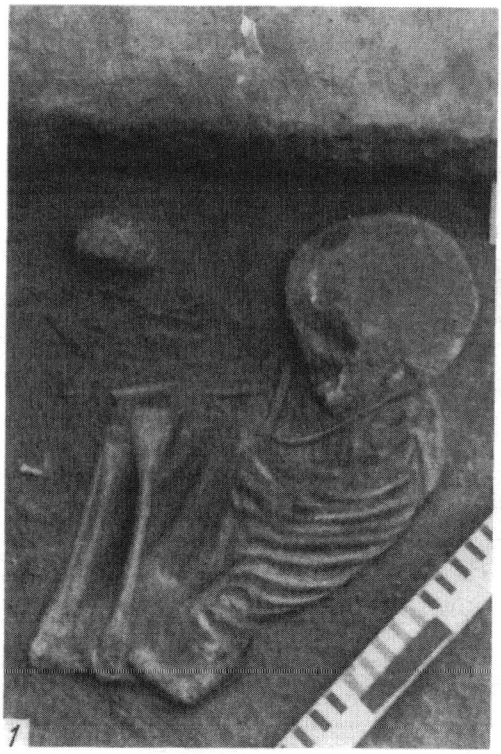

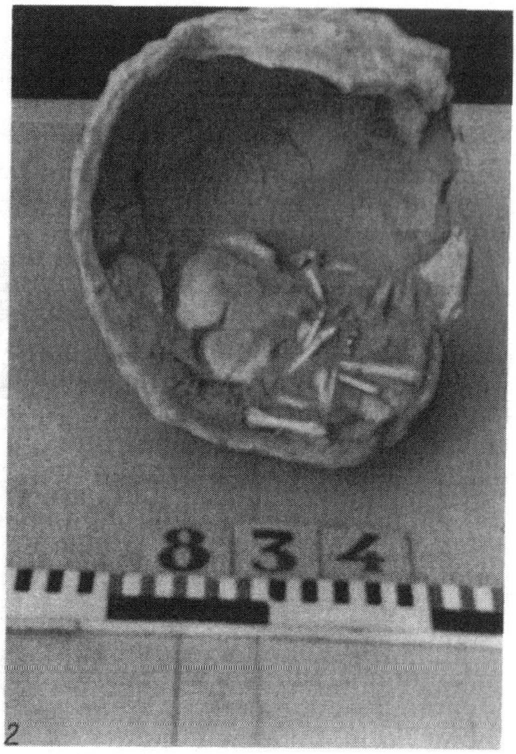

Figure 6.17:1–4. Burials from the lower levels. *1.* No. 102, Level 8. *2.* No. 103, Level 8.

forms were also built for such special structures as granaries and grain-drying platforms (such a platform was also found in the third level at Tell Hassuna, Lloyd and Safar 1945, 272, n.7). The walls are substantial with an average thickness of 30–35 cm, widening at the bottom to 50–60 cm. Some walls are more than 10 m long. The technique of construction is well developed and is for the most part identical throughout all levels. All the walls are made of clay lumps, either massive blocks or comparatively thin slabs bound together with clay mortar. The clay contained chopped straw or twigs, and some gypsum to cement the individual blocks. As is common in pisé technique, the blocks were not predried before use. The blocks and slabs were not of uniform size, but in a few instances some standardization of form and size is observable, a feature which increases over time. Thus, for example, in Level 12 the walls of room 382 are made of yellow clay slabs 50 cm × 24 × 5 cm; in Level 11 the outer walls of room 381 are made of slabs measuring 40 cm × 30 cm × 5 cm, whereas the slabs on the inner sides are considerably smaller, measuring 18 cm × 17 cm × 6 cm. In Level 10 the grain-drying platform is made of rectangular slabs of different sizes (23–25 cm × 20 cm × 3–4 cm). By contrast, Level 9 has yielded a wall made of blocks that were of standard rectangular form and uniform size (35 cm × 25 cm × 6 cm). Unusually narrow and long oval blocks of red clay were found among the remains of a large round structure (oven?) in Level 9. These approached 50 cm in length and 7 cm in breadth and height. The double walls of room 234, Level 8, consisted of different components. The internal ones were made of common brown clay slabs, measuring 35 cm × 25 cm × 5 cm, whereas the outer walls were especially thick, consisting of gray-green

3. No. 102, Level 8.

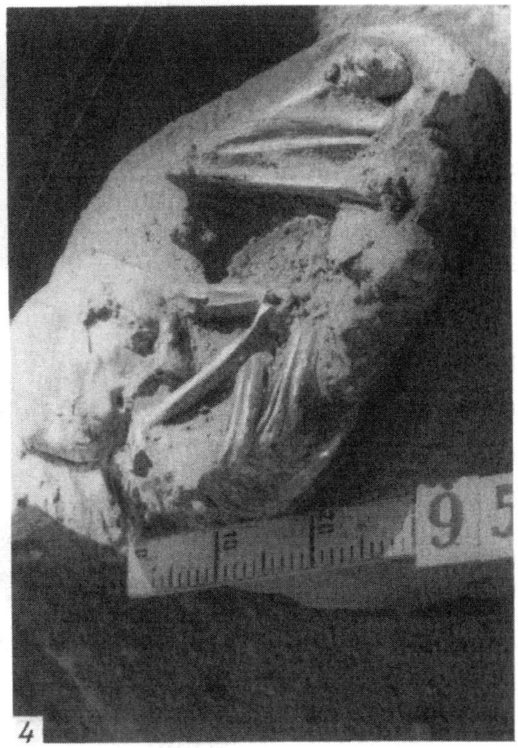

4. No. 132, Level 9.

blocks of clay, 80 cm × 35 cm × 40 cm. From Level 12 onwards the walls were plastered on both sides, the plaster being up to 1.5 cm thick. Such plastering has been recorded on the walls of all three original complexes (31, 35, 41).

The "buttresses," which were later to play such an important part in Sumerian architecture, are a very early architectural feature at Yarim Tepe I. A long wall, which crossed square 57 and was connected with complex 40 of Level II, was reinforced on its western side with large buttresses which projected 30–50 cm from the wall's surface. In all subsequent levels such buttresses become an almost obligatory element of their buildings. As we have already noted, the floors sometimes had special foundations made of potsherds or small stones which were covered with reed flooring and then plastered with clay that contained an admixture of chopped straw and gypsum, the presence of the latter in the plaster was sometimes considerable. In some instances there were platforms strengthened with numerous posts and intended, probably, for grain storage; in both Levels 11 (square 47) and 8 (room 272 of complex 29) these were associated with large stone mortars. It should be pointed out that gypsum plaster has been recorded not only in the floors of houses but also on the surfaces of the passages or "lanes" between them.

The roofs of the houses had a three-layered structure consisting of matting, clay and gypsum, a common feature in the area to this day. Such a roof was supported by wooden poles that rested, in turn, on the walls and a ridge beam. Few rooms had doorways; presumably they were accessible from the roof, whereas auxiliary structures had low walls. Primitive hearths and sophisticated ovens coexist from the twelfth level. The ovens are round or rectangular in plan, with openings on the side or on the top. In Level 10 and later, two-story pottery kilns with a system of holes connecting the fire chamber with the baking chamber were encountered. From the earliest stages of its development the settlement displays a characteristically Hassuna complex; baking ovens, "husking trays," platforms for drying, and grain grinders or mortars.

Burials

The lower levels have yielded over twenty burials (Fig. 6.17). These are mostly infants (17) found under the walls and corners of houses or beneath their floors, thresholds, hearths, and sometimes in large vessels or under large potsherds. The skeletons lie contracted on the right or left side or, but only occasionally, on the back. The orientation of the skeletons varies. Occasionally, the remains of the dead are accompanied by archaic clay vessels and animal bones (sheep, goats, horns of oxen). With the single exception of contracted burial 126 (in a narrow, oval pit sunk from Level 11), which belongs to a female under twenty-five years of age, the remains of adult persons provide evidence of some very unusual ritual practices (e.g., dismembered corpses found in special round or rectangular structures). In Level 12, inside a tholos (diameter 2.5 m) miscellaneous bones of two or three adults accompanied by sheep bones and fragments of not less than four vessels were discovered (burial 131).

In Level 11, burial 105 was found inside room 282. The floor measured 1.35 m × 1.05 m, and its walls and floor were gypsum plastered. The floor was traversed by a trough 50 cm wide, within and by the sides of which lay the bones of two feet in anatomical order, the head of a femur, and some fragments of other bones of an adult person. Similarly, room 346 (1.70 m × 1.25 m) of the same building complex (30) yielded dismembered remains of another adult person: arranged in anatomical order were bones of part of the spine, thorax and limbs and the skull. In Level 7 two dismembered skeletons were found inside the central room of complex 17.

The discoveries at Yarim Tepe I show that intramural burials are present in all levels and that they were associated with specific burial ritual. These are mainly infant burials and for the most part those of children under 1 year of age (over ten examples); two burials belonged to children 1.5 and 3 years old; two others, to children aged 4 and 6, and two more to children of 10–12 years. The tradition of burying children inside the settlement was very consistent and has also been recorded at pre-Hassuna sites; from the earliest levels of Umm Dabaghiyah and Tell Sotto (Bader 1975, chapter 3 in this volume) infant burials were frequently found under the floors and walls of the houses, under hearths and inside large vessels. Later, during the Hassuna (Lloyd and Safar 1945, 267, n.7; el-Wailly and Abu es-Soof 1965; al-Adami 1968; and Mortensen 1970) and post-Hassuna periods this tradition became widespread among a number of early agrarian cultures both in the Near East and southeastern, eastern, and central Europe. Everywhere the predominant type of burial ritual was inhumation, with the dead lying on the right or left side or, rarely, on the back. Accompanying grave goods and sacrificial food are encountered, but only occasionally.

As for the bones of adults from the lower levels of Yarim Tepe I, all are related to some unusual ritual. The small number of such burials does not allow us to categorize them as customary; in an area of 400 square meters the remains of some ten or more dismembered skeletons have been discovered in the lowest six levels. In this connection it should be noted that dismemberment rituals are an extremely ancient tradition in the Near East (Cauvin 1978). However, the question of the usual method of Hassuna burial must remain open, though we can postulate with a high degree of probability that there were extramural necropoli (Kirkbride 1973b), similar to those of the Halaf culture that we know to have existed together with unusual and individual intramural burials (Merpert and Munchaev 1982a, chapter 10 in this volume).

Pottery

At Yarim Tepe I all the previously known groups of Hassuna ceramics are represented, i.e., coarse, archaic painted, standard painted, incised, and painted-and-incised. A considerable amount of archaic painted pottery appears in Level 12 and coexists with the coarse ceramics, the two constituting a single early Hassuna ceramic complex. Level 7 is transitional between the "archaic" and the "standard" wares. Some of the forms are extremely conservative and persist throughout all levels. Among them are thick-walled oval basins, including those with carinated bases, open cups, and hemispherical or spherical jars with pointed rims and cylindrical necks. The occurrence of large storage jars with pointed rims, abruptly carinated bodies, and small flat bases is more limited; they and some specific forms of cups and jars are found in the lower levels only.

The classification of the pottery from Yarim Tepe I is based on shape, whereas technological differences within each group were noted separately. Five basic kinds of vessels have been distinguished, each with several distinct types.

Type I. Large storage vessels, represented by three varieties: (1) large jars with a wide, low neck, an abruptly flaring bell-shaped body with a sharp carination, concave lower sides, and a flat base; (2) medium-sized vessels with a low, broad neck, rounded body and flat base; and (3) medium-sized vessels with a high cylindrical neck, and body similar to (1) (Fig. 6.18:a). All these vessels are coarse. The clay has an organic temper. Ornamentation is rare and is represented exclusively by simple applied decoration (Fig. 6.18:b). Such vessels are found for the most part in Levels 10–12, with some isolated examples up to Level 7.

Type II. Jars, represented by four varieties: (1) wide-necked, with a rounded, slightly flared rim, low cylindrical neck, a spherical and somewhat flattened body, and a wide flat base; (2) jars with tall, slightly everted necks, oval bodies, and small flat bases; (3) jars with a flared neck, a spherical body, and a rounded or pointed base; and (4) jars with a minimal neck, rounded rim and body, and a flat or rounded base (Fig. 6.18:c). These vessels are both coarse and dark, or thin with a light slip. The ornamentation is dominated by the archaic painting techniques (eight variations); applied decoration has been recorded on five large jars from Levels 10–12, and incising on one vessel from Level 8 (Fig. 6.18:d). These vessels are common in all the early levels; variety (1) predominates in Levels 10–12, varieties (2) and (3) in Levels 10 to 8, and variety (4) is typical of all six of the early levels.

Type III. Open, flat-based cups or bowls, represented by six varieties, among all of which there are coarse and fine examples: (1) with straight or sharply everted, slanting sides and a rounded or pointed rim, (2) rounded sides and a tapering rim, (3) sharp carination and straight pointed rim, (4) rounded sides, sharp concave rim, (5) rounded sides, slightly flared rim, (6) straight walls and a flattened everted rim (Fig. 6.19:a). The archaic painting makes use of eleven patterns (Fig. 6.19:b). From Levels 7–12 geometrical compositions gradually grow more complex. Predominant in the lower levels are cups or bowls with a straight rim, and in the upper ones, with the concave variety.

Type IV. Tall, coarse, and fine pots with rounded or straight sides, represented by five varieties that range from those with

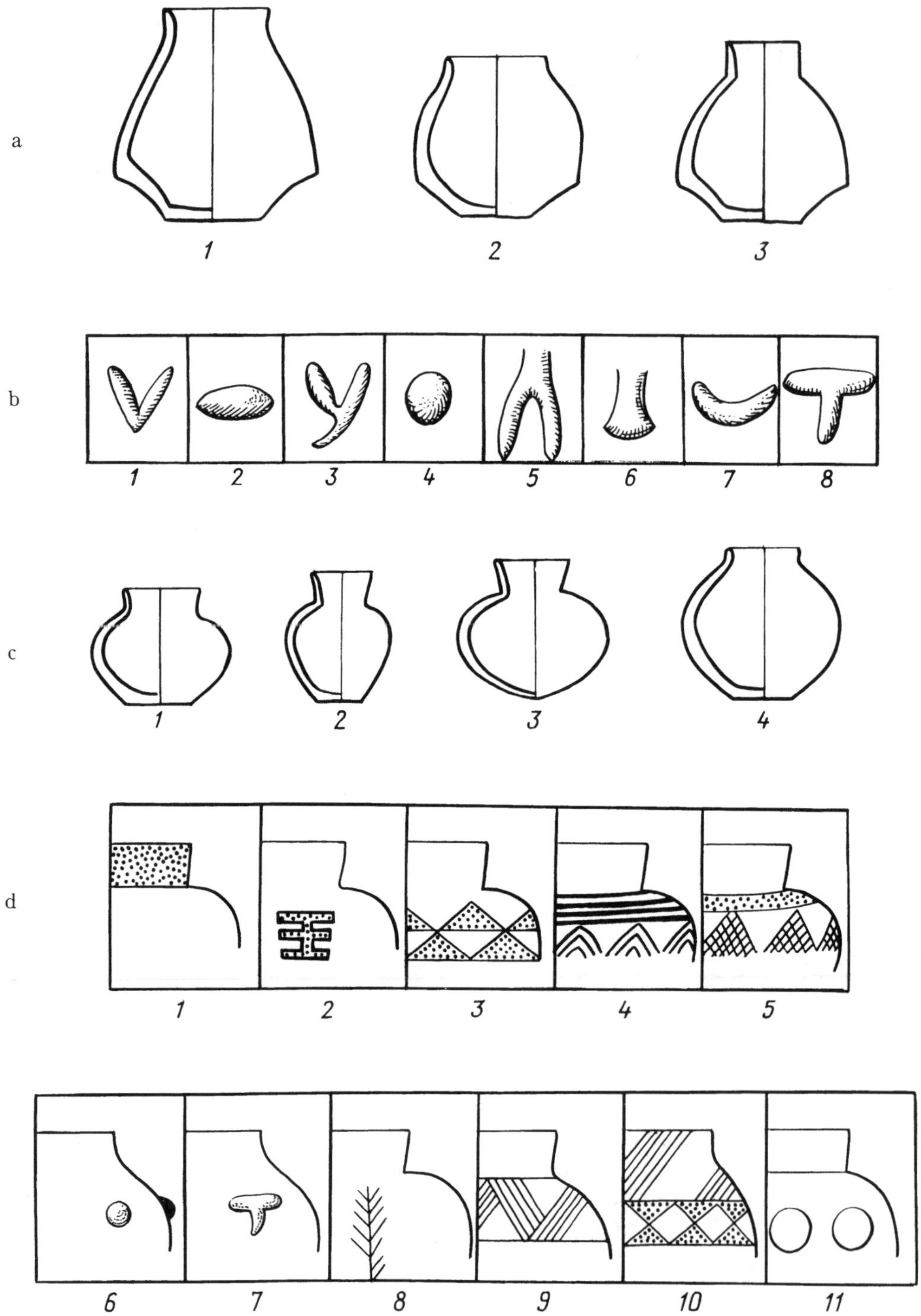

Figure 6.18:a–d. Vessel types and decorative styles: *a,* Type I; *b,* 1–8 appliqué; *c,* Type II; *d,* 1–5, 9–11 painted; 6–7 appliqué; 8 incised.

a

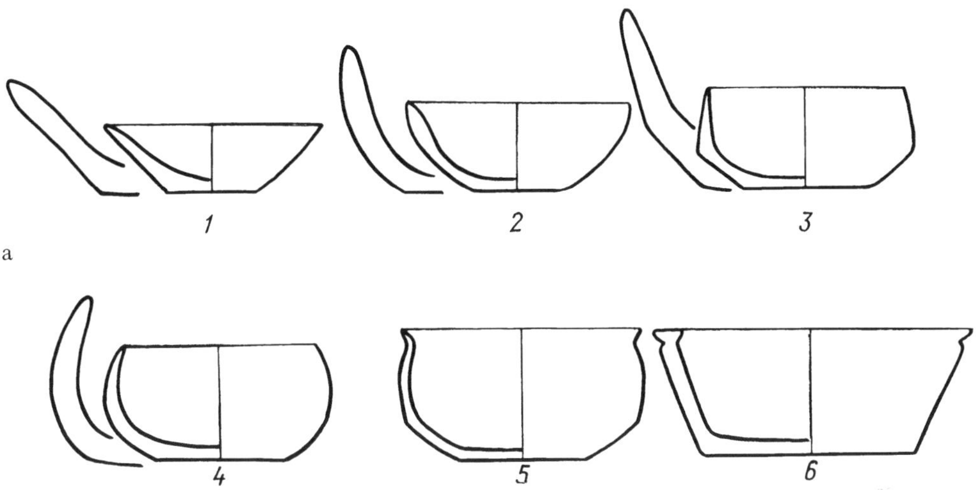

b

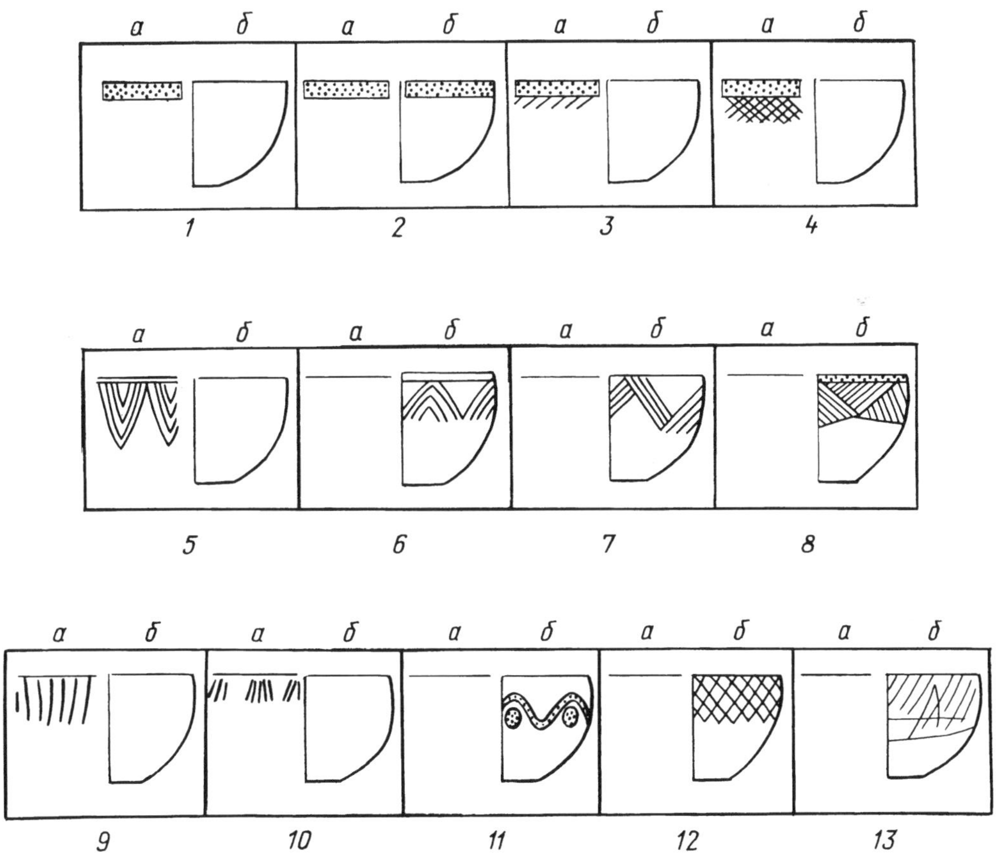

Figure 6.19. Vessel types and decorative styles: *a*, Type III. *b*, 1–11 painted; 12–13 incised decoration. The interior of the vessel is shown on the left and the exterior on the right in each case.

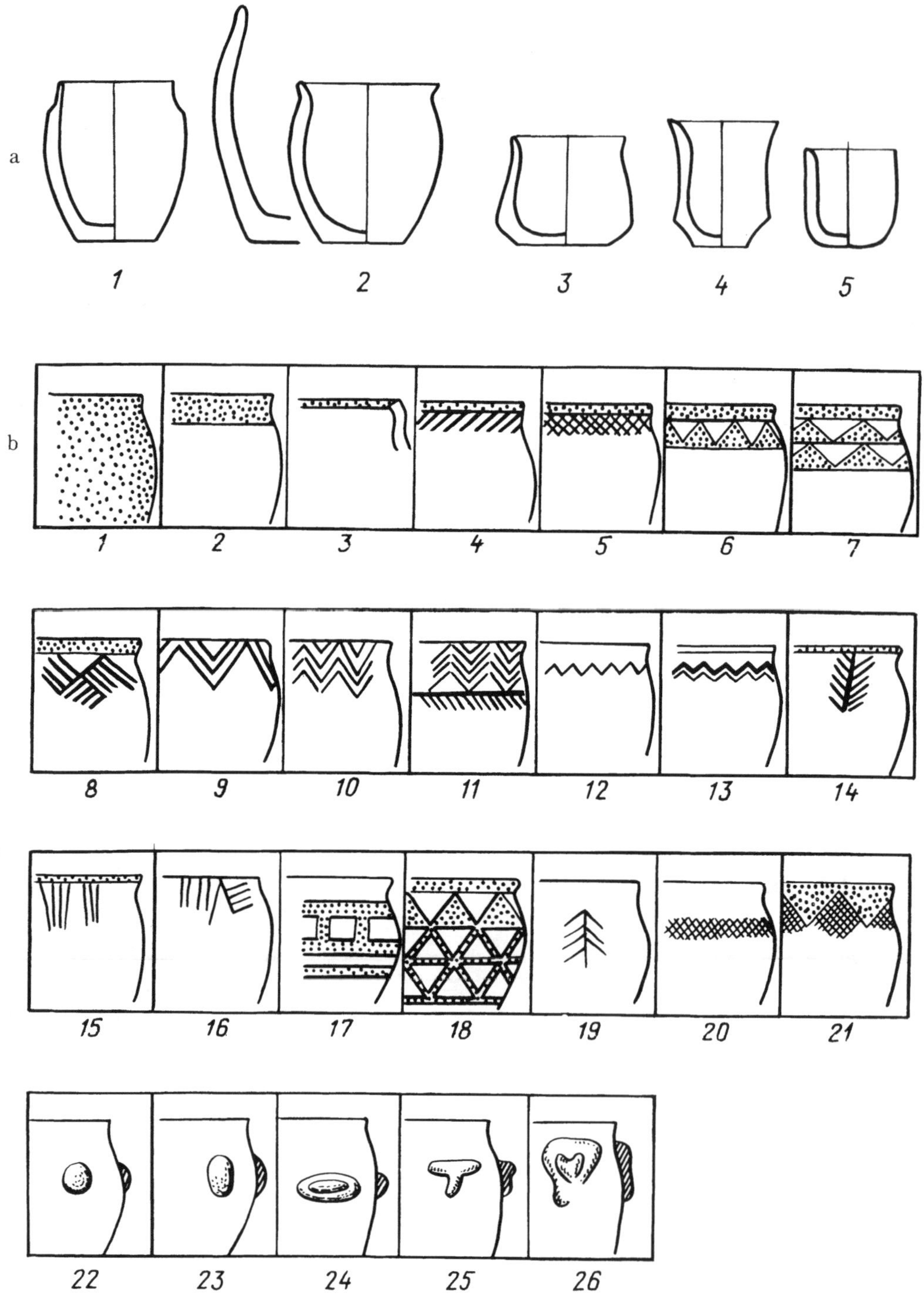

Figure 6.20 Vessel types and decorative styles: *a,* Type IV; *b,* 1–18 painted; 19–20 incised; 21 painted and incised; 22–26 appliqué.

a very distinctive profile to simple cylindrical ones: (1) medium-sized pots with rounded sides, a sharply thinning rim and a wide flat base, (2) large and medium-sized pots with rounded sides and wide flat bases, (3) medium-sized wide pots with straight walls, widening at the flat base, (4) tall pots of medium size with concave walls and flared rim, (5) simple cups of small and medium size (Fig. 6.20:a). There are eighteen variations in the painted designs, which become more complex on the cups and bowls (Fig.6.20:b). There are seven examples of applied ornament, all from Levels 11 and 12, and two examples of incised decoration in Levels 8 and 9. Incised decoration becomes common in Level 7, where the standard painted and painted-and-incised types also appear.

Type V (Fig. 6.21). Basins are represented by two varieties, round or oval with smooth bases, and oval "husking trays." The first variety is common in all the early levels. The second is divided into two subtypes: the first, with a cellular pattern in the base, prevails in Levels 10–12, and the second, with "ribbing," in all subsequent levels. "Husking trays" are, of course, known also from Tell Sotto and Umm Dabaghiyah (Bader 1975, chapter 3 in this volume; Kirkbride 1972, 9, Pls. XII, XV; Kirkbride 1973b, 5, Pls. II, X; Merpert 1978, 140, 148).

Already in Level 12 there appear some unusual vessels with pedestal or ring bases. There occurs also a small and distinctive group of dark red, painted and burnished, thin-walled vessels of which the source has not yet been identified (Fig. 6:22:1–4; Bader 1975, chapter 3 in this volume; Kirkbride 1972, 5; Mellaart 1975, 139). (See Figs. 6.23–6.25 for photographs of ceramic vessels from all levels.)

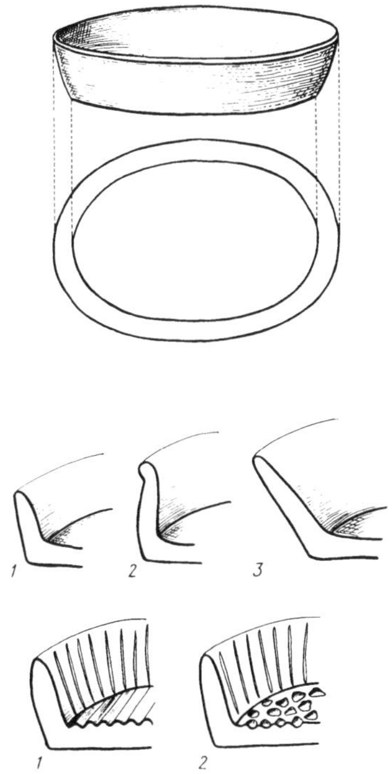

Figure 6.21. Vessel types: Type V

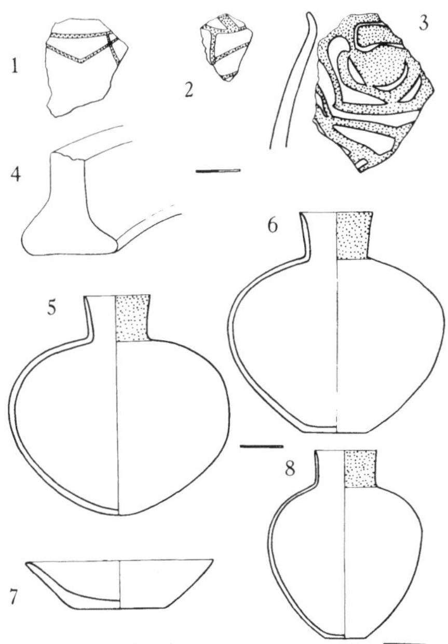

Figure 6.22:1–8. Types of vessels and decorative styles from the lower levels. 1–4 from pit 71 in the original ground surface; 5–8 from level 10.

Figure 6.23:1–5. Vessels from levels 12 (1,3,5) and 11 (2,4)

Figure 6.24:1–5. Vessels and example of applied decoration from levels 10 and 9

Figure 6.25:1–6. Vessels from levels 9 (1–3,5) and 8 (4,6)

Stone and Bone Artifacts

The total number of pieces of flint and obsidian from the lower six levels at Yarim Tepe I is not great (not more than six hundred from the area of 400 square meters), far less than the numbers recovered from the earlier settlements in the area. Flint constitutes 77 percent and obsidian 23 percent of the finds. The obsidian originates from some eastern Anatolian source or sources, and an overwhelming majority of the pieces of obsidian are finished tools. The majority of flint objects are cores, blades, flakes, and debitage. The variety of tools is extremely limited (Fig. 6.26). Among the flint artifacts it is possible to distinguish endscrapers on rectangular or oval flakes and blades, and side-scrapers on oval flakes, all with a retouched edge (Types A1 and A4 after Mortensen 1973, Pl. XX); also it is possible to distinguish borers on flakes and blades, large knives on blades (only in the lowest levels), and sickle blades. Standardized sickle blades, including those with a denticulated working edge, are found for the most part in the later levels of the settlement, a fact which corresponds well with the data from Choga Mami where the proportion of sickle blades increases from 25.4 percent in the Early Samarra phase to 44.2 percent in the so-called Late Samarra. Four projectile points have been found only in Level 12, all made from blades. Three are stemmed and leaf-shaped, the fourth is shaped like an elongated rhomboid (Fig. 6.27:8–11). Obsidian occurs as blades, both very large knives for dressing animal carcasses (established by trace analysis) and, commonly, sickle blades and scrapers. Although the blade tools from Yarim Tepe I represent a widespread ancient tradition, they find their closest analogies and indeed their ancestral forms, not in Jarmo or other Zagros sites, but above all in the corresponding complexes of the lower levels of Tell

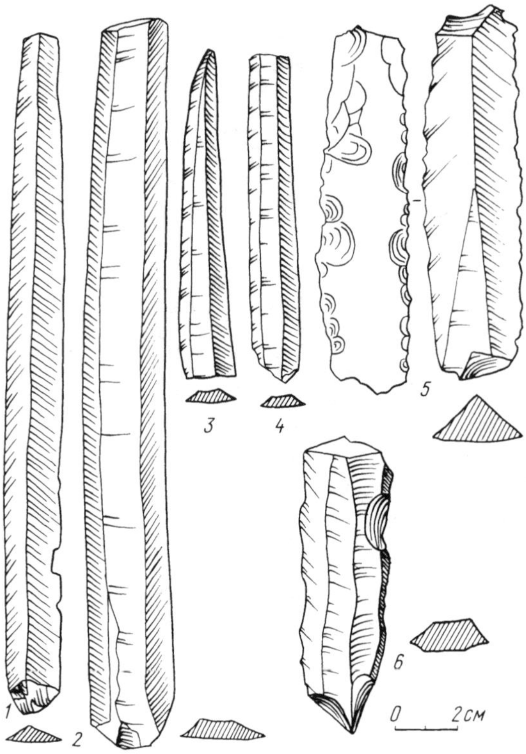

Figure 6.26:1–6. Obsidian tools from lower levels

Sotto and Umm Dabaghiyah (Mortensen 1973, 53; Kirkbride 1972, 11, Pl. XVII). The same is true of the arrowheads, of which the earliest examples are to be seen in a series from Tell Maghzaliyah (Kirkbride 1972, 6–7, Pl. XVII; Kirkbride 1973b, 6; Munchaev and Bader 1979, Figs. 3 and 9).

Levels 12 to 10 have yielded seven wedge-shaped axes (Fig. 6.27:1–7). Two maceheads have been found: a coarse basalt example in Level 12, and a polished, pear-shaped one of marble in Level 8. Grinding and pounding tools are represented by numerous mortars, flat and saddle-shaped querns, as well as by handstones and pestles of different types. Only two hoes have been found, one with a hole for a handle. In the lower levels we have found fragments of ten polished marble or alabaster bowls and jars, and seven palettes.

The variety of bone artifacts is limited, mostly awls made of the leg bones of small herbivorous animals. There are also needles, spatulas and polishers. In Level 10 an ox shoulder blade with uniform incisions on it was found, possibly indicating some form of reckoning.

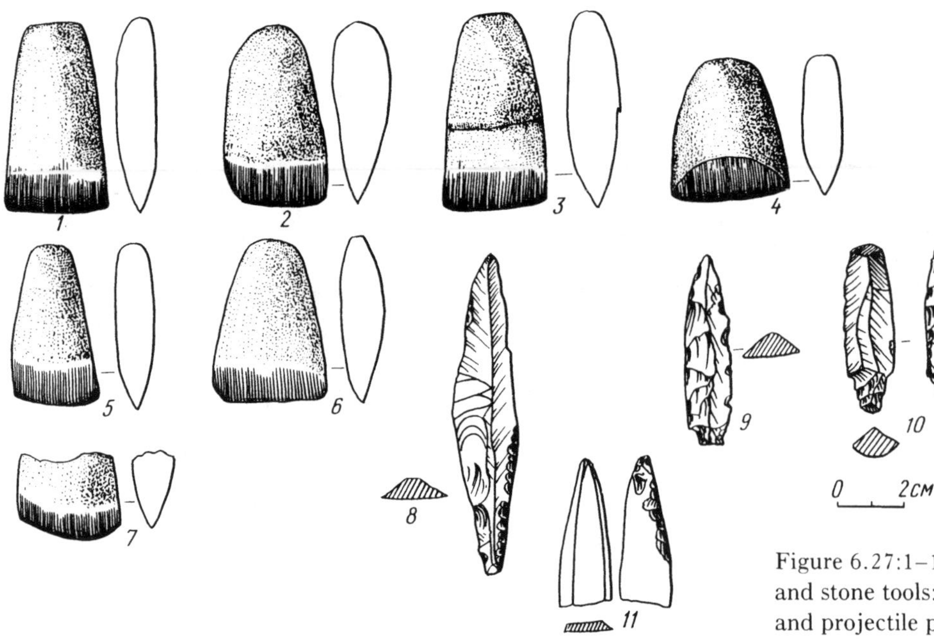

Figure 6.27:1–11. Flint and stone tools: axes and projectile points

Other Finds

In the lower levels of Yarim Tepe I the excavations have revealed twenty-seven conical and biconical spindle whorls made of clay, alabaster, marble, and gypsum. Sling bullets of clay, numerous in the upper levels, are rare in the lower ones. There are many beads of different stones (marble, serpentine, limestone, azurite, carnelian, chalcedony, alabaster, turquoise, rock crystal) as well as of obsidian, bone and shell. Fourteen types have been distinguished among them: rhombic flat, oval flat, cylindrical, barrel-shaped, tubular, spindle- and ring-shaped, spherical, triangular flat, trapezoid, rectangular flat, parallelepiped, as well as three- and two-component beads. The most colorful and varied necklaces come from the lower levels. For example, Level 9 has yielded a necklace of fourteen turquoise beads (nine cylindrical and five tripartite ones); Level 10, a necklace of ten disc-shaped carnelian beads; and lastly, and most interesting, from Level 12, a necklace of fifteen beads of different sizes, colors (gray, green, yellow, cherry, red, blue, etc.) and materials (including rock crystal and shell) (Fig. 6.28), found in the center of the round ritual structure 319. This necklace is the most striking and interesting so far found in Hassuna levels at any site. The only fragment of a polished stone bracelet comes from Level 9.

Figure 6.28. Bead necklace from Level 12

Stone pendants are rare: four examples in simple oval shapes. Stone seal-pendants are of particular interest. Seven were found in Yarim Tepe I, of two types; round ones are characteristic of the upper levels, and rectangular specimens, of the lower ones. The five rectangular examples come from Levels 9–11; they are made of quartzite and are of uniform shape, with a cross-hatched pattern on the obverse and an eye for suspension on the reverse (Fig. 6.29).

A particularly important item was metal, encountered twenty-six times in the settlement as a whole, with twenty-one examples coming from the lower levels discussed here:

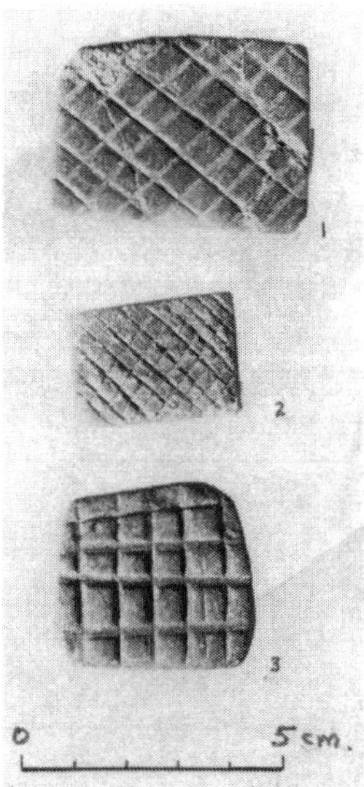

Figure 6.29:1–3. Stone seals from lower levels (1–2, Level 8; 3, Level 9)

thirteen pieces of copper ore (two from Level 7, four from Level 8, three from Level 10, and four from Level 12), which suggest the smelting of metal in the settlement itself, and four metal artifacts; a copper sheet bead (Level 7), a copper ring (Level 10), a ring-shaped pendant made of a copper plait (Level 11), and, perhaps most important, a lead bracelet (Level 12). The latter lay on virgin soil under the corner of room 345, the earliest in the central section of the excavations. It seems to have been buried there as a foundation deposit when construction began. The bracelet is 9.2–10 cm in diameter, and is made of a lead rod 7–8 mm thick. The ends are bud-shaped and thickened (Munchaev and Merpert 1981, 309, chapter 12 in this volume). Copper artifacts have also been found both in pre-Hassuna Tell Sotto (Munchaev and Bader 1979, 110) and, even earlier, in the preceramic settlement of Tell Maghzaliyah.

Note

1. The authors wish to express their gratitude to Prof. David Oates and Dr. Joan Oates for information about the unpublished results of the survey and access to the material. For a reference to the earliest sites, see J. Oates (1968:4, n. 9) and for Hassuna-Samarra settlement, J. Oates (1972:300).

7

THE ARCHAIC PHASE OF THE HASSUNA CULTURE

N. Ya. Merpert

The site of Yarim Tepe I was excavated by the Soviet archaeological expedition in Iraq from 1969 to 1976 inclusively. It is only the second Hassuna site to be excavated systematically on a broad scale, the first being Tell Hassuna, where large-scale excavations were conducted by S. Lloyd and F. Safar in 1944 (Lloyd and Safar 1945).[1] In this chapter, several questions concerning the early periods of occupation at the site will be discussed.

The upper levels of the site (1–7) are characterized by the abundance of "standard" Hassuna pottery, and imported Samarra vessels begin to appear with Level 6. The lower levels (8–12) are characterized by an overwhelming preponderance of "archaic" Hassuna pottery, including large, sharply carinated pots with supporting bases, and material culture reminiscent of the preceding Tell Sotto culture, including vessels with decorations in relief. The upper levels were exposed over an area of 1700 square meters, and the lower levels down to sterile soil over an area of 400 square meters. This scale of excavation allowed us to examine the entire cultural deposit over a broad area, the specific character of its individual building phases, and their association with each other.

In its entirety, the cultural deposit at Yarim Tepe I, which exceeds 6 m in depth, spans the entire Hassuna period as it was determined at the site of Tell Hassuna. Yarim Tepe I was occupied until the very end of the Hassuna period, and like Tell Hassuna, it lacks pre-Hassuna levels. By the material culture assemblage of the deposit, its thickness, and the period of its accumulation, Yarim Tepe I can be correlated with Tell Hassuna and can be differentiated from Umm Dabaghiyah and Tell Sotto. The Hassuna settlements at these latter sites were relatively short-lived, existing only in the archaic and the very beginning of the standard Hassuna pottery phases which lie directly atop pre-Hassuna occupations. The final phases of occupation at Yarim Tepe I contain relatively high frequencies of Samarra ware. However, these phases, beyond question, precede the initial occupation of the neighboring Halaf settlement of Yarim Tepe II, and, thus, the appearance of Halaf material culture in the Sinjar region. The analysis of the material culture from the upper levels at Yarim Tepe I and a detailed comparison of these levels with the Halaf sites is beyond the scope of this article. However, two conclusions related to this topic reached as a result of the Soviet expedition's work in the region are as follows:

1. The seven-meter Halaf deposit at Yarim Tepe II directly overlies the remains of a Hassuna settlement, located in the southern part of the mound. The latter existed both in the archaic and in the standard Hassuna periods, which is indicated not only by ceramic finds, but also by sculpture with close parallels at Yarim Tepe I. Thus, at Yarim Tepe II, one observes the second example (Tell Hassuna being the first) of a direct stratigraphic relationship between the Hassuna and the Halaf strata. At Tell Hassuna, the strata containing Halaf material, 6, 7, and 8 are thin and mixed with Hassuna elements according to A. Perkins (1957, 25), and only material culture from the late phase of the Halaf is present. However, at Yarim Tepe II, the Hassuna settlement is overlain by a Halaf stratum with a thickness of seven meters. This stratum contains no fewer than nine building phases, reflecting an extended period of Halaf occupation, by no means confined to the later part of the period.

2. The cemetery for the Halaf settlement of Yarim Tepe II is located on the slope of Yarim Tepe I, which had long since been abandoned. This stratigraphic association can also be viewed as extremely significant. One can add to this the complete absence of any Halaf material culture in the Hassuna deposit at Yarim Tepe I, including its uppermost levels.

The above facts indicate that the Hassuna settlement at Yarim Tepe I and the Halaf occupation at Yarim Tepe II did not overlap temporally in the Sinjar plain, nor did the cultures that these sites represent.

Comparative Stratigraphy

Turning first to the earliest building level (12) at Yarim Tepe I and also to the material recovered from slightly earlier pits dug into sterile soil, these can be related to Tell Hassuna Stratum 1a through a comparative analysis of the recovered ceramic assemblages from these strata, especially the decorated sherds. Comparison with materials not only from Tell Hassuna, but also from the earlier sites of Tell Sotto and Umm Dabaghiyah, is very important in establishing this relationship. Such a comparison firmly indicates, on one hand, continuity in the development of a number of vessel forms and decorative styles. On the other hand, distinctive changes, particularly in terms of the latter, can be discerned. Sophisticated decorative reliefs are found on many large vessels in the pre-Hassuna strata (Kirkbride 1972, 1973a, 1973b). These are present, both in Stratum 1a at Tell Hassuna and in Level 12 at Yarim Tepe I, but are degraded and vestigial in nature. In both settlements, only a few of the simplest molded decorations are found, and variation among them is relatively insignificant. A similar statement can be made for the marble and alabaster vessels. The stone vessel tradition, which was previously so well developed, was clearly ending as well, since it is only represented by isolated finds. Furthermore, painted ceramic vessels, which were commonplace at Umm Dabaghiyah (Kirkbride 1972, 13, Pl. X), have almost completely disappeared by the time of the initial occupation of Tell Hassuna and Yarim Tepe I, though a few isolated examples were found in the pits dug into sterile soil at the latter. With the founding of Yarim Tepe I, the origins of a new and distinctively Hassuna decorative style, primarily represented by painted designs on vessels, can be observed. This style is characterized by extremely simple geometrical motifs, especially in the early phases of the Hassuna, including simple wide ribbon designs, and straight-lined figures. These painted patterns are markedly different from the complex curvilinear compositions noted on vessels from Umm Dabaghiyah. The first stages in the development of this new decorative style are represented in the lower strata of both Tell Hassuna and Yarim Tepe I.

The basal levels of both sites are remarkably similar in material culture assemblages, and little can be found that would suggest that Yarim Tepe I, Level 12 is significantly earlier than Tell Hassuna Stratum 1a. However, it should be noted that in Level 12, finds of painted pottery, which number in the dozens, are significantly more numerous than in Stratum 1a of Tell Hassuna where only a single painted vessel was found. A subsequent increase in the number of painted vessels and in the complexity and variety of the painted designs is indicated in the stratigraphic sequences of both sites (Tell Hassuna Strata 1a–1c and Yarim Tepe I Levels 12–7). Incised ornamentation, which appears first in Level 10 of the latter, has also been found in Stratum 1a of Tell Hassuna (Perkins 1957, 2). Similar parallels can be found in other categories of artifacts as well, including large flint and obsidian flakes, flint projectile points, and stone axes. The available data indicate that Stratum 1a at Tell Hassuna and Level 12 at Yarim Tepe I are related and contemporaneous. Thus, one cannot assign Level 12 to a period preceding the original occupation of Tell Hassuna (Bader 1975, 111, chapter 3 this volume).

At both sites, the lower phases of occupation are virtually identical, reflecting a single stage in the development of early agricultural settlements within northern Mesopotamia with the onset of the Hassuna. The Hassuna assemblage was subsequently modified, but nevertheless, a remarkable degree of continuity can be traced over the course of a number of centuries.

The nature of the linkage between the Hassuna and the preceding early agricultural settlements, and the Hassuna's position within the developmental sequence of northern Mesopotamia have recently taken on special significance with the discovery of new sites such as Tell Sotto and Umm Dabaghiyah, representing new cultures that preceded the Hassuna in the region. N. O. Bader, excavator of Tell Sotto, stresses the connections between the latter and Tell Hassuna, to the point of concluding that a genetic connection could be traced between them (1975, 107–108, chapter 3 this volume). In this linkage, the results of the excavations at Yarim Tepe I are especially important.

First, it must be emphasized that substantial architectural elements were discovered in Level 12 of this site. These elements suggest the early presence of a well-developed building tradition which included a variety of structural types and preconstruction planning of complexes and of the settlement in general. Results from the Yarim Tepe I excavations not only supplement the Hassuna period database, but radically alter the previous picture presented by the excavators of Tell Hassuna that was based on the absence of permanent dwellings in the earliest stratum of the latter. At Tell Hassuna, only traces of ephemeral "tent camps" of hunters and herdsmen unfamiliar with plant cultivation were encountered (Lloyd and Safar 1945, 262, 271). The assertion that the earliest single room houses appeared in Stratum 1b of Tell Hassuna, overlying this "camp," is now largely discredited (Lloyd and Safar 1945, 272). The complex multiroomed structures of Level 12 at Yarim Tepe I, constructed using sophisticated building techniques, testify to an already well-established tradition that must have been developed in preceding periods. An analysis of the pre-Hassuna settlements in northern Mesopotamia, including Tell Sotto, Umm Dabaghiyah, and Kültepe, has led to the identification of numerous structural floorplans and building techniques that can be linked beyond question with the architectural remains of the earliest phase at Yarim Tepe I. Thus, both can be encompassed within a single tradition. N. O. Bader was entirely justified in writing: "The large multiroomed houses, so typical of Tell Sotto and Umm Dabaghiyah, have now been discovered in the lower strata of Yarim Tepe I. They must also be present in the lowest stratum of Tell Hassuna" (1975, 107, chapter 3 this volume).

The building tradition observed in the earliest Hassuna levels is already quite developed. Its ultimate origin must predate the settlements at Tell Sotto and Umm Dabaghiyah since both already possess relatively complex architecture. The roots of this tradition can only be speculated upon, and empirical evidence must be sought further back in time. Until recently, the dwellings at Jarmo dating from the seventh millennium B.C., with their walls set on stone foundations, were considered the earliest rectilinear structures known from Mesopotamia. Now, equally ancient and considerably more distinctive examples have been discovered at the margins of the Sinjar plain in the foothills of the Jebel Sinjar, several kilometers from Yarim Tepe I. A well-established tradition of rectangular structures is clearly represented at the multicomponent, preceramic site of Tell Maghzaliyah, which was also excavated by the Soviet expedition. The entire occupation sequence at this site dates from the seventh millennium B.C. at the latest. At this time, if not earlier, similar structures were constructed in southern Anatolia, stretching from Çayönü, where cellular and grill plan structures were found (Braidwood 1973; Braidwood et al. 1974) to Çatal Hüyük and Hacilar. One observes a similar picture in southwestern Iran adjacent to Mesopotamia (Hole 1977, 44ff.). This tradition has early parallels in the eastern Mediterranean as well. In certain areas of the latter, including sites within Palestine and Jordan, the tradition of rectangular structures overlies and is intermingled with an earlier tradition of circular dwellings, the latter possibly traceable back to the Mesolithic dwellings of the Natufian culture. This combination is encountered at Beidha and the transition between the two traditions is found in the preceramic levels at Jericho, where the shift from round to rectangular dwellings is connected with northern influences.

The above evidence supports the position that several architectural traditions within Southwest Asia can be traced to a period long before the beginning of the Hassuna, in which rather sophisticated building techniques and a variety of structural forms can be observed even at this early date. The possibility exists for a combination of these traditions within the Hassuna culture itself, suggested by the presence of both rectangular and round structures in the lower levels of Yarim Tepe I. As a general rule, in the construction of dwellings and associated outbuildings, rectangular structures prevail. This architectural tradition developed directly out of the Tell Sotto and Umm Dabaghiyah building traditions. It is interesting to note that this tradition was passed on during the Tell Sotto period to more southerly regions within Mesopotamia, where it appears in highly developed form in the lowest stratum of Tell es-Sawwan as an integral part of the Samarra material culture assemblage.

However, by no means should it be construed that architecture forms and building techniques are the only links between the Hassuna and preceding culture represented by Tell Sotto and Umm Dabaghiyah. One can list entire categories of artifacts recovered from Yarim Tepe I in which the basic prototypes can be found in these earlier settlements. This applies to flint and obsidian artifacts, and, indeed, to the entire stone tool and weapon assemblage, including adzes and pins. It also applies, although to a lesser extent, to the ceramic tradition. N. O. Bader, touching briefly on this question, recently stated that "the clay vessels from Tell Sotto and the lowest stratum of Tell Hassuna are very similar in ware and manufacturing technique. . . . The crude cooking ware and the thin-walled vessels of the second, third, and fourth phases of Tell Sotto include a number of transitional forms with the latter, such as ribbed vessels and oval tubs with corrugated bottoms" (1975, 107; chapter 3 this volume). J. Mellaart also noted that in the Hassuna culture, particularly its initial stage, there was "a great deal of crude pottery forms, including the husking tray, borrowed from the Umm Dabaghiyah culture" (1975, 148). It must be emphasized that the large vessels with corrugated bases found at Tell Sotto and Umm Dabaghiyah were the prototypes of the most diagnostic vessel found in the lowest level of Yarim Tepe I. P. Mortensen, in discussing the possibility of establishing a continuous line of ceramic development from Jarmo to Hassuna 1a, has stressed the complete absence of this vessel type at Jarmo which, as noted above, dominated the early Hassuna ceramic assemblage (1970, 148).

The origins of this diagnostic form and other basic forms of crudely manufactured pottery of the archaic Hassuna can now be securely traced. In our discussion along these lines, the previously mentioned hypothesis put forth by N. O. Bader that postulates a genetic connection between Tell Sotto and the archaic Hassuna deserves further attention. In contrast to Bader's position (1975, 107–108, chapter 3 this volume), this author does not consider the evidence at hand sufficient to establish a genetic link between the two sites such that they represent two phases within the development of a single culture. Perhaps it would be more accurate to differentiate two distinct developmental stages within the Neolithic of the Sinjar Valley, with each stage represented by related, yet independent cultures.[2]

Various differences between the Tell Sotto and Hassuna material culture assemblages have already been mentioned. These can be expressed in terms of artifact traits, including decorative styles and ceramic vessel forms, which are often quite sensitive cultural markers both spatially and temporally. One can also adopt the position that the differences between the assemblages of the two sites are more fundamental in nature, which is only explainable by the movement of new groups into the Sinjar region, even to the extent of completely divorcing the Umm Dabaghiyah/Tell Sotto culture from the later Hassuna. This position is taken by J. Mellaart, who concludes that the common features between Umm Dabaghiyah and Hassuna "are outweighed by the differences between the cultures" (1975, 141). However, his conclusion that "the origin of the Hassuna culture is not clear to any degree," and that "it certainly did not develop out of the Umm Dabaghiyah culture" (1975, 149), also seems overly extreme on the other end of the spectrum from Bader's position.

The material differences noted above do not obscure the fundamental role that the Tell Sotto/Umm Dabaghiyah culture played in the formation of the Hassuna culture. Therefore, the origins of the Hassuna can no longer be considered ambiguous. It is now untenable to argue that Hassuna culture was derived from a source external to northern Mesopotamia in an already developed form that replaced a local and considerably less developed culture of hunters and pastoralists (Lloyd and Safar 1945, 262). It should be noted that even before the discovery of sites of the Tell Sotto type, P. Mortensen, commenting on the origin of Hassuna settlements, surmised that it was "not at all impossible that they developed from still earlier settlements in the Tigris drainage" and that

> these sites remain to be found. Their discovery is problematic since in many cases they may be underneath later cultural accumulations or meters of alluvial deposit. The possibility that they might appear one fine day is confirmed by the recent unexpected discoveries, under ecological conditions not very different from those along the Tigris, of two settlements on the Euphrates, Mureybet and Bouqras, which date between the ninth and seventh millennia B.C. (1970, 132)

The earlier sites, whose discovery Mortensen foretold, have now been located and excavated. Settlements of the Tell Sotto/Umm Dabaghiyah type permit us to speak with confidence of the existence of a local substrate from which Hassuna culture developed, thus providing an answer to the question of its origin that can be supported with concrete evidence. We can assert with some confidence that in northern Mesopotamia, Hassuna culture did not arrive or develop in a vacuum and was not, as had once been thought, the earliest agricultural society in the Sinjar Valley. Furthermore, there is now little basis for considering the Hassuna a direct descendant of Jarmo or settlements of the Jarmo type

(Braidwood and Howe 1960, 161; Masson 1964, 51) or for a scenario placing the first agricultural commununities in the Zagros, from which they subsequently expanded into the fertile areas of the Jezira. Tell Sotto and Umm Dabaghiyah represent an essential link between cultures of the Jarmo and the Hassuna types (Bader 1975, 108–109). Extrapolating along this line of reasoning, one could speculate that specific antecedents of other traditions await discovery, particularly the Samarra and Halaf. Thus, the Hassuna cannot be considered the only early agricultural society or even the first in the Tigris/Euphrates drainage (Mellaart 1975, 141ff).

In reference to the Sinjar plain, the concept of an earlier substrate may be extended further back into time. Empirical evidence exists for the presence of sites that precede Tell Sotto and reflect even more ancient stages in the development of sedentary settlements. These sites are identical to Jarmo, in an evolutionary sense, in their role in the development of early agricultural societies. However, they cannot be considered as derived from Jarmo-type settlements, but represent an independent line of development in the foothills and plain of the Sinjar, a region that is comparable from an ecological perspective to the corresponding areas of Zagros.

As early as 1968, J. Oates noted the presence of material culture similar to the Jarmo assemblage at the site of Tell Ghassul, near Tell Afar. In 1977, within the same general area, in the border zone between the plain and the foothills, N. O. Bader discovered the multicomponent site of Tell Maghzaliyah mentioned above, which has an 8 m cultural deposit, essentially aceramic, that is contemporaneous with the settlement at Jarmo, if not earlier. Further excavation of this and other sites from the same period is required to establish the degree to which these sites were connected with the development of later cultures in the region, especially that represented by Tell Sotto. Also, a distinct possibility exists that sites which will provide new links with the past are yet to be found in the region. However, even given the limited evidence to date, we can confidently state that other sedentary settlements connected with the origins of a food-producing economy were present in the Sinjar plain. In light of recent discoveries, the comparison made by P. Mortensen, mentioned above, between the regions in the upper Tigris drainage and the area along the upper Euphrates where such early agricultural settlements as Mureybet and Bouqras are located, seems remarkably accurate. The origins of early agricultural societies in northern Mesopotamia can apparently be traced throughout a vast territory and to numerous sources.

Economy

One can now speak of considerable time depth in the development of agricultural settlements in northern Mesopotamia, particularly in the Sinjar plain. A discussion of the controversy centered on the findings of domestic-sheep remains at Shanidar is beyond the scope of this article, although a similar and undisputed discovery made in the lowest stratum at Ali Kosh in the Deh Luran Plain (Bus Mordeh phase) (Hole, Flannery, and Neely 1969, 344) lends support for animal domestication at a very early date. The Jarmo data pertaining to early animal husbandry are limited, though domesticated goat is certainly present, but evidence for other domesticated species is less clear. However, the continuation of the animal domestication process at this time is not in doubt. As for the domestication of plants, primarily cereals, the critical evidence from Jarmo is well known.[3] The corresponding data from Tell Sotto are only now being processed, although the early agricultural economy of the settlement has been firmly established.

J. Mellaart notes that Umm Dabagiyah

> is located on the margin of the dry farming zone and the high concentration of gypsum and other mineral salts in the subsoil makes irrigation impossible here. For this reason, the evi-

> dence for plant cultivation is limited at best. Although there is sketchy evidence for domesticated emmer and single-row wheat and barley, peas, lentils and standard bread wheat must have been imported, since they do not grow well in open steppe. (Mellaart 1975, 137; Helbaek 1972, 17–19)

However, given the fact that the Tell Sotto culture is found further to the north in the heart of the dry farming zone on the edge of the piedmont, the setting at Umm Dabaghiyah can be considered atypical of the culture as a whole. Among the osteological material obtained in the first field seasons at Umm Dabaghiyah, five types of domesticated animals were identified: sheep, goat, cow, pig, and dog (Bökönyi 1973, 9–12). Sheep and goats predominate among the domesticates, and cattle are present only in small numbers, indicating they are of little economic significance at this early date (ibid., Mellaart 1975, 137). Hunting played an essential role in the subsistence economy, especially of onager, which comprised 68 percent of the osteological remains, but also of gazelles, wild pigs, and in certain instances bison and badgers. On the whole, hunting still played a more important role than animal husbandry, since the bones of domesticated animals make up only 10 percent of the faunal assemblage. Nevertheless, the latter indicate that settlements of the Tell Sotto type possessed the complete Southwest Asian complex of domesticated animals with the exception of the horse and camel. Thus, they were well beyond the initial stages of animal domestication found at Jarmo.

The next stage in the development of a mixed economy based on plant cultivation and animal husbandry can be observed at Yarim Tepe I. The evidence from the lower levels, beginning with Level 12, does not significantly differ from the upper strata. The fully sedentary and long-term occupation of the site and its location in the Sinjar plain, currently one of the most fertile and actively farmed regions of Mesopotamia, indicate a fully developed food-producing economy. Only agriculturally based settlements can exist in the region for extended periods. The established character of the agricultural economy is suggested by numerous related structures such as grain-storage bins and grain-drying platforms, encountered in nearly all of the levels, beginning with Level 10. Similarly, agricultural tools, including grain-grinders, mortars, and pestles are found in each residential complex. In contrast to Tell Sotto, stone hoes are found only in small numbers, which is typical, with a few notable exceptions, of all known settlements of the Hassuna and Samarra cultures. Sickle blades have been identified by G. F. Korobkova through use-wear analysis. They were encountered in the earliest level at Yarim Tepe I, and their relatively small numbers can be attributed to the general paucity of the stone inventory and also, quite probably, to manual methods of harvesting that are even today widely practiced in this region that do not require the use of special tools.

Finally, direct evidence of plant cultivation comes from carbonized remains of grain that have been identified in all strata beginning with Level 12, where numerous remains were found inside a jar associated with burial 131 in tholos 333, and also within the ash pits near the base of the walls of complex 41, one of the buildings of the original settlement. These paleobotanical remains are now in the process of being analyzed. For this reason, this article will be limited only to general comments on the recovered plant materials from Yarim Tepe I.

The preliminary analysis of the Level 1–7 paleobotanical remains was undertaken by Z. V. Yanushevich, F. Kh. Bakhteev, and L. V. Prishchepenko in 1972. The analysis identified two species of wheat, hard (*Tr. durum Desf.*) and soft (*Tr. aestivum L.*), barley (*Hordeum* species), wild cereals of the family Gramineae, and chickpeas (*Cicer* species).

In 1978, G. N. Lisitsyna used flotation techniques to recover more than thirty charred remains of seeds belonging to various cereal and leguminous species from samples within the cultural deposit. It should be emphasized that most of these remains were found in the aforementioned ash pit within Level 12. Preliminary analysis of this material has yielded the identification of two forms of wheat, hard (*Tr. durum Desf.*) and spelt (*Tr. spelta L.*), naked barley (*Hordeum vulgare ssp. nudum*), wild cereals of the family Gramineae, and wild pulses of the Leguminosea family.[4] Thus, the most important cereal plants in the Southwest Asian complex of cultigens are represented in morphologically domesticated forms from the very beginning of the occupation of the site. Furthermore, they more clearly exhibit domesticated traits than the paleobotanical remains from the preceding periods.

A similar statement can be made for the osteological remains that suggest the presence of a developed form of animal husbandry. The initial inspection, carried out during the course of the excavation by S. Bökönyi, indicated the presence of domesticated sheep, goats, pigs, and cattle, along with hunted wild species. These preliminary observations were later verified by the eminent Soviet paleozoologist, the late Professor V. I. Tsalkin. The faunal assemblage is now being further processed by V. I. Bibikova. Without trying to anticipate the results of her study, I will address, with Bibikova's generous permission, only the most general data, relating to the cultural deposit as a whole, noting beforehand that the lower phases do not differ fundamentally in their faunal assemblages from the subsequent layers.

First, we find a marked difference between the faunal record at Yarim Tepe I and the osteological remains from Umm Dabaghiyah. At the former, the wild animals are limited in number, significantly lower than the corresponding indices for domesticated animals. The portion of the faunal assemblage analyzed to date by V. I. Bibikova and the others noted above, can be broken down into the following species:[5]

cattle (*Bos taurus*): 143 bones
sheep (*Ovis aries*): 171 bones
goat (*Capra hircus*): 110 bones
mixed sheep and goat (*Ovis aries* and *Capra hircus*): 380 bones
pigs, both domesticated and wild (*Sus domestica* and *Sus scrofa*): 355 bones
dogs (*Canis familiaris*): 49 bones
Mesopotamian fallow deer (*Dama mesopotamica*): 4 bones
mouflon (*Ovis orientalis*): 18 bones at least
bezoar goat (*Capra aegagrus*): 8 bones
gazelle (*Gazella subgutturosa*): 14 bones
onager (*Asinus onager*): 10 bones
leopard (*Pantera pardus*): 1 bone
jackal (*Canis aureus*)?: 1 bone

Even if we attribute half of the pig remains, which are difficult to identify as domesticated, to hunted animals, 82 percent of the faunal assemblage is still composed of domesticated species compared with the 10 percent noted above from Umm Dabaghiyah. Increasing reliance on domesticated animals and the corresponding decline in the role of hunting in the subsistence of the settlement are clearly evident from this data. Additionally, domesticated cattle already made up at least 14 percent of the herded species, and when included with the pigs, this number rises to 30 percent and probably considerably more, since the classification of only half of the pig remains as domesticated is probably conservative. A clear trend toward the increased exploitation of domesticated animals emerges, a direct consequence of the fully sedentary life of the early agricultural settlements.

Results analogous to those cited above were obtained from the faunal assemblages at Tell Hassuna, Matarrah, and other contemporaneous settlements. This article will be limited to a comparison of the evidence for the development of a food-producing economy at Yarim Tepe I and Tell es-Sawwan. The latter is a Samarra settlement at a similar level of development and contemporaneous with Yarim Tepe I from the earliest phases onward, though Yarim Tepe I was settled somewhat earlier. At Tell es-Sawwan carbonized plant remains were found in the ditch dug during the initial occupation of the site and later covered by the the remains of this occupation (Stratum I). An analysis of these remains showed that the inhabitants of the settlement cultivated emmer, single-row wheat, standard bread wheat, six-row barley, with and without awns, and flax. Certain morphological peculiarities of the standard bread wheat, the six-row barley and the large-seed flax indicate the practice of simple irrigation techniques.[6]

The numerous remains of carbonized grain, granaries, and agricultural tools in the subsequent strata demonstrate the continued development of an already-established agricultural tradition. A similar statement can be made for animal husbandry practices. The presence, from the initial occupation at Tell es-Sawwan onward, of domesticated goats, sheep, and dogs is clearly indicated in the faunal record, as is the modest role played by hunting in the subsistence economy, on the basis of the isolated finds of gazelle, red deer, bison, and onager, comparable to Yarim Tepe I (Flannery and Wheeler 1967).

Overall, the evidence for the presence of a food-producing economy from both settlements at comparable stages of development is remarkably similar. It conclusively demonstrates that the basic aspects of this economy were fully developed in the preceding period, and prevailed in the subsistence strategies of both the Hassuna and the Samarra cultures from the beginnning.

Trade

The surplus generated by this highly developed agricultural base created numerous opportunities for the inhabitants of the region to develop active and large-scale exchange systems. The necessity of this exchange is emphasized by D. Kirkbride in connection with the limited presence of natural resources locally, especially wood and high-quality stone (Kirkbride 1972, 1973a, 1973b, 1974, 1975). At Umm Dabaghiyah, as the excavator notes, many types of stone, lumber for building, and even certain plant species, such as lentils and peas, were brought in from outside. This statement is equally applicable to Yarim Tepe I. The high level of exchange activity and the variety of directions from which it was carried out are demonstrated at the latter by the remains of wooden structures, the presence of obsidian tools, and the discovery of large numbers of beads made from various types of imported stone, including chalcedony, carnelian, marble, and other types. Nonlocal raw materials, such as obsidian, marble, turquoise, carnelian, and copper, suggest relations, quite possibly mediated by intervening groups, with rather distant regions, including Iran, Anatolia, southern Arabia, and the Mediterranean littoral. To obtain the abundant natural resources of these regions, the early agriculturalists of northern Iraq could only offer the fruits of their highly productive lands in return.

In summarizing what has been presented thus far, it should be reemphasized that the Hassuna culture developed out of an already established and continually developing local food-producing tradition in northern Mesopotamia. The latter not only occupied the foothills and "hilly flanks" of the region, but also extended far into the adjacent plain. It is found both in the northern part of the Jezira and in adjacent areas to the west (the Euphrates plain) and to the east (Khuzistan). Even the earliest agricultural societies were involved in complex exchange interaction, and newly arriving groups were quickly drawn into local cultural and economic systems, further expanding them. These exogenous stimuli, together with internal developments, determined the course later societies would follow in forming new cultural configurations within Mesopotamia.

The Hassuna culture was one of these configurations. It can be viewed as a regional manifestation of a particular stage in the development of early agricultural settlements in northern Mesopotamia. The Hassuna is closely linked with the preceding local cultural stage represented by Tell Sotto, but can also be differentiated from it by certain material culture peculiarities that were influenced by the factors mentioned above. In the developmental sequence from the Tell Sotto to the Hassuna culture, influences from adjacent regions to the east and northeast (the Zagros), and the south and southwest, due to local displacements of relatively large groups, cannot be discounted. In any case, this is a complex, multivariate process, and it would be overly reductionist to search for a single variable responsible for the development of the Hassuna culture.

Conclusion

In concluding, I will touch very briefly upon questions concerning the chronological placement of the lower strata of Yarim Tepe I, since a more detailed consideration of these matters will be appropriate only after the sequence is published in detail. The material recovered from the lower strata can be grouped within the chronological divisions for the Hassuna culture previously proposed by the excavators of Tell Hassuna (Lloyd and Safar 1945, 161ff.), and also by Perkins (1957, 2ff.) and Mortensen (1970, 126ff.). However, the more recently excavated materials from Yarim Tepe I have also generated a need for further "fine-tuning" of this chronology. First, Lloyd and Safar's position that Stratum 1a of Tell Hassuna should be separated from the Hassuna culture proper, since the former is characterized by "crude Neolithic" rather than painted pottery more Chalcolithic in character found throughout the rest of the Tell Hassuna deposit, is contradicted by recent evidence. The presence of painted pottery in Level 12 at Yarim Tepe I, otherwise identical to Hassuna 1a, and the crude Neolithic pottery characteristic of the latter found through Level 8, permit us to replace a sequence that designates stratum 1a at Tell Hassuna as a "special complex" with one that stresses a continuous line of cultural development both at Yarim Tepe I and at Tell Hassuna from their initial occupations. This unity is further confirmed by all other aspects of the material culture, including burial customs and the use of metal.

On the other hand, Braidwood and Mortensen, who have always argued against the artificial separation of Stratum 1a from the rest of the Tell Hassuna occupation, have been proven correct in this respect (Braidwood 1945; Mortensen 1970, 103ff.). However, Mortensen concurred with Lloyd and Safar on the temporary nature of the earliest settlement at Tell Hassuna in the absence of substantial structures and the prevalence of crudely manufactured plain wares. Like the excavators of the site, he concluded that the original inhabitants were primarily hunters, a conclusion substantiated by the finds of projectile points and projectiles for slings. Thus, he doubted that the first settlers possessed a highly developed food-producing economy, although he did admit that they may have originated from sedentary settlements that were unknown at the time (Mortensen 1970, 130–31). The above cultural and economic traits formed the basis for Mortensen's phase "A" in his tripartite periodization that includes Tell Hassuna Stratum 1a, Ali Aga, and Operation VI:5 at Matarrah.

The sites mentioned above are roughly contemporaneous with the lower levels of Yarim Tepe I. It must be emphasized that the economic aspects that once distinguished Mortensen's Phase "A" have now been refuted, whereas the cultural differences remain to a considerable extent. Thus, the very basis for isolating Phase "A" needs to be reconsidered, since it does not differ from Phase "B" to the extent it once did (Mortensen 1970, 130–31) and, indeed, can only be distinguished by ceramic trait frequencies. In fact, the basic discerning marker between Phases "A" and "B" that also can be found at Yarim Tepe I, is the gradual transition from "archaic" to "standard" Hassuna pottery, noting that they coexisted over a significant time interval. This transition includes a number of changes in other aspects of the material culture, caused both by internal development and by external influences, the latter originating primarily from the Samarra region further to the south. In this respect, viewing differences in material culture as separate phases within a unified cultural sequence seems more appropriate. Mortensen's tripartite periodization, in our view, remains useful if Phase "A" is defined by the prevalence of "archaic" Hassuna pottery, Phase "B," by the appearance and subsequent proliferation of "standard Hassuna pottery"; and phase "C," by the continuation of the latter with the introduction of significant quantities of Samarra pottery and other aspects of its material culture.

The relative chronology of the Hassuna period was established by the excavators of the eponymous site. Further comparison has confirmed the basic positions of these investigators, while "fine-tuning" them. The chronological framework from which this sequence was developed encompassed early agricultural settlements over a broad expanse of territory, including the eastern Mediterranean (Jericho IX, Megiddo XX, Ras Shamra IVA–B, Amuq A–B, Carchemish), southern Anatolia (Mersin XXIV–XX, Sakçe Gözü), and northern Mesopotamia (Ninevite 1). However, many of these comparisons, particularly with the western sites, are based mainly on isolated and not particularly diagnostic finds (Lloyd and Safar 1945, 265ff.; Mortensen 1970, 131) as has been emphasized by J. Mellaart (1975, 149). However, over the past thirty years this temporal framework has been noticeably refined. Among the factors responsible for this, the following seem especially significant:

1. The discovery and excavation of sites of the Tell Sotto and Umm Dabaghiyah type, which more precisely established the position of the Hassuna culture in the northern Mesopotamian sequence.

2. The excavations at Tell es-Sawwan and Choga Mami, which illuminated the Samarra culture and the stages of its development. This, in turn, has allowed a detailed comparison with the Hassuna culture and its developmental sequence.

3. The excavation of Tell Shimshara, which permitted a more detailed comparison of the Hassuna culture with the Zagros group of sites.

4. Excavations in the Deh Luran Plain and the subsequent development of a new and extremely important chronological framework, from the ninth to the beginning of the fourth millennium B.C., in a region intimately connected with Mesopotamia. This framework made it possible to compare not only specific phases, but entire chronological epochs as well (Hole, Flannery, and Neely 1969; Hole 1977, 24–37).

5. The excavation of the Yarim Tepe group, which has made it possible to establish the direct relationship of the Hassuna period within the Sinjar plain, with the Tell Sotto culture at one end and the Halaf on the other.

In light of the data from these studies, the Hassuna culture follows directly upon the Tell Sotto culture, which, in turn, follows cultures of the Jarmo type chronologically, and in its early phases retained definite elements in common with them (Bader 1975, 108, chapter 3 this volume). The beginning of the Hassuna culture is close in time to the origins of the Samarra culture, e.g., the lowest stratum at Mattarah and Stratum I at Tell es-Sawwan, and the beginning of both may be synchronous with the Tell Sotto–Hassuna transition. In the Deh Luran Plain, the Hassuna culture can be synchronized with the Sefid and Surkh phases (Hole 1977, Figs. 43–49), which lie between the Mohammed Jaffar phase, contemporaneous with late Jarmo, and the Choga Mami Transitional phase contemporaneous with the Halaf (Hole 1977, 24–37).[7] Thus, the end of the Tell Sotto culture and the Mohammed Jaffar Phase yield the *terminus post quem* for the Hassuna culture; whereas Tell es-Sawwan and the Sefid and Surkh phases provide the *terminus ad quem* for the Hassuna and late Samarra cultures; and finally the Choga Mami Transitional Phase and the Halaf (at least within the Sinjar Valley) provide the *terminus ante quem*. Within this framework, the lower strata at Yarim Tepe I are contemporaneous with the Sefid phase in the Deh Luran Plain and with the two lowest strata at Tell es-Sawwan, preceding the appearance of Samarra painted pottery.

Let us now briefly review the corresponding absolute chronological data. The extremely poor preservation of the organic remains from Tell Sotto and Umm Dabaghiyah has prevented us from obtaining radiocarbon dates for these sites. The only date from this early period was obtained from a sample found in stratum XV at the site of Telul eth-Thalathat, located 40 km to the east of Tell Sotto, excavated by a Japanese expedition (Fukai, Hariuchi, and Matsutani 1970). This sample (TK-24) was dated at 5570 ± 120 B.C. (J. Sato, T. Sato, Otomori, and Suzuki 1969, 513). Kirkbride correctly associates stratum XV at Telul eth-Thalathat with Umm Dabaghiyah and consequently with Tell Sotto, but doubts the correctness of the date adduced, considering it too late for the end of that culture (Kirkbride 1973a, 1973b, 1974, 1975).

Mellaart considers the TK-24 date reasonable, but only for the Tell Sotto/Umm Dabaghiyah culture which is confined to the first half of the sixth millennium B.C. and possibly the very end of the seventh millennium B.C. (1975, 140–41). In this case, the Tell Sotto culture would logically occupy a chronological position between the end of Jarmo and the beginning of Hassuna periods.

The accuracy of the TK-24 radiocarbon date is confirmed, in our opinion, by the entire group of radiocarbon dates from sixth millennium B.C. Mesopotamia which has now been systematically analyzed. Based on stratigraphic and typological criteria alone, the final phase of the Tell Sotto culture cannot be chronologically separated to any significant extent from the initial phases of the Hassuna, and consequently, of the Samarra cultures (Bader 1975, 108, chapter 3 this volume). The lowest stratum of Tell es-Sawwan, marking the beginning of the Samarra culture, yields a date of 5506 ± 73 B.C. (P-855), close in time to TK-24 (Stuckenrath and Ralph 1965, 190). A similar result was obtained from a sample from the lowest stratum of another Samarra site, Matarrah: 5610 ± 250 B.C. (W-623) (Rubin and Alexander 1960, 182). Consequently, the middle and upper strata of the Hassuna and Samarra settlements, including Yarim Tepe I, have radiocarbon dates that span the second half of the sixth millennium B.C., up to its end. A date of 5349 ± 200 B.C. was obtained from a sample (P-856) from Stratum III of Tell es-Sawwan, a layer containing Samarra pottery (Stuckenrath and Ralph 1965, 190). The sole date obtained for Tell Hassuna comes from Stratum 5 and is close to the boundary between the sixth and fifth millennia B.C. (5080 ± 200 B.C.) (W-660) (Rubin and Alexander 1960, 180). Finally, two dates were obtained for Level 7 of Yarim Tepe I: 5090 ± 100 B.C. (LE-1070) and 5200 ± 90 B.C. (LE-1086). Thus far, there are no radiocarbon dates for the final phases of the Hassuna settlement. An analysis of a number of samples from the final Samarra stratum at Choga Mami yielded an average date of 4900 B.C. (VM-483) (J. Oates 1972, 49–50).

It must be stressed that the above dates were given without dendrochronological calibrations. Such calibrations would increase the antiquity of these dates by an average of five to six hundred years, making them roughly contemporaneous with the calibrated dates from Chagha Sefid in the Deh Luran region, especially the Sefid and Surkh phases (6300–5400 B.C.), for which F. Hole has proposed an absolute chronology (1977, 27). However, the author does not consider the issue of date calibration to be conclusively resolved, and thus, the uncalibrated dates were retained in this article. They should primarily be considered in a relative sense and only as gross chronological markers.[8]

In accordance with these data, the most probable date range for the final phase of the Tell Sotto culture is the second quarter of the sixth millennium B.C., perhaps closer to its end.[9] Consequently, the early phases of Yarim Tepe I and the Hassuna culture in general date from the end of the second quarter into the third quarter of the sixth millennium. The radiocarbon dates mentioned above from stratum VII at Yarim Tepe I give the approximate *terminus ante quem* for the early Hassuna. The end of the occupation, however, as was also noted above, precedes the origin of the Halaf settlement at Yarim Tepe II, which can be placed at the very beginning of the fifth millennium B.C. (the date of the penultimate occupation of that site is SOAN-1291: 4710 ± 40 B.C.).

Notes

1. Two other sites, Matarra and Tell es-Sawwan, which were previously classified as Hassuna settlements, must now be placed within the related, but certainly independent, Samarra culture.

2. This author's disagreement with N. O. Bader's position stems, not so much from the relationship between the sites under consideration, but from our differing conceptions of "archaeological cultures," a problematic issue at best both theoretically and as it is applied specifically to northern Mesopotamia.

3. Additionally, G. N. Lisitsyna's analysis in 1978 of the Tell Maghzaliyah data established beyond question the presence of domesticated grain at the site.

4. The author expresses deep gratitude to G. N. Lisitsyna, who kindly allowed him to use the results of her research, and provided the data mentioned here.

5. S. Bökönyi's field observations are listed in the publications of the preliminary results of the first excavation seasons at Yarim Tepe I (Merpert and Munchaev 1971a, 160).

6. This distinguishes Tell es-Sawwan from Yarim Tepe I, with the latter located in a region more favorable for dry farming. In northern Mesopotamia, unlike the middle Tigris region, irrigation systems were not necessary (see Helbaek 1964, 45–48).

7. Hole 1977, 24–37. Originally, we compared the Hassuna culture with the developmental sequence in Khuzistan based on the results obtained from the excavated material at Ali Kosh, from the Mohammed Jaffar phase to the Sabz phase (see Merpert and Munchaev 1971a, 161). Subsequent excavations at Chagha Sefid (Hole 1977, 24–37) made it possible to review this comparison and further refine it.

8. The dates obtained from Tell es-Sawwan and Tell Shimshara have not been presented because they were obtained from problematic samples (see Mortensen 1970, 136).

9. The Mohammed Jaffar phase of the Deh Luran region with an associated radiocarbon date of 5870 B.C. should also be synchronized with this period (Hole, Flannery, and Neely 1969, 334).

8

YARIM TEPE II: THE HALAF LEVELS

N. Ya. Merpert and R. M. Munchaev

a. Yarim Tepe II: The Upper Halaf Levels

Mention has already been made of the cemetery dug into the abandoned settlement on Yarim Tepe I (see also chapter 10). Typical Halaf pottery was found in the graves (see Fig. 6.6:7), as well as some stone vessels and mace-heads (see Fig. 6.9:3,4). Its most interesting feature is the form of some of the graves with a lateral chamber at the bottom of a vertical shaft, the earliest known occurrence of this method of interment. The settlement to which the cemetery belonged was located at the mound of Yarim Tepe II, about 250 m west of Yarim Tepe I and overlooking the brook, Joubara Diariasi, which has cut deeply into its eastern flank; the "half-mound" that remains has given its name to the whole site. Yarim Tepe II appears originally to have been roughly circular in plan, with a diameter of 100–140 m; its surviving length is 120 m from north to south, with a width of 25–40 m and a height of almost 9 m. In order to discover the character and thickness of the occupation deposit, we began in 1969 by excavating a step trench for stratigraphic purposes from the highest, central part of the mound down to the level of the brook.

The Halaf is one of the most fascinating of the cultures associated with early agriculturalists in the Old World. It was named after the site of Tell Halaf in Syria, excavated by Baron M. von Oppenheim (1943) almost fifty years ago, and since then, a considerable number of Halaf sites have been investigated. It now appears that the culture extended from northern Mesopotamia, including parts of Iraq and Syria, into southeastern Turkey, and that its influence was actively felt westward into Palestine and the Levant (Kaplan 1960, 32–36; Braidwood and Braidwood 1960; and Garstang 1953), and northeast to the region of Lake Van (Reilly 1940, 145–78) and even Transcaucasia (where a typical Halaf pot has been found at Kültepe, Nakhichevan, by O. A. Abiboulayev, and is on exhibition at the Museum of the History of Azerbaijan, Baku). At present two main groups may be distinguished, a western one including among others Tell Halaf, Brak, and Chagar Bazar (Mallowan 1947), and an eastern group characterized by Arpachiyah (Mallowan and Cruikshank Rose 1935) and Tepe Gawra (Tobler 1950). None of these settlements has been completely excavated, the best known at present being Arpachiyah, just east of Mosul, and many essential questions remain unresolved, notably the problems of the origin and early development of the culture and the interrelation between its various regions. In this connection the excavation of

Yarim Tepe I, occupying as it does an intermediate position between the western and eastern groups of sites, must prove of the greatest importance, and this was a prime motive in our decision to begin work there in 1969.

Architecture

An area of some 400 square meters adjoining the original step trench is now in the course of excavation. The upper levels have proved to be badly disturbed by numerous graves, of which some forty have already been excavated, and by many pits evidently dug for domestic purposes; about ten pits so far opened have yielded material of Late Assyrian aspect. Very little remains of the two uppermost building levels, and Level 1 was represented only by isolated patches of a hard-packed earth floor up to 5 cm thick. The remains of similar floors of Level 2 were found in many places at a depth of 50–80 cm below the surface datum, with, in square 23b, a curved pisé wall that appears to have belonged to a round structure; a large stone found nearby may have been part of a buttress supporting the same building. Level 3, averaging 1.1 m, but in places up to 1.5 m, below datum, (hereafter mbd) was also extensively damaged, but has yielded isolated dwellings with auxiliary structures and hearths as well as extensive remains of flooring, and a further two levels have been exposed at depths from 1.7 mbd to 2.8 mbd. We now have, in addition to a considerable quantity of artifacts of various types, information that contributes to the study of houseplans and building techniques both at Yarim Tepe and in Halaf communities in general in the fifth millennium B.C.

Yarim Tepe II shows no evidence of deliberate planning and, like other known Halaf sites, such as Arpachiyah (Mallowan and Cruikshank Rose 1935), was intensively built up. Almost all the dwellings are one-room mud brick houses, with the round tholos plan and a diameter of 3–4 m. In rare cases the walls were lined externally and internally with a thin gypsum plaster, painted in red. The best-preserved example proved to be a tholos in square 23b at a depth of 1.2 mbd–2.05 mbd, which was 4.5 m in diameter. Its walls stood to a height of not less than 80 cm and were on average 25 cm thick, reinforced in places by buttresses. Two doorways were found, on the northwest and southeast sides, of which the latter was evidently the original entrance, later blocked and replaced by an opening in the opposite wall. One of the stone door sockets remained in situ. A number of successive floors were cleared, each composed of fine, hard-packed greenish gray clay (at Yarim Tepe II this was obtained in the nearby stream and was used to coat floors and sometimes walls of domestic houses) and gypsum, and in the north wall of the house was an intriguing feature, a rectangular recess 55 cm high, 40 cm wide and 20 cm deep. Adjoining this tholos on the south, and slightly later in date, were the remains of another less well-preserved but interesting round house. It was divided internally into two unequal parts by a pisé partition, and the smaller section was in turn subdivided into two compartments that were undoubtedly used for domestic purposes, probably the storage of produce. The finds were informative, for they included considerable quantities of charred cereal grains as well as crushed storage jars, a large pestle and other objects.

Comparable structures designed for habitation and general domestic purposes have been unearthed in the lower building levels. In Level 4, squares 19 and 24, the remains of three adjoining houses were discovered at a depth of 2.0–2.8 mbd (tholoi 5, 6, 9). They had a regular, round plan and the walls stood to a height of 60 cm with an average thickness of 25 cm. Number 5, the largest, was 4.15 m in diameter and the smallest, number 9, 2.75 m. On the northwest side of number 6 lay a small pisé annex, 1.4 m × 1.2 m, opening directly off the round room. In the center of the latter were found traces of the hearth, while scattered around the floor were two stone querns, a pestle, a grinding stone and a quantity of sherds. The annex was used for storage, and yielded two jars, a sling bullet, one flint and one obsidian sickle blade, and a number of animal bones. A similar pisé house, with a round room 3.8 m across and an adjoining rectangular annex 1.55 m × 1.1 m, was found in square 29 at a depth of 2.7–3.0 mbd. Here the rooms were connected by an opening 45 cm wide, with a surviving sill 10 cm high. We may assume that this combination of a circular main room with a rectangular subsidiary annex is one of the standard house plans in the settlement (see Mallowan and Cruikshank Rose 1935, Fig. 13), but the most popular type of dwelling was still the single-roomed round house. To date, some thirty tholoi, including the large houses and smaller subsidiary rooms, have been found at Yarim Tepe II, more than in all other Halaf sites combined. At Arpachiyah there were ten; at Tepe Gawra, one; at Nuzi, two; and at Çayönü, in southeast Turkey, nine (Braidwood, Braidwood, Tulane, and Perkins 1969, 1275–76). The largest example of this type so far, tholos number 7, was also excavated in 1971; it was 5 m in diameter and had walls 60 cm thick.

It must be added, however, that rectangular buildings have been found at Yarim Tepe II. One was discovered in 1971 in square 19 at a depth of 3.1–3.4 mbd. It was approximately 3.5 m square with walls up to 20 cm thick, and in the east corner was a space 40–50 cm wide separated by a clay partition from the rest of the room, perhaps for cooking, since a small round portable hearth made of clay was found in it. Other groups of two or three rectangular rooms of different sizes have been found, and in the southeastern sector of the settlement was a particularly interesting complex consisting of several small rectangular rooms, 1.65 m × 1.2 m, 1.4 m × 1.1 m, and the like, one round structure 1.5 m in diameter, and two clay ovens.

There has hitherto been no generally accepted interpretation of the function of the tholos-type buildings in Halaf settlements, and their use as dwellings seemed particularly open to question. But the excavations at Yarim Tepe II, and also at Çayönü, in southeast Turkey, where a number of these round pisé buildings have come to light, have established their domestic character beyond doubt, and in Yarim Tepe II they clearly represent the basic type of house. Pisé buildings of the tholos type were also found in the earliest level of Yarim Tepe II, in the step trench dug in 1969. Thus, we may now confidently conclude that the circular pisé building is a typical form of residential architecture among the fifth millennium population of northern Mesopotamia, and it is in sharp contrast to the forms characteristic of Yarim Tepe I, although, as we have observed, several round structures were found in the earlier settlement. The archaeological evidence in general, and that of Yarim Tepe II in particular, demonstrates that the cultural and historical development of northern Mesopotamia in the Halaf period is characterized not only by a change in house plans, a further development of the agricultural and stock-breeding economy and a growth of population, but also by the appearance of some new elements in the material culture. It is not our purpose in the present article to establish the causes of these very considerable changes in the development of the region at the dawn of the Metal Age, but we should like to call attention to one very interesting fact. The latest archaeological research in Transcaucasian territories has established that, at about the same time, the round pisé building of the tholos type made its appearance in the Kur-Araxes valley as the basic form of domestic architecture (Kushnareva and Chubinshvili; Djaparidze and Djavakhishvili 1971). This can hardly be dismissed as a coincidence, especially when we bear in mind that imported specimens of characteristic Halaf painted pottery are first found in Transcaucasia at the same date. Questions arise, of the widest importance and interest, concerning the cultural and historical links between the ancient Near East and the Caucasus in the epoch of the formation and development of early agricultural communities, a subject which has hitherto been largely neglected.

We cannot conclude this summary discussion of the architecture of the Halaf period at Yarim Tepe II without a reference to the only building so far discovered that appears to be other than domestic in character. Unfortunately, this important pisé structure, in squares 23 and 28 at a depth of 1.6–2.35 mbd, was severely damaged by later buildings and pits, and we cannot yet present a restoration of its plan. The principal feature was a wall 8.4 m long, aligned approximately from northwest to southeast and standing to a height of between 40 and 80 cm. It was built in two phases; the earlier southeastern sector, about 4 m long with two adjoining chambers 6.0 m × 1.75 m and 3.5 m × 2.0 m, was constructed in the usual manner, with layers of light yellow clay 5–6 cm thick alternating with 2–3 cm thick layers of greenish gray clay, and was coated with plaster up to 4 cm thick. A 30 cm deposit of occupation debris accumulated against this structure before the building of the second phase, which extended the principal wall 4.4 m to the northwest. The extension, a solid mass of reddish clay, was buttressed at some points on both faces and varied in thickness from 0.4–1.0 m, and a small room, 1.75 m × 1.62 m, adjoined its south side at the far end. Unfortunately its damaged state leaves us in ignorance of the overall plan and many structural features that might have afforded clues to its purpose, but it is noteworthy that none of the usual debris of domestic occupation, including pottery, stone tools, and bones, was found within the building or in its immediate vicinity; we may conclude that it was some public edifice, perhaps related to cultic activities. Cult buildings are known to have appeared in an early phase of the development of agricultural communities (Mellaart 1965), but none has yet been discovered in northern Iraq as early as the fifth millennium, and in general only one building has so far been identified as a shrine in the context of a Halaf settlement. This is the remains of a rectangular pisé structure of several rooms, 6.25 m × 5.1 m overall, which was found by M. E. L. Mallowan at Tell Aswad in northern Syria and plausibly interpreted by him as religious in character (Mallowan 1946, 1956).

Ovens and Kilns

The remains of a considerable number of hearths and ovens of different construction, form, and purpose have been discovered in the various levels of Yarim Tepe II. These include small, round clay hearths, dome-shaped *tannur* ovens up to 1 m in diameter for baking bread, two chamber furnaces, and large kilns for firing pottery. The ovens were sometimes built into the walls of houses but were more often located close by, in the open air. In the northeastern sector of the settlement a closely packed group of round ovens of the tannur type was uncovered, all built of clay and up to 80 cm in diameter.

Stone and Bone Objects; Evidence for the Economy

Among the considerable range of objects recovered from the excavation of Yarim Tepe II were stone and obsidian implements, including sickle blades, querns, grinding stones, pestles, and so on, that reflect all the operations of harvesting and preparation of agricultural produce. About two hundred flint and, more commonly, obsidian blades originally formed part of composite sickles, and some still bore traces of the bitumen that was used to fix them in wooden or bone hafts. The corners of many blades show traces of sheen, indicating that they were set at an angle in the haft to give the sickle a serrated cutting edge. The stone material also included a number of ornamental pendants, and among small terracotta objects were a few female figurines and a stamp seal (see Figs. 6.10:12 and 6.9:6 respectively). A considerable quantity of car-

Figure 8.1:1-9. Halaf pottery

bonized grain was found and has been submitted for specialist study and identification, together with a rich faunal assemblage. The animal bones from the 1969 season were studied by the late Professor V. I. Tsalkin, who concluded that all the main domesticated species, including cattle, sheep, goat and pig were represented among the herds and flocks of the Yarim Tepe II settlement (see Bibikova 1981).

Pottery

The pottery includes all the known Halaf repertoire, displaying the standard range of techniques (Dabbagh 1965). It is handmade, tempered with a mixture of finely crushed stone, sand, and shell and commonly slipped and burnished. It falls into two main groups, painted and unpainted, of which the second is more numerous; it is noteworthy that the highest standard of modeling is displayed in the painted wares, especially the smaller vessels, among which there are some thin-walled examples of very fine quality. The absence of handles is a typical Halaf characteristic. The most frequent forms are bowls, cups (see Fig. 6.8:5,6,9; Figs. 8.1:5 and 8.2:4), and a variety of jars with bulbous bodies and low undifferentiated or everted rims (Figs. 8.1: 1,2,4 and 8.2:1–3). Many large jars were up to 50 cm or more in height, and some of these have been completely restored; they are of pear-shaped or globular form and constitute a type not previously reported from other Halaf

Figure 8.2:1–4. Levels 3–4. Halaf pottery.

not previously reported from other Halaf sites. As can be seen from the examples illustrated (Figs. 8.1 and 8.2), the decoration extends over the whole vessel from rim to base. Of very common occurrence at Yarim Tepe II are small oval basins with vertical sides, roughly made and with thick walls; only some of these are painted.

The extremely attractive Halaf painted pottery is richly represented. It is mainly monochrome, but there are a considerable number of bichrome and polychrome specimens; various shades of red and brown are the dominant colors. The bowls and cups are usually covered with painted decoration inside and out. Up to the present over 350 different motifs have been recorded from Halaf pottery. Various geometrical patterns prevail, such as simple hatched triangles and rhomboids, straight and undulating lines and zigzags, together with the more elaborate "chessboard", four-petalled flowers, the Maltese cross and the bucranium (Fig. 8.3:2). Of especial interest from Yarim Tepe II are the naturalistic motifs, such as the fragments of a cup with a representation in black paint of a flock of feeding aquatic birds (Fig. 8.3:5–6), sherds with realistic portrayals of goats and deer (Fig. 8.3:1,3), and even an entire scene with birds of prey attacking a flock of deer (Fig. 8.3:4); parallels for these motifs may be found in particular on the pottery from Tell Halaf (Dabbagh 1966, Pls. XIV–XXIX). Attention should also be called to two pieces for which no known parallels exist in the Halaf repertoire; the first is a bowl with its light yellow outer surface ornamented with large circles of solid brown paint (Fig. 8.2:4), and the second a sherd with a realistic representation, in dark brown paint on a light ground, of a rampant cheetah, *Acinonyx jubatus* (Fig. 8.4 [Oppenheim 1943]).

Since only upper levels of the Yarim Tepe II settlement have so far been partially excavated (however, see chapter 8b for subsequent work), the authors have deliberately refrained in this article from attempting any detailed comparative analysis of the pottery and formulating conclusions from it. At present we may confidently assert that several characteristic features of our material point to a close association with the western group of Halaf sites, while there are at the same time obvious traits in common with the eastern group. It is equally clear that in our pottery we have both shapes and motifs that are unparalleled in either group, and Yarim Tepe II occupies a special place among the sites of the Halaf culture in the fifth millennium B.C. It has also contributed to the establishment of its chronology, for the examination of carbonized grain found at a depth of 1.7–1.9 mbd yielded a radiocarbon date of 3210 ± 130 B.C. (Porada 1965, 177).

Figure 8.3:1–6. Levels 3–4. Painted Halaf pottery with naturalistic motifs.

Figure 8.4. Level 3. Painted Halaf sherd showing cheetah.

b. Yarim Tepe II: The Lower Halaf Levels

Excavation was initiated in the center of the surviving tell over an area of 600 square meters, whereas the lowest levels were exposed over an area of only 500 square meters (Figs. 8.5 and 8.6). In total, nine building levels have been identified (Fig. 8.7). The two latest are greatly disturbed by graves and pits of As-

Figure 8.5:1–3. Photographs from the first excavation season (1969)

syrian and Hellenistic dates. In the southeastern part of the tell, remnants of a small Hassuna settlement have been found, underlying the earliest Halaf level (about two hundred fragments of typically Hassuna pottery were found here).

In all the Yarim Tepe II levels the main form of house architecture was a single-room, domed tholos, 3–5 m in diameter; rarely, there is evidence of a flat roof. A considerable number of small tholoi were excavated, with diameters ranging from 1.5 m to 2.0 m, which served some domestic purpose. Rectangular structures occur, but only one, in Level 5, can be regarded as a dwelling (a single room house, measuring 4.4 m × 2.2 m); the others, sometimes divided by transverse partitions into smaller compartments, must have served some other purpose.

From the earliest level (9) all the structures were made of clay layers of varying lengths and breadths, averaging 5–6 cm thick. However, mud brick was also known to the inhabitants of the settlement, though they used it only rarely. In section 23b of Level 5 a part of a wall of a rectangular structure was built of rectangular bricks (23–25 cm × 14 cm × 7–9 cm) mortared with gray clay.

Of considerable interest are the ruins of a large building complex of cruciform shape uncovered in squares 18–19 at a depth of 3.80–4.85 mbd (Level 6). In the center of this unique building was a round structure (tholos 31) with massive walls 60–70 cm thick and an inner diameter of 2.60 m. The tholos was divided into two parts, of which the southern was the smaller. The northern portion was further subdivided by two transverse walls into three compartments (0.5 m × 1.0 m, 1.5 m × 0.7 m, and 1.4 m × 0.65 m). On the inside of the central compartment there have survived two flat alabaster slabs, and it seems probable that the upper part of the inner walls of the tholos was also finished with similar stone slabs. On its northern and southern sides the tholos was directly adjoined by rectangular structures; on the eastern and western sides, there was a corridor (0.5–0.9 m

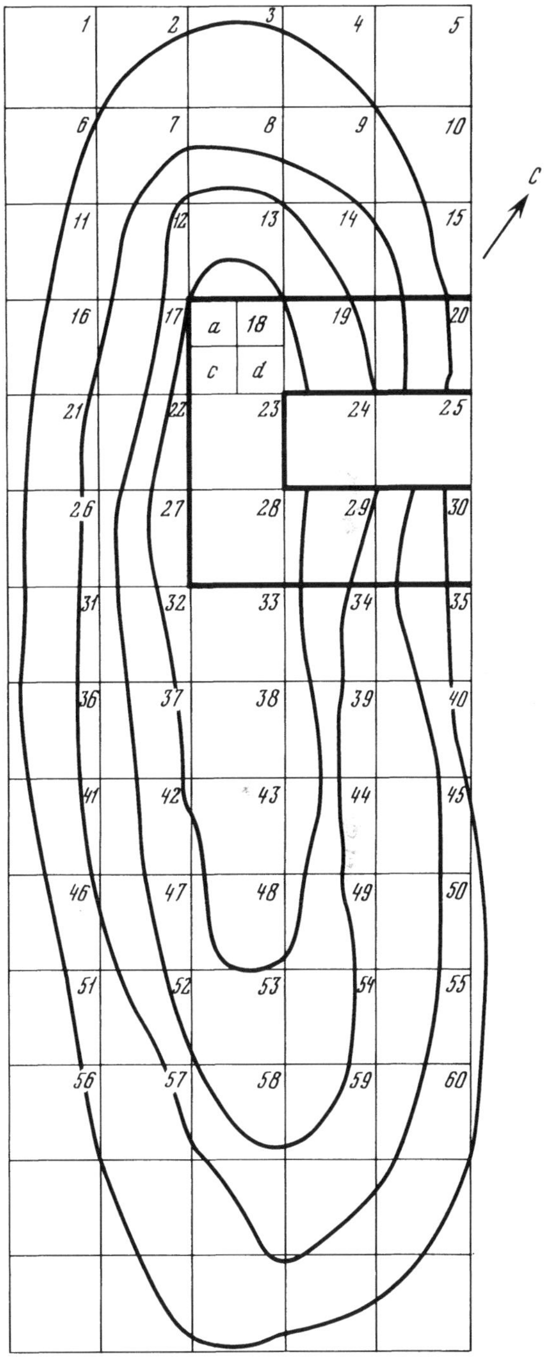

Figure 8.6. Plan of the grid over the site and excavated area

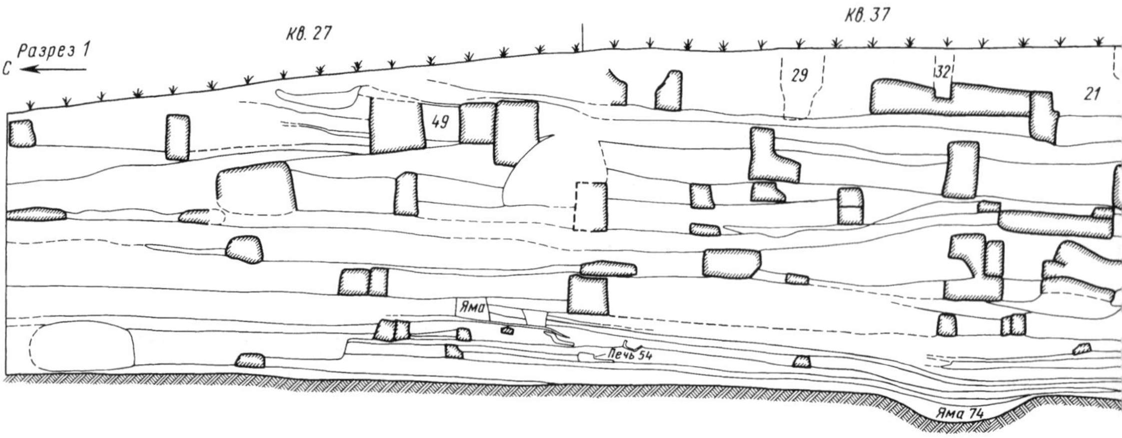

Figure 8.7. North-south section. Key: С = North; Яма = pit; Печь = oven; Разрез = cross-section; Кв. = room; Погр. = burial; Полы и границы ям = floors and borders of pits; Остатки стен = remainders of walls; Номера помещений печи = location number of ovens; *thin solid line,* floor; *line-and-hatch-mark,* walls; *number,* number of the room; *heavy solid line,* oven.

wide) between the tholos and its ancillary rooms. The tholos was further enclosed by rectangular multiroomed structures, comprising some twenty-two rooms in all.

What is the purpose of this structure? Undoubtedly, it is not a living complex. This is indicated not only by the very small size of the rooms but also by the lack of pottery and other domestic debris. Nonetheless, we do not believe that it was a "ritual" structure. Its most likely purpose would seem to have been a large public storehouse, standing in the very center of the settlement. It is important to stress that this complex was built on a site previously occupied by ordinary structures and that after its destruction, it was built over once again with ordinary buildings.

All the lower and middle levels of Yarim Tepe II yielded a considerable number of hearths and ovens. These were for the most part circular and oval bread ovens (of the tannur type) and, occasionally, large pottery kilns.

The Earliest Level of Yarim Tepe II

It is of particular importance to note the buildings and objects which were discovered in this earliest level and therefore characterize this original stage of the Halaf settlement (Fig. 8.8). Our description will begin in the southern section. Here, at a depth of 6.65–6.80 mbd in square 29, the remains of three narrow, rectangular, multiroomed structures were discovered, running along the north-south axis (355, 358, 365). The walls (25–35 cm wide) were made of clay slabs (5–10 cm thick), coated with a layer of mud plaster (in one case it was possible to determine the exact dimensions of the slabs: 85–87 cm × 22–24 cm × 8–10 cm). All three structures were built at the same time; their foundations rest on virgin soil at a depth of 6.76–6.80 mbd. They were probably granaries, but they did not function long and were destroyed during the time of this same earliest level. To the north, and contemporary with these structures, there was a further rectangular structure (307), of which only the transverse and some parts of the longitudinal walls have survived. Ruins of a clay oven (308, length 1 m), consisting of a combustion chamber and an ash pit, were found nearby, together with the remains of a further oven, sunk into the virgin soil.

To the west of this industrial area there stood several dwellings, including the remains of six houses of tholos type (Fig. 8.9). Four were excavated almost completely (61,

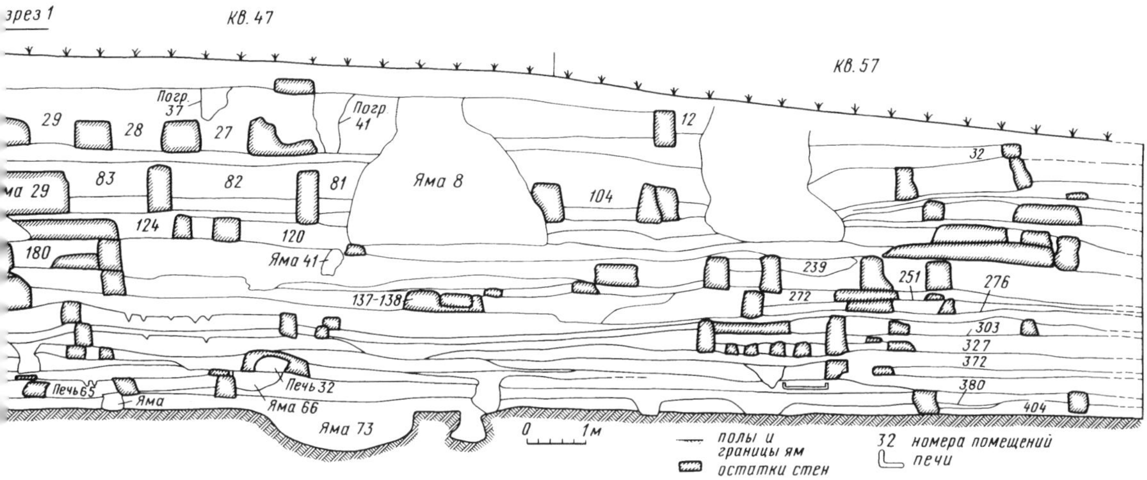

67, 71, and 73). Their walls, which survive in places to a height of a meter, were made of blocks of clay with an admixture of gypsum, coated with mud plaster. Southwest of tholos 71 there stood another, only partially excavated (75), with a round clay oven in its center. Tholos 61 contained two ovens, the southern of tannur type, the northern, apparently a hearth; the tholos itself had a flat roof. Tholos 67, the largest in this level, some 5.3 m in diameter, was reinforced on its eastern side by additional walls, inside and out. The walls of this tholos rested on a specially prepared clay platform. Beneath the floor plaster was a small depression containing fragments of an intentionally broken painted vessel with a corrugated surface (Fig. 8.10:7) and three microlithic obsidian tools (Fig. 8.10:1,2), covered with the ash of a fire. In addition, animal bones, fragments of vessels, five clay and two stone spindle whorls, a pendant of gray stone with circular incisions (Fig. 8.10:6), a pierced flat stone pendant (Fig. 8.10:5), a fragment of a clay figurine (Fig. 8.10:4), and a unique, triangular, copper seal-pendant (Fig. 8.10:3) were found under the floor just north of the depression. It seems possible that all these items were placed under the foundations when the house was built, and that they have some ritual significance.

The remains of two structures unearthed in the southern corner of the excavations belong to the earliest period of the settlement. They are a rectangular structure and a potter's kiln. They were built at the same time, and earlier than tholoi 67 and 75. (Tholoi 61 and 66 were later still.) Only the inner platform of the kiln (diameter 1.85 m to 1.98 m), with six flues some 15 cm in diameter, and the lower part of the fire chamber have survived (Figs. 8.11 and 8.12). In the center of the platform was a larger hole (diam. 35 cm to 40 cm); the fire chamber was connected to the upper kiln oven not only by the six smaller channels, up to 1 meter deep, but by the larger central hole as well. The entrance to the kiln, on the south, was oval (40 cm × 40 cm × 18 cm; the remains of the kiln have been transferred to the Mosul National Museum and are exhibited there).

No dwellings were found in square 19, only the remains of a few ovens and pits like those found elsewhere at this level. Some of these pits were quarries for clay, but a few, such as 60 and 61, were intentional constructions. These were sunk some 2.7 m into virgin soil and appear to have been for water storage.

Thus, we can see that Yarim Tepe II was densely built upon from its inception. Indeed, such dense construction was characteristic of

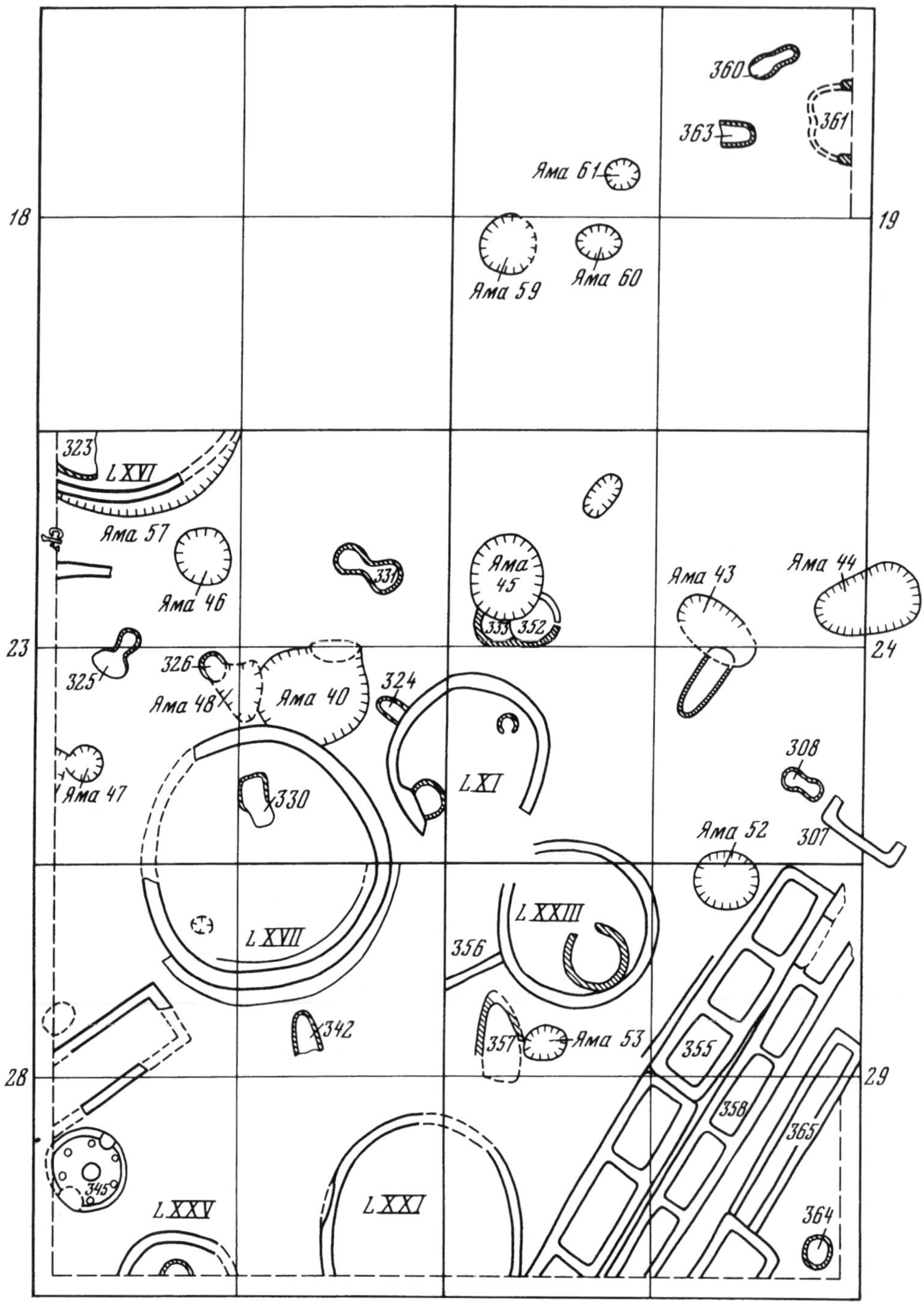

Figure 8.8. Plan of the earliest level (9)

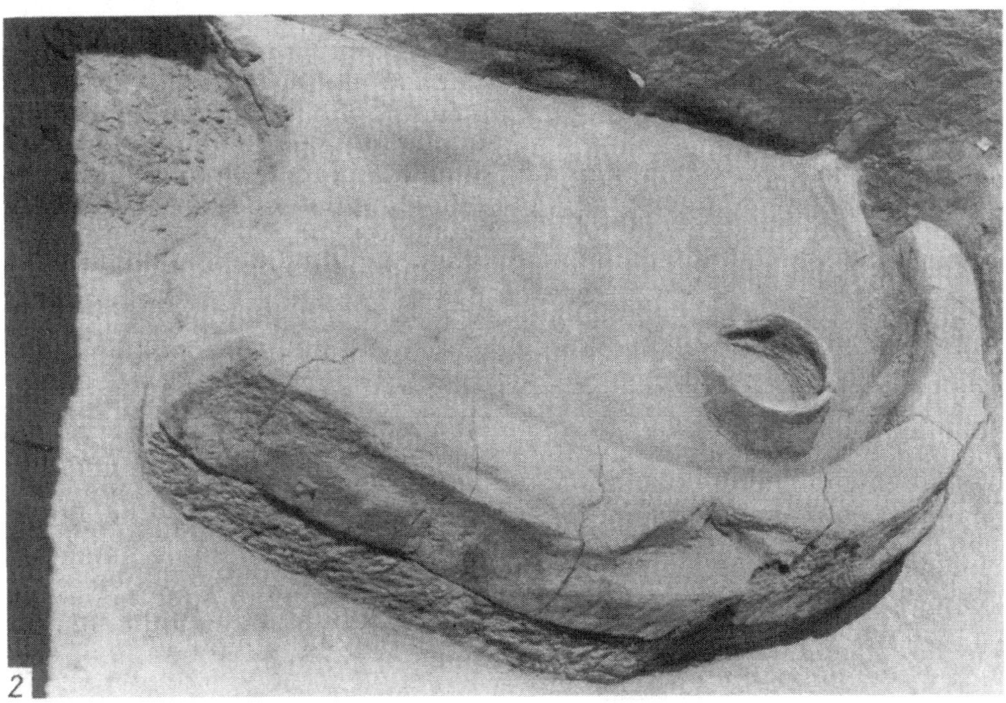

Figure 8.9:1–2. Remains of domestic auxiliary structures (tholoi) 1.73, 2.61

Figure 8.9:3. Remains of domestic auxiliary structures: tholos 67

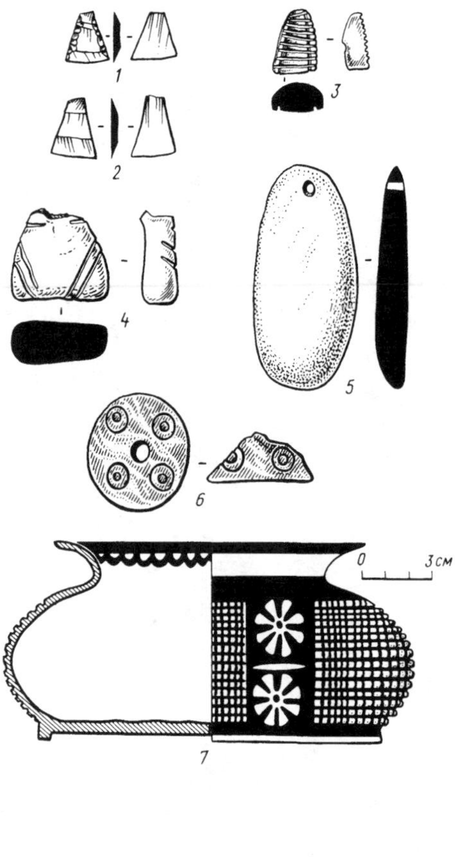

the settlement throughout its existence. To summarize, from the earliest period of the settlement its inhabitants built not only houses of a specific architectural form but also various structures for other domestic purposes, such as grain drying and storage, kilns for baking pottery, and reservoirs. Moreover, all the structures in the southern half of the excavated area were oriented parallel to the north-south line. On the whole, the same pattern of planning and orientation of buildings is observable also in the later stages of the settlement's life.

During the period of Halaf development in the Sinjar plain the traditional Mesopotamian techniques of construction remained the same as those in use at Yarim Tepe I. Apart

Figure 8.10:1–7. Objects from tholos 67: *1–2*, Obsidian tools; *3*, Copper seal-pendant; *4*, Fragment of a clay figurine; *5*, Stone pendant; *6*, Stone spindle whorl; *7*, Painted vessel.

Figure 8.11. Pottery kiln (structure 345)

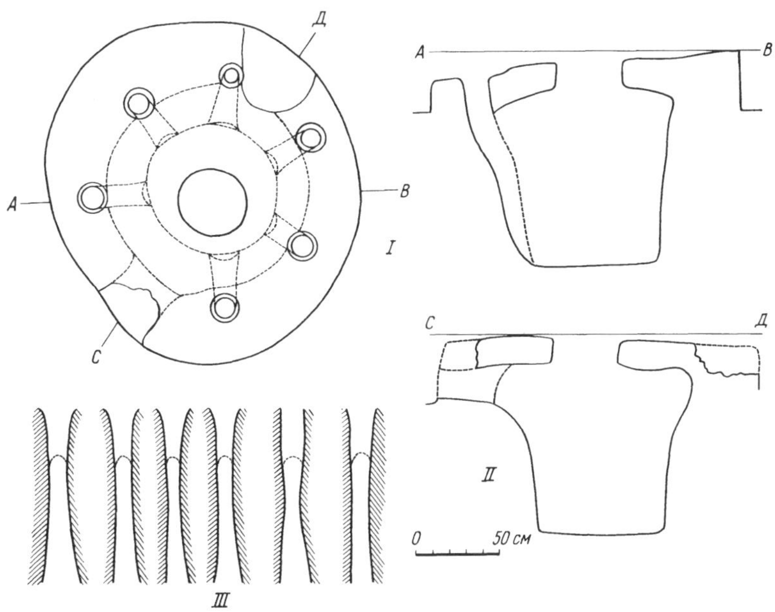

Figure 8.12. Pottery kiln 345 plan and sections

from the common yellowish brown clay, gray-green loam was widely used; this is found in almost every area of North Mesopotamia. This loam has important properties such as good resistance to moisture and quick drying after rain, properties that were, probably, identified from the very earliest stages of settlement here.

In the earliest level of Yarim Tepe II almost all of the structures are built of clay layers. Such a technique is typical for settlements of the Halaf culture as a whole as well as for those of the preceding Hassuna and subsequent Ubaid cultures. The width of the layers depended on the thickness of the walls, their thickness varied from 6–12 cm, with the length sometimes reaching 1 m. For instance, the walls of an auxiliary structure found in the sections of square 29 were made of layers measuring 85–87 cm × 20–22 cm × 8–10 cm. From the earliest level the houses of Yarim Tepe II are surprisingly uniform. They were single-room round structures with flat or dome-shaped roofs. The largest among them is tholos 67 with a diameter of 5.3 m (Fig. 8.9:3). It has been established that the construction of the tholos had been preceded by a ritual ceremony. The question arises whether this large building was a cult structure, since the tradition of burying ritual objects under public buildings is well known in later Mesopotamian tradition.

Thus, in the earliest period of the Halaf settlement a round single-room house was the basic form of dwelling. However, in the earliest level no tholoi were connected to any auxiliary rooms, whereas in the later levels there were usually small rectangular structures attached to one or, often, two, or even three sides. But does this observation apply to house forms in other Halaf settlements? Round structures of the tholos type have been discovered, as is well known, in Halaf levels of settlements situated in the Tigris and Euphrates valleys as well as in southeastern Turkey (Mallowan and Cruikshank Rose 1935; Tobler 1950; Mellaart 1966; Mellink 1964; Braidwood, Çambel, Lawrence, Redman, Stewart, and Stewart 1974), i.e., over the entire territory occupied by the Halaf culture. On this basis, the round house should be the most typical, but is this true for all stages of Halaf development? At present, it is not possible to provide a simple answer to this question. In particular, the two best-known Halaf settlements, Arpachiyah and Yarim Tepe II, provide apparently different pictures. On the basis of the new excavations (1976) I. Hijara believes that at Arpachiyah the early Halaf levels (XI–IX, following Hijara) are characterized by rectangular architecture (Hijara 1978, 1980). Round houses are present in Levels VIII–VI and, especially, in V–II (which correspond to Level TT10 of Mallowan). According to Hijara, in the final Halaf stage (Level I or TT6) the architecture becomes again rectangular; at Yarim Tepe II, however, we have demonstrated that the architecture remains essentially the same throughout all the building levels. The conclusions at Yarim Tepe II are based on extensive excavation of the site, whereas Hijara's reconstruction of the architectural history of Arpachiyah reflects the evidence of a single trench, which can hardly be regarded as convincing. We obviously need new data from a number of sites to resolve such questions finally. Such data should also help to solve the problem of Halaf origins.

Cult Objects

The mortuary data from Yarim Tepe II is discussed in detail in chapter 10. These data illustrate the diversity of Halaf burial ritual and the general complexity of ideology of these early agriculturalists in North Mesopotamia. In addition to the burials, there are other examples of ritual at Yarim Tepe II: special pits were found with remains of fire, the burnt bones of animals, and intentionally broken vessels. One such pit (0.6 m × 0.5 m × 0.2 m, section 19a, depth 5.45–5.64 mbd) contained scattered fragments of a unique painted anthropomorphic vessel in the shape

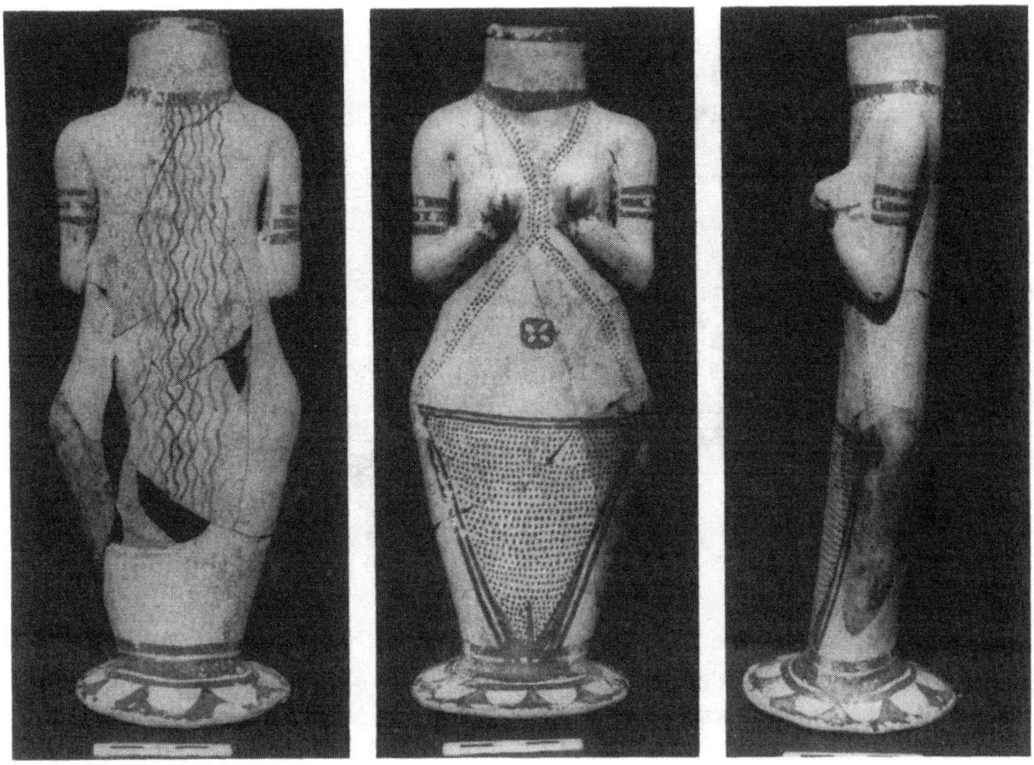

Figure 8.13:1–3. Views of a clay anthropomorphic vessel

of a female figure (Fig. 8.13), a broken alabaster plate (see Merpert, Munchaev, and Bader 1981: 45, Fig. XV, bottom right), a quadrangular stone seal with a carved design and an eyehole on the back for suspension, and the fragments of a simple clay bowl. Another pit (0.8 m × 0.5 m, section 28a, depth 6.6 to 6.8 mbd) contained fragments of several vessels, including a zoomorphic vessel which resembles a pig (Figs. 8.14 and 8.15). Although zoomorphic vessels are known at other Halaf sites, these two specimens are unique.

In the absence of similar cult objects at other Halaf settlements, no detailed interpretation can yet be given. But a few observations can be made. In the first place, the evidence for ritual ceremonies described above was found in the same area of the settlement as the remains of inhumations.

Moreover, the presence of unique objects in such contexts would appear to be characteristic. The copper seal-pendant found in the deliberate deposit under the floor of tholos 67, the earliest metallic seal so far discovered, belongs to this same group of unique objects. The trapezoidal microlithic obsidian tools found under the floor of the same tholos are also unusual. It should be emphasized that this evidence for ritual is found only in the early levels at Yarim Tepe II; moreover, it seems likely that the inhumations and the other evidence for ritual are related.

Stone and Bone Artifacts

By far the largest group of stone tools from both the early and the later levels at Yarim Tepe II are for agricultural purposes. They include grinding stones for grain made of local limestone (twenty-five from the lower two levels), pestles (ten), mortars (ten), and a few

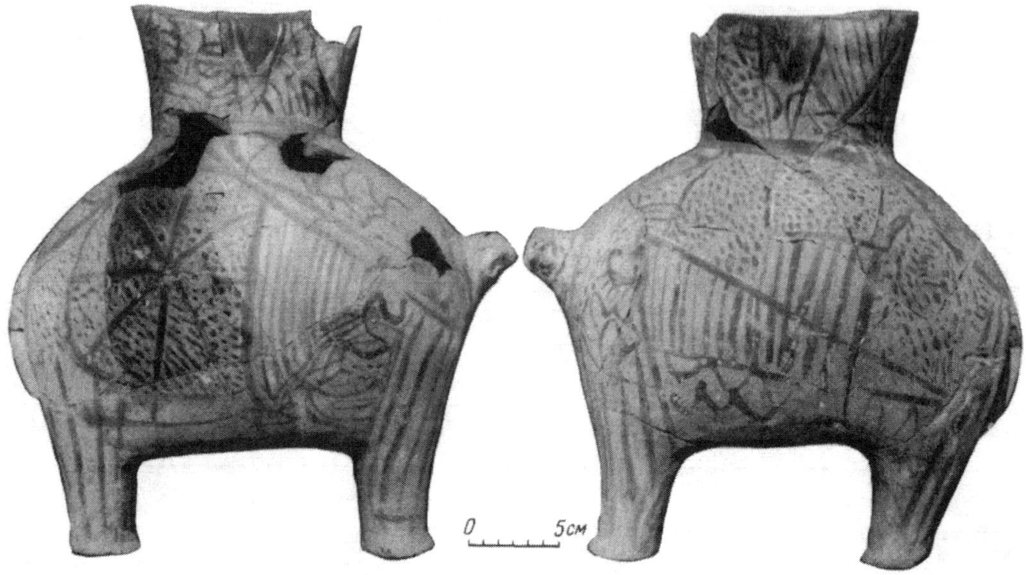

Figure 8.14:1,2. Both sides of a clay zoomorphic vessel

rounded stones that are possibly digging stick weights. Two basalt mortars are particularly highly polished and have a very distinctive shape. Flint and obsidian tools are plentiful, especially in the early levels (Figs. 8.16–8.18). Flint is most common as semifinished products and production waste (350 specimens altogether from Levels 8 and 9), whereas obsidian occurs almost exclusively as finished tools (180 objects). Obsidian tools are three times more numerous than flint. As at other Halaf sites the most common tools are narrow retouched blades, which were hafted to form knives or sickles (cf. Merpert, Munchaev, and Bader 1981: 43–44, Figs. XIII–XIV). Flint was also used to make scrapers, drills, and sharp points. As in the earlier Hassuna phase at Yarim Tepe, the obsidian was imported from Anatolia, whereas the flint was local.

Among the other stone objects from Yarim Tepe II (Levels 8 and 9) are one diorite and two alabaster conical spindle whorls, and five complete small vessels (Fig. 8.19) as well as fragments of ten others. The stone bowls were made of alabaster, limestone, marble, and other types, and do not differ, on the whole, from vessels found in the later Halaf levels or on other Halaf sites. Their fine quality attests to the skill of the Halaf stone workers, as does that of other objects, in particular the pendants and beads found at Yarim Tepe II. The former are of two types: flat, pierced oval ones (Fig. 8.20:1,2) and seal-pendants (Fig. 8.20:3–11). Nineteen of the latter have been found at Yarim Tepe II, seventeen of which come from the lower half of the deposits. However, such objects are not yet attested in the earliest level and only four come from Level 8. The seal-pendants are small and occur in a variety of forms. They are made for the most part of steatite and serpentine, but there are some of marble, sandstone, diorite, and rock crystal. Such stone artifacts are very characteristic of the Halaf culture, the most significant collections coming from Arpachiyah and Tepe Gawra (Mallowan and Cruikshank Rose 1935, 90–96, Fig. 50, Pl. VII; Tobler 1950, Pls. XCIIb, 1–4, CLXII, 17–19). Indeed it would appear that the Mosul area was the center of production of such objects. We have already noted the carnelian,

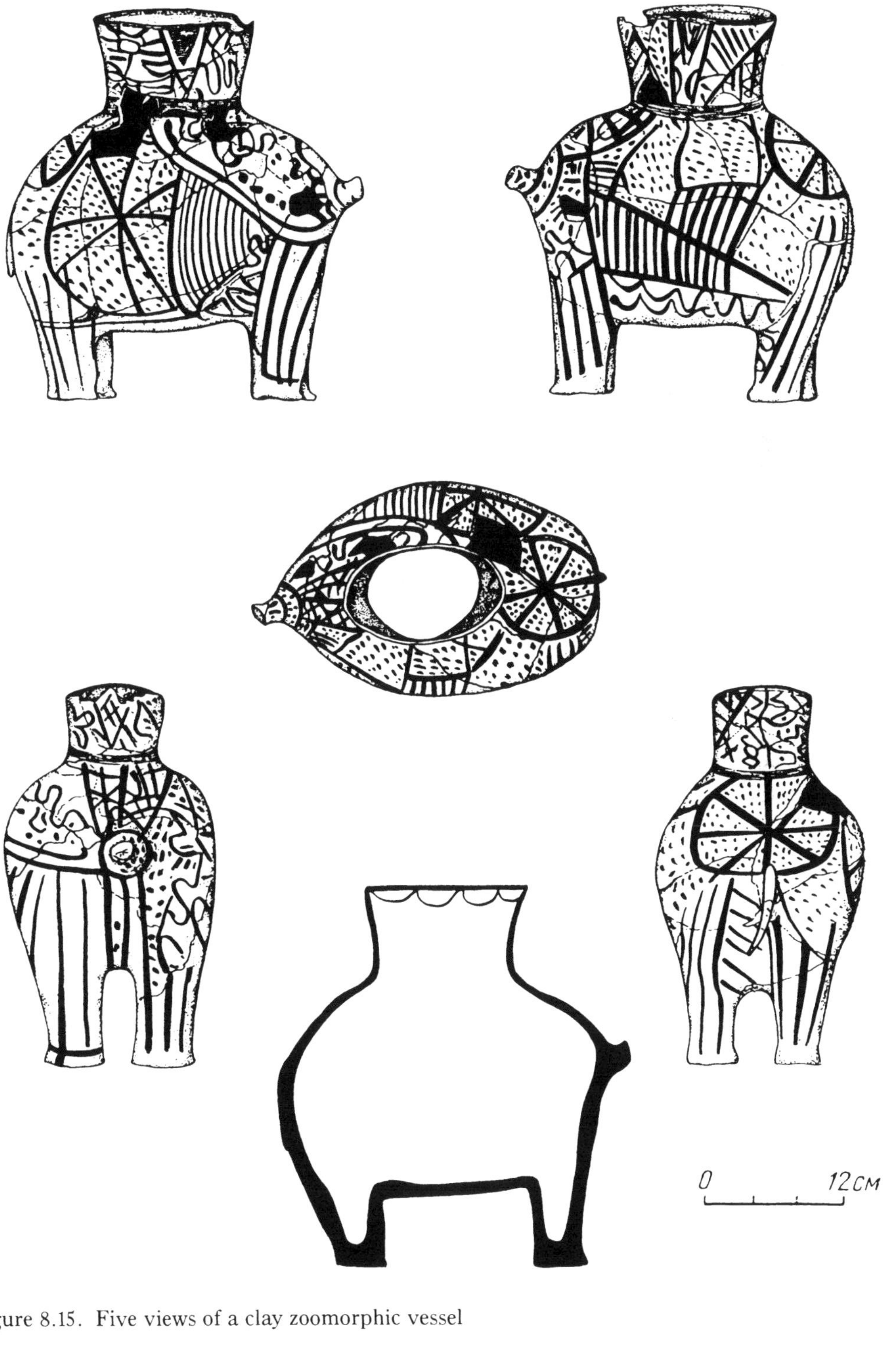

Figure 8.15. Five views of a clay zoomorphic vessel

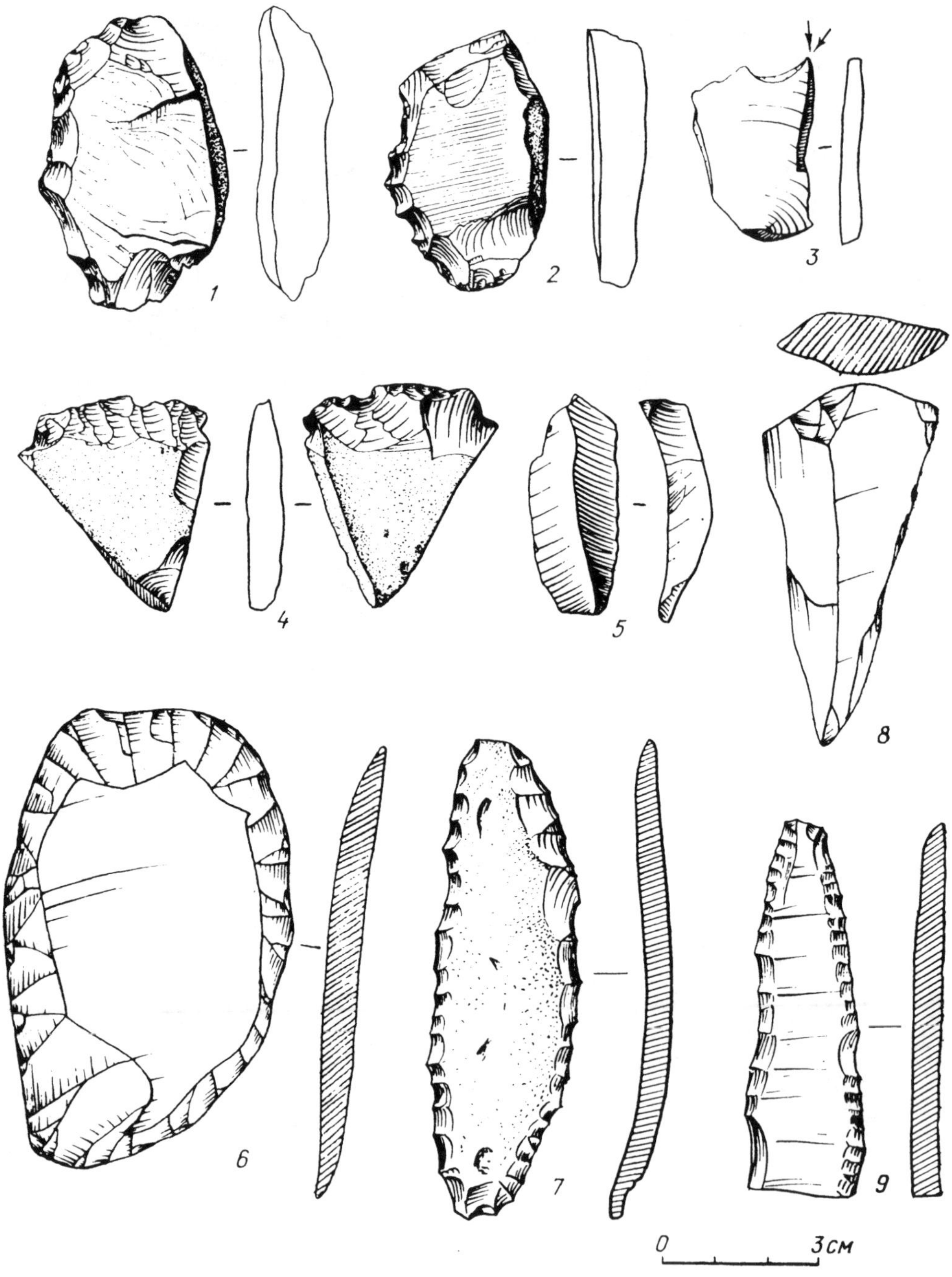

Figure 8.16 Flint tools

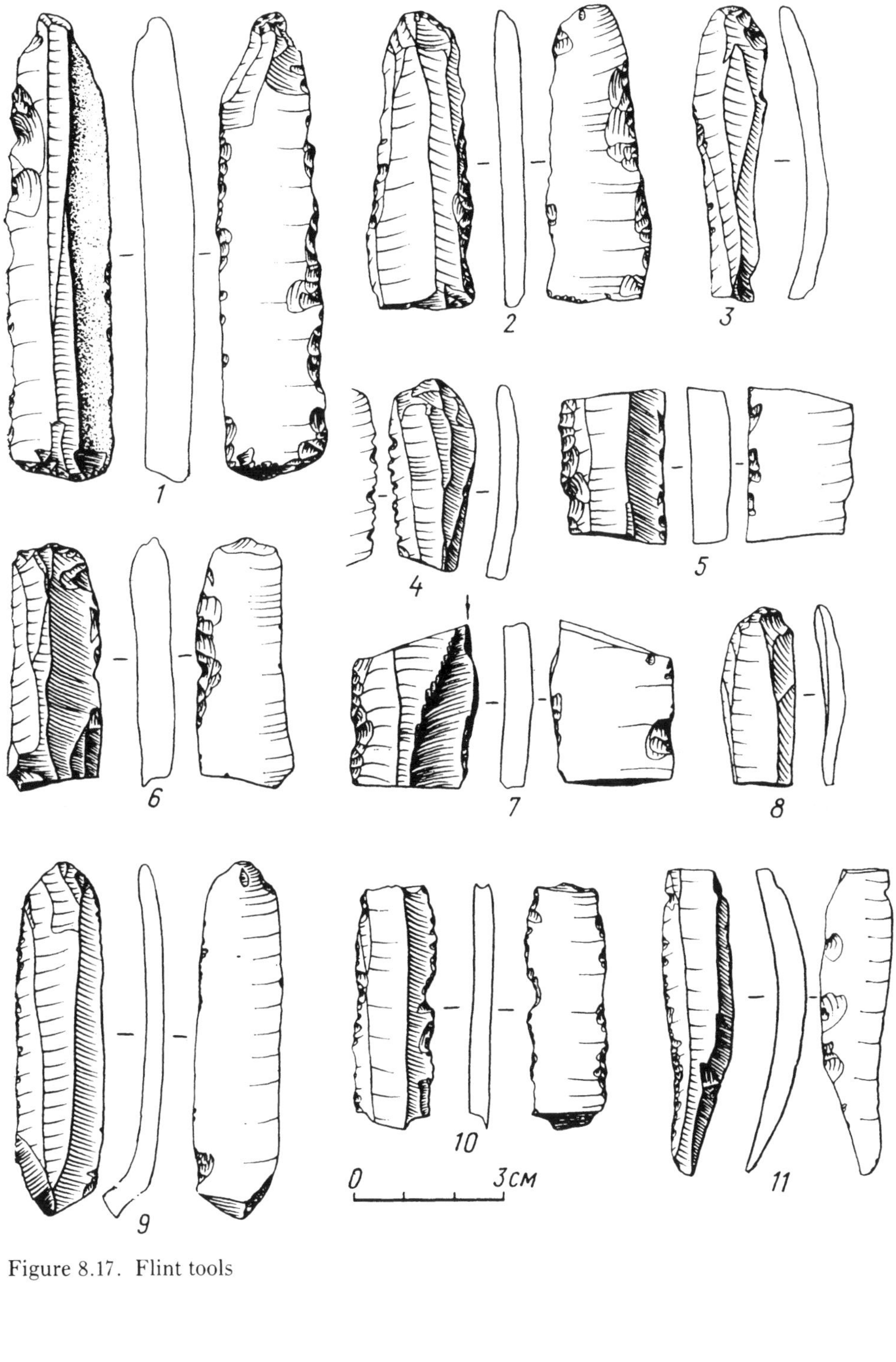

Figure 8.17. Flint tools

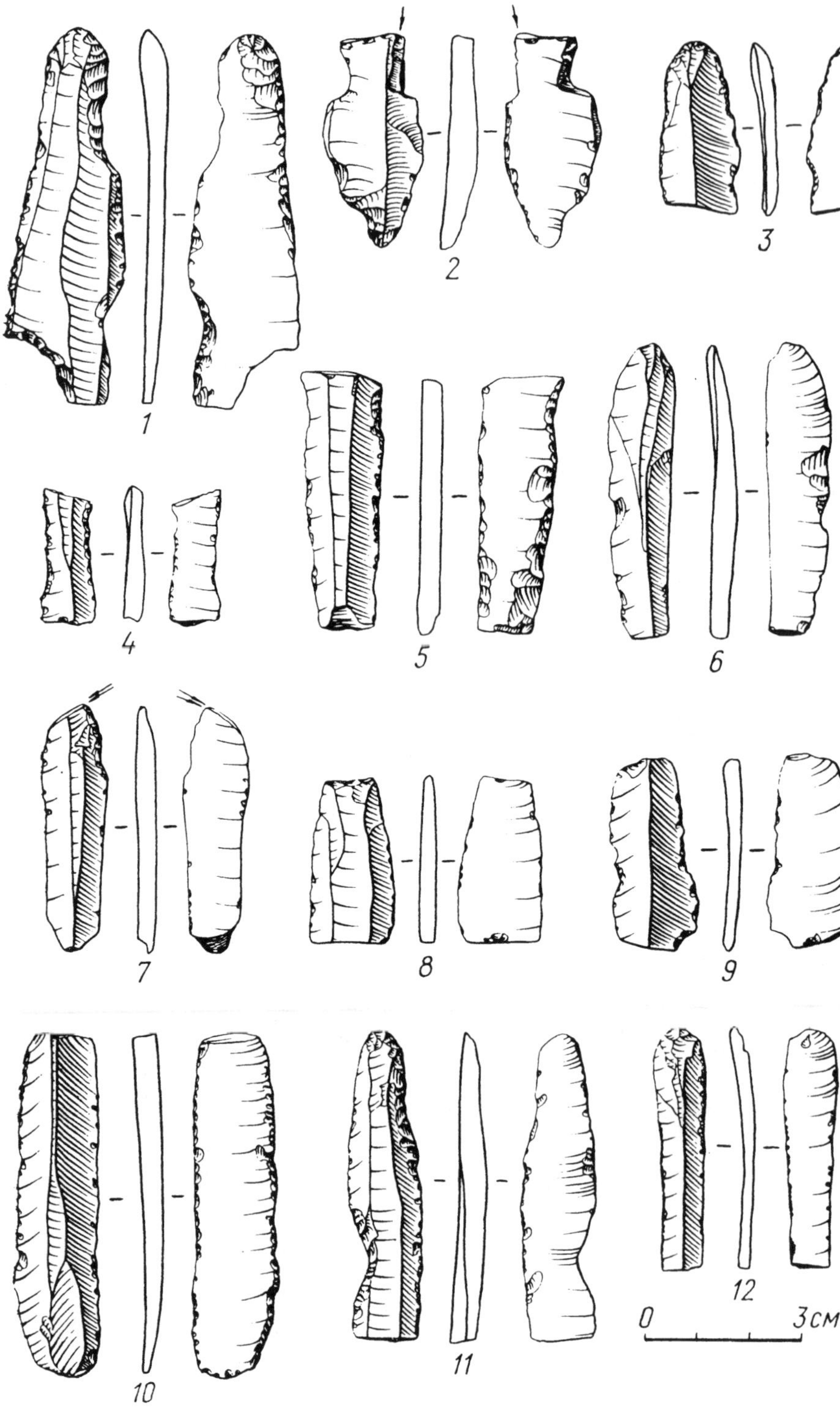

Figure 8.18. Obsidian tools

Figure 8.19:1–6. Stone vessels

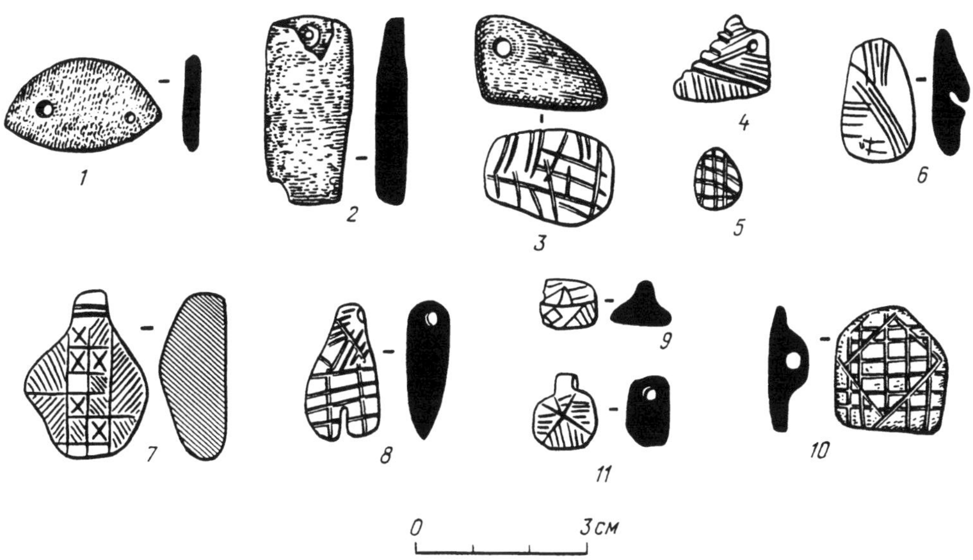

Figure 8.20:1–11. Stone seal-pendants

rock crystal, and steatite beads found in burial 58. Bone objects are standardized but not common: eleven awls and two spatulas only were found in the lower levels.

Pottery and Other Clay Objects

Clay objects occur in enormous numbers in all levels of Yarim Tepe II. Such artifacts include scrapers and spindle whorls. The former are made of potsherds and are oval or triangular in form; the inside working edge is coarsely chipped (see Fig. 8.21). Such objects occur in all levels (more than a hundred in Level 9 alone) and at other Halaf sites. Presumably they were used for preparing animal skins. In Levels 8 and 9 more than thirty spindle whorls were found, mostly biconical. "Sling bullets" were also found in all levels and are typical of the early agricultural settlements of Mesopotamia from the pre-Hassuna period onwards.

The pottery vessels are divided into three main groups: painted, simple smoothed, and coarse. In the lower levels of Yarim Tepe II (depth 5.2–6.95 mbd) 25,397 fragments of different vessels were found, including 10,707 painted sherds. This figure, however, does not take into account the complete vessels from the level or pottery from the burials and from the 1969 trench. Of these, 12,175 sherds were recovered from Level 9 (4,997 painted, 6,524 smoothed, and 654 coarse). The unpainted smoothed pottery predominates, providing approximately 55–60 percent of all pottery recovered throughout all levels at Yarim Tepe II. Roughly 40 percent of the sherds were painted, but the proportion of coarse ware is insignificant. All the vessels are handmade. Among the coarse pottery, thick-walled and flat-based jars or bowl shapes predominate (Fig. 8.22:2,3). Their form, color (usually dark brown), and the poor quality of clay and firing make them very conspicuous among the rest of the ceramic repertoire; the

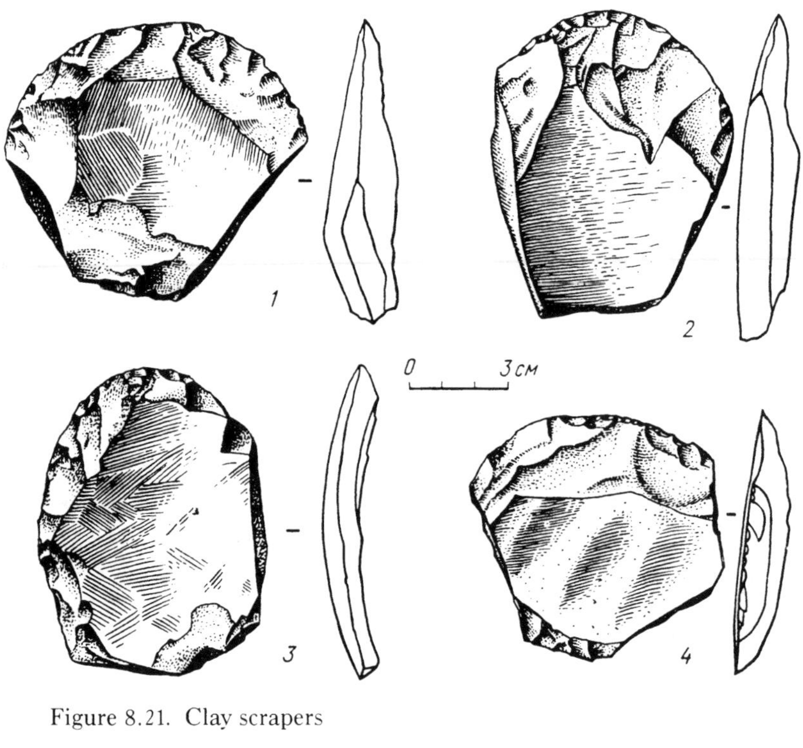

Figure 8.21. Clay scrapers

Figure 8.22:1–8. Coarse and thin-walled vessels

clay contains a heavy proportion of organic temper. Such vessels are found in all levels and show no change throughout the whole period of Halaf occupation.

The technical quality of the smoothed pottery is higher, though still inferior to the painted variety. The clay is of a better quality with some sand and grit temper. The firing is uneven. The surface of the vessels is slipped (light brown, yellow, or white) and sometimes burnished. Both thin-walled (0.5– 0.7 cm) and thick-walled (1.0–1.3 cm) vessels are found. The latter include large deep bowls and flat-based pots (up to 40–50 cm in diameter and height) with no handles or ornamentation (Fig. 8.22:7,8). The thin-walled vessels are relatively diverse in form and in some instances have spouts (Fig. 8.22:4) and projecting lugs for suspension (Fig. 8.22:1,6).

The painted pottery is of a very high quality with respect to both clay and firing. These vessels are slipped and burnished and are of different colors (brown, cream, rose, white). Their forms repeat those of the simple smoothed ceramics, but each form occurs in a considerable variety of types (Figs. 8.23–8.25). Five bowl types have been established (Figs.8.26–8.29). All are richly painted, frequently on both interior and exterior surfaces. Some types are common also in the later levels, but types 4 and 5 (Figs. 8.26:2,3 and 8.27:3,4) are typical at Yarim Tepe II of

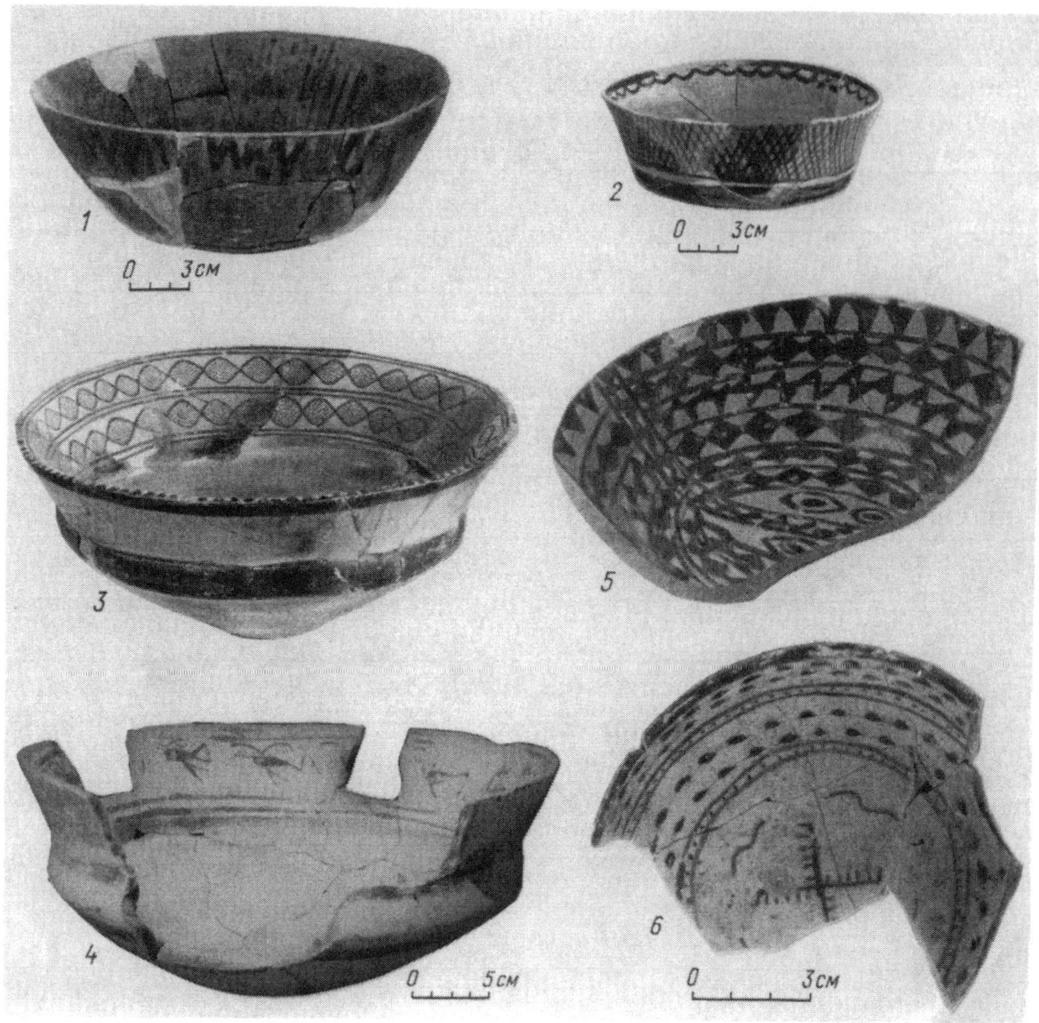

Figure 8.23:1–6. Painted bowls

the earliest levels only. Painted pottery is common at all Halaf sites, but so far, no other site has produced the great diversity seen at Yarim Tepe. Comparative analysis has shown that a considerable number of the painted bowls from the early Halaf levels of Arpachiyah resemble those from the lower levels of Yarim Tepe II.

The shallow basins and plates are similar in form to the bowls (Fig. 8.29), but only a few have been found in Levels 8 and 9, with not a single polychrome plate of the Arpachiyah type among them. The latter, of course, occur for the most part only in the latest Arpachiyah level (TT6). Another form of painted vessel typical of the early levels of Yarim Tepe II and second in quantity only to the bowls is the jar (Merpert, Munchaev, and Bader 1981: 47, Fig. XVII, top left). This general pattern continues throughout all levels, but in the latest stage of the settlement some finer jars with high necks are found, similar to those known at Arpachiyah (Mallowan and Cruikshank Rose 1935, Fig. 59:7, 67). In the opinion of some researchers jars constitute nearly 50 percent of all Halaf vessels (Le Blanc and Watson 1973). This view is not sustained by the evidence from the western Halaf sites where jars are relatively rare, though they seem to be more common in the Tigris valley, as at Arpachiyah. Low-necked jars are also known from Yarim Tepe II.

The Yarim Tepe Halaf pottery finds its closest parallels at Arpachiyah. However, it includes the most typical forms from both western and eastern Halaf sites as well as some unique examples, in particular the zoomorphic and anthropomorphic vessels described above. A smaller fragment of a vessel similar to the elephant- or pig-like example from Yarim Tepe II (cf. Merpert, Munchaev, and Bader 1981: 37–38, Figs. VII–VIII) was found at Arpachiyah (Mallowan and Cruikshank Rose 1935, Pl. Va), but we know of no parallel for the very striking female vessel.

The paint colors include brown, red, red-brown, black-brown and, rarely, white, against a background of pink, white, yellow-pink, cream, and light green. The painting is usually monochrome; only in a few instances are there combinations of brown and white. Geometric ornament is predominant and frequently makes up complex and graceful compositions. Depictive motifs are also common. We have found fragments of not less than twenty vessels decorated with the typically Halaf bucrania, generally in the realistic style known at western Halaf settlements such as Chagar Bazar rather than the more stylized Arpachiyah examples (Mallowan and Cruik-

Figure 8.24:1–5. Painted pots

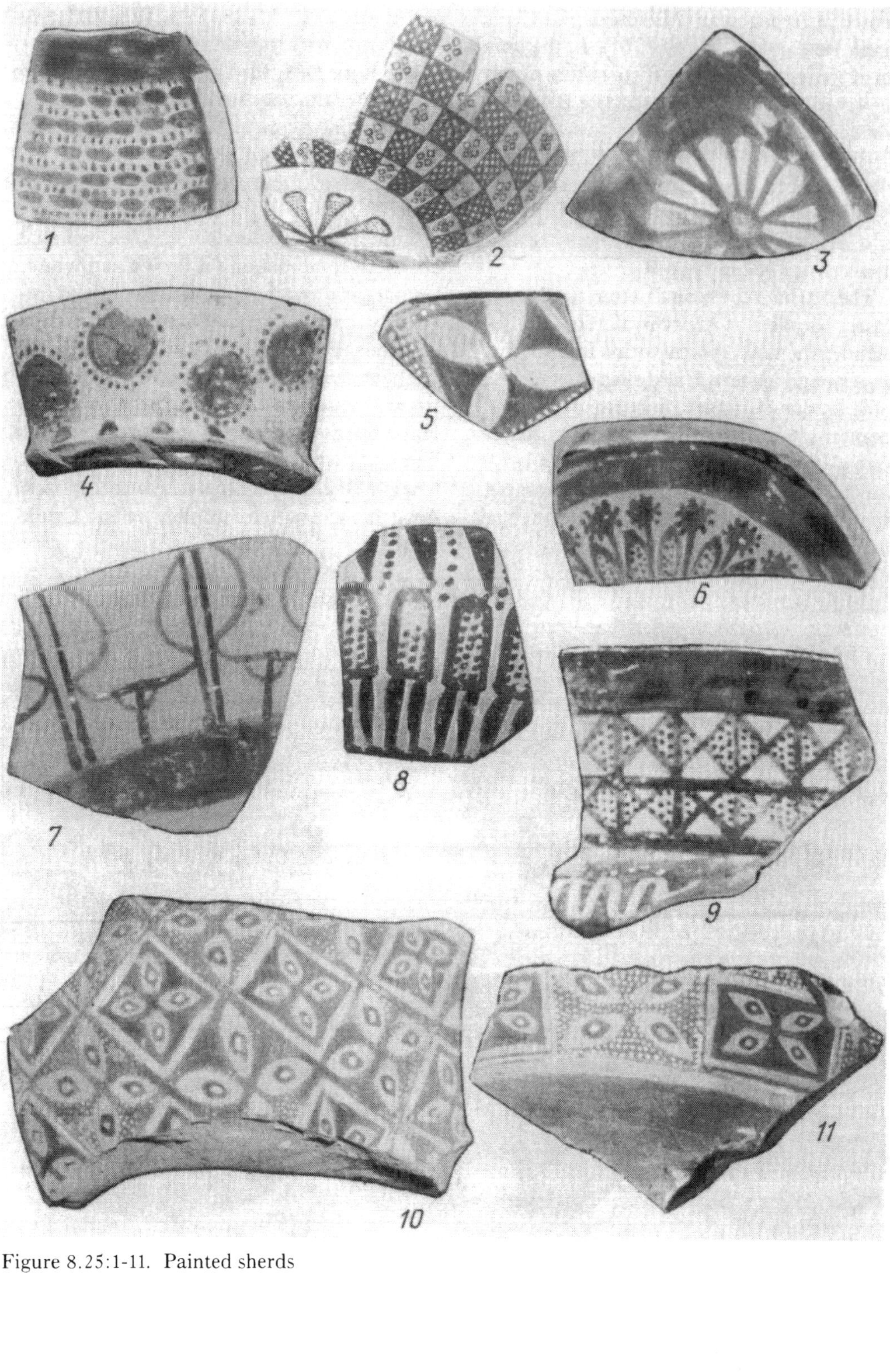

Figure 8.25:1-11. Painted sherds

Figure 8.26:1–3. Drawings of painted bowls

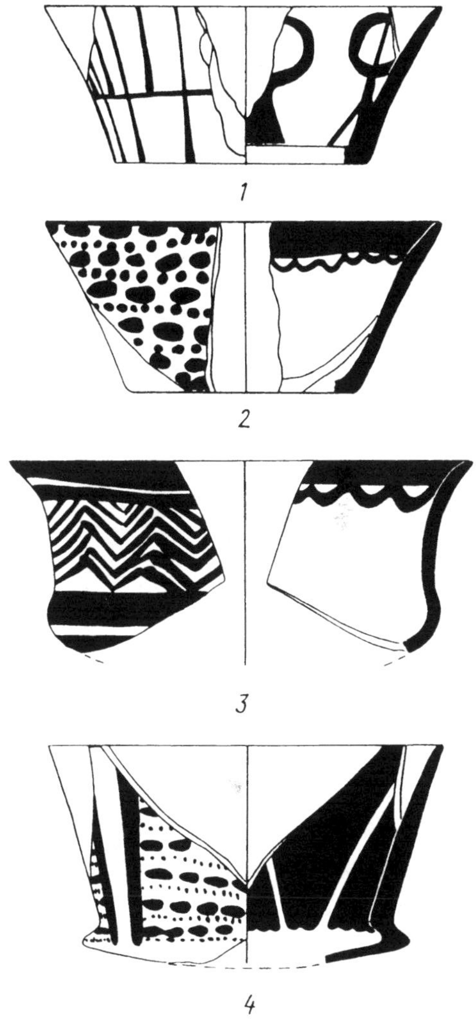

Figure 8.27:1–4. Drawings of painted bowls

shank Rose 1935, Figs. 74, 75). Although the similarities to Arpachiyah are close, depictive motifs like those from Yarim Tepe II are more common on the western sites. Prominent among these is the "huts and flowers" motif, of which over a hundred examples were found in Levels 9 to 8 (Fig. 8.28). This design, as is well known, is also attested at Brak, Chagar Bazar, and other Halafian sites, including Arpachiyah (Mallowan and Cruikshank Rose 1935, Fig. 77; Mallowan 1936, Fig. 25, 4, 5, etc.), though by far its most frequent occurrence is at Yarim Tepe II.

Yarim Tepe has also yielded several sherds with schematic depictions of a tree or bush and some anthropomorphic motifs. Again, these motifs are known on western Halafian ceramics and, occasionally, at Arpachiyah. Mallowan has classed them among the Samarran repertoire. In the earlier levels of Yarim Tepe II a bird motif is common, sometimes painted in white. Most common is the depiction of a single row of birds (Fig. 8.30). These differ from similar motifs found in the later levels of Yarim Tepe II, a situation found also at Chagar Bazar. Also known at Yarim Tepe are an unidentified animal (Fig. 8.31:8), a fish (Fig. 8.31:1), and a snake (Fig. 8.31: 3–5).

Figurines constitute the final group of clay artifacts. Two complete figurines and fragments of seven others have been found in Levels 8 and 9 (Fig. 8.32:1–9). All are stylized and one type is cone-shaped (Fig. 8.32:5,7,8). Such figurines have been encountered in all

Figure 8.28:1–9. Drawings of painted bowls

levels at Yarim Tepe II and are known also from other Halaf sites (Mallowan and Cruikshank Rose 1935, Fig. 47:9, 13, 16). A second type is represented by a small seated figure with exaggerated hips and thighs. The lower part of one such figure has survived (Fig. 8.32:9). No other Halaf site has produced an example precisely similar to this Yarim Tepe II piece, but it is in general comparable to a well-attested Halaf type. The remainder of the figurines constitute a third type. They resemble some figurines from Arpachiyah (clay and one stone example) and especially, from Tepe Gawra (Mallowan and Cruikshank Rose 1935, Fig. 47-19, Pl. Xa, 920, Fig. 52-3; Tobler 1950, Pl. CLIII,2). The origin of the Halaf figurines of this type is as yet unclear. They cannot be convincingly compared with any earlier group of anthropomorphic figures, but to a considerable extent this is characteristic of Halaf plastic art as a whole which, as has been correctly stressed by some authorities, stands by itself in prehistoric Mesopotamia, its differences from Hassuna and Samarra being greater than any similarity (Antonova 1974, 54).

Figure 8.29:1–12. Drawings of shallow bowls and platters

Lower Levels of Yarim Tepe II: Conclusion

The question of the origin and the character of the earliest stages of the Halaf remain unanswered. Such traits as are known have been established largely on the basis of the ceramic materials from Arpachiyah. However, the stratigraphy of that settlement cannot be considered to be absolutely certain. The traits of the early complex defined there are only approximations since they were established without statistical analysis. In the lower levels of Yarim Tepe II all of these are present in considerable numbers but in combination with some other features that are thought to be characteristic of subsequent stages of Halaf development. Therefore, the question whether these Yarim Tepe levels belong to the earliest or to the beginning of a middle stage of Halaf must remain open for the present. There can be no doubt, however, that the 7-meter-thick deposits of Yarim Tepe II reflect a very long period of development. That this development had begun relatively early is confirmed by radiocarbon determina-

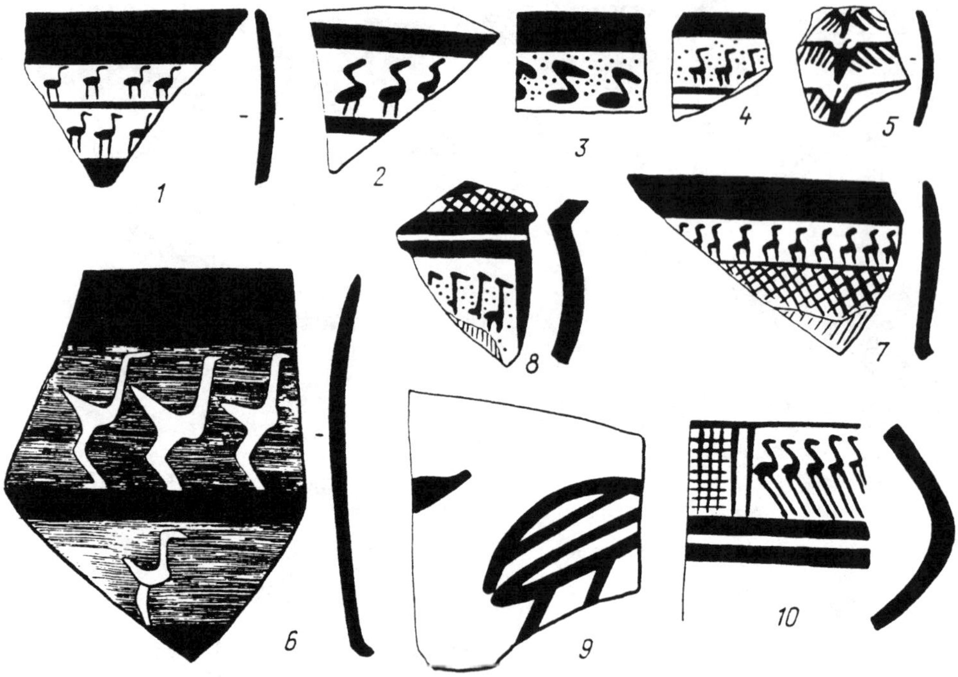

Figure 8.30:1–10. Drawings of painted sherds

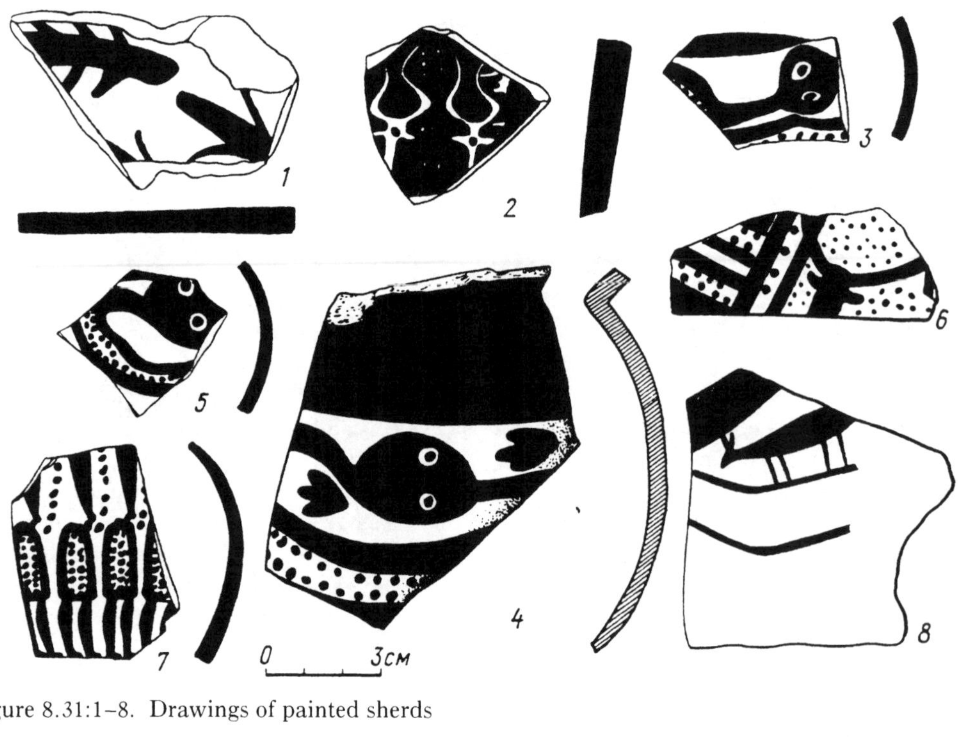

Figure 8.31:1–8. Drawings of painted sherds

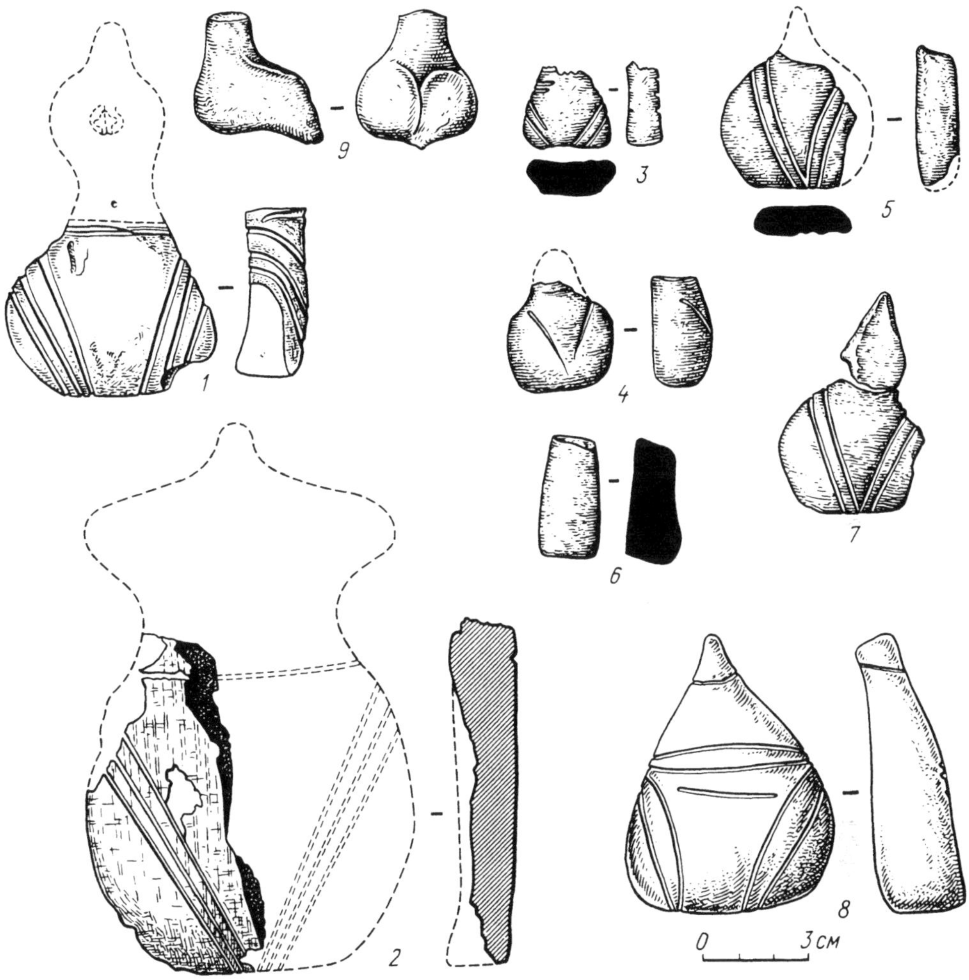

Figure 8.32:1–9. Drawings of clay anthropomorphic figurines

tions from various levels of the settlement (4840 ± 180 B.C. to 4210 ± 130 B.C. [Munchaev and Merpert 1981, 266]).

All of the Yarim Tepe II deposits are demonstrably later in time than the final stage of the Hassuna culture, as can be seen from the fact that all these levels lie directly over the ruins of a small Hassuna settlement. Further evidence for this relationship can be seen in the situation of the Halaf necropolis on the abandoned tell of Yarim Tepe I. It cannot be questioned that the necropolis emerged here after the lifetime of the Hassuna settlement.

We can conclude that in the late sixth or at the beginning of the fifth millennium B.C. in the Sinjar plain the Hassuna culture was replaced by a new and genetically alien Halaf culture. The latter was related to a new ethnic group that established cultural unity over a considerable territory in northern Mesopotamia and eastern Turkey. The questions of the genesis, original homeland, and paths of diffusion or migration of the Halaf culture remain open. It is significant, however, that Yarim Tepe II occupies a geographically central place in the Halaf area and combines traits of both western and eastern groups of Halaf sites. The economy of Yarim Tepe II was based on highly developed forms of food production. In this respect the Halaf culture inherited and developed the traditions of the preceding cultures and itself constituted an important stage in the process of general development that resulted in the formation of one of the world's earliest civilizations and turned Mesopotamia into a vital center in the early cultural history of mankind.

9

YARIM TEPE III: THE HALAF LEVELS

N. Ya. Merpert and R. M. Munchaev

Yarim Tepe III is the largest in a group of six mounds bearing the same name. It is located approximately 7–8 km west of Tell Afar, on the right bank of the Djubara-Diariasi stream about 250 m from the site of Yarim Tepe I. The southern edge of the mound cannot be separated from the northwestern edge of Yarim Tepe II. On the north, south, and west, Yarim Tepe III is bounded by regularly plowed fields, but the greater part of the mound itself has not been disturbed by agricultural activities.

The average diameter of the mound is 225 m (Map 9.1). Its eastern and southern slopes are noticeably steeper than the rest. The highest point on the mound lies in its southeast quarter. Its average height is 10 m above the level of the surrounding fields and 12 m above the stream. The disturbances that are visible on the surface of the mound can be attributed to a small settlement that existed there during the 1930s.

The abundant cultural material on the surface of the mound includes typical examples of Northern Ubaid and Halaf pottery, and also isolated fragments of later ware, notably glazed pottery. Additionally, obsidian and flint flakes have been found in considerable numbers.

The excavations at Yarim Tepe III were conducted by the Soviet expedition from 1977 to 1980. On the southern slope of the mound adjacent to its central part, an area of more than 500 square meters has been cleared. In addition to this, trenches of 25 square meters each have been cut into the base of the mound on the northern, southern, and western sides. The major excavation has reached a depth of 7 m, and the trenches have been dug down to sterile soil. As a result, it has been established that the depth of the cultural deposit averages 12 m. The lower 8 m of the deposit are associated with the Halaf culture, and the upper 4 m belong to the Ubaid.

a. Results of the 1978–1979 Field Season

The primary objective guiding our investigation of Yarim Tepe III was to examine the stratigraphic sequence at the site, especially the correlation between the Ubaid and Halaf layers. Additionally, the relationship between Yarim Tepe III and the previously excavated Halaf site of Yarim Tepe II was to be studied. Toward these ends, excavations at Yarim Tepe III began in 1977 in the southeastern part of

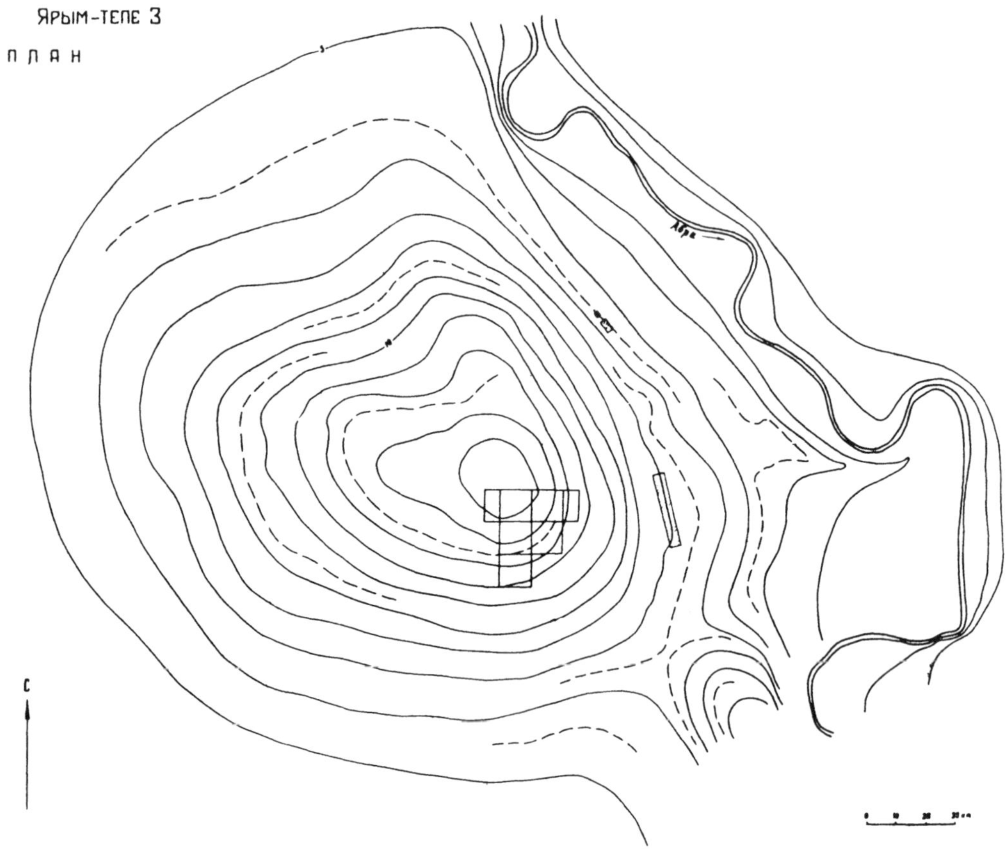

Map 9.1 Topographic map of Yarim Tepe III and excavated areas.

the tell where the remains of an Ubaid settlement were exposed over an area of 600 square meters (Bader, Merpert, and Munchaev 1981). There are at least three Ubaid building levels which are interrelated through common building plans and construction techniques. The most prominent structures either remained unchanged or were only slightly modified throughout the entire Ubaid period at the site (e.g., complexes 4 and 8; rooms 15, 16, 27, and 89). Other structures were entirely rebuilt with their ground plans substantially altered. The Ubaid stratum at Yarim Tepe III reaches the modern surface of the tell and its total depth exceeds 3 meters. No strata later than the Ubaid have been discovered to date. As for the earlier periods, a 5 m × 5 m test pit, begun in 1977 on the southern slope of the tell, has cut through the Halaf stratum, which is thicker than the overlying Ubaid stratum.

Two additional test pits were started in the 1978–79 season. Test pit 2 was placed 60 m away from the center of the mound on its northern slope, 6 m below its apex. The cultural deposit in this area continued down to a depth of 12.4 m, with the material recovered belonging entirely to the Halaf period. At least four distinct building levels could be differentiated in this test pit at the levels of 7.40, 7.60, 8.80, and 9.60 m below datum (mbd). The remnants of a large tholos were exposed at approximately 8.80 mbd, indicating that during this period substantial structures extended to the northern limits of the settlement. This tholos was over 5 m in diameter and its internal buttresses formed a cross-shaped internal space. Architectural remains were not encountered in the levels directly beneath the tholos (c. 9.60 mbd). Instead, fifteen compacted and often clay-covered surfaces were exposed, alternating with debris layers. The greatest concentration of material culture remains was encountered in the fill and on the floors of the tholos and related buildings. At lower depths the artifact density gradually decreased.

A similar stratigraphic sequence was exposed in test pit 3, which is located on the western slope of the mound, 90 m from its center, and roughly 6 m below its apex. In this test pit, sterile soil was reached at approximately 11 mbd. It is important to bear in mind that the excavation in the central part of the tell has reached the level of 6 mbd, and that all three test pits began at roughly 6 mbd and were excavated down to sterile soil. Thus, the entire stratigraphic section of the cultural sequence of Yarim Tepe III has been exposed.

Test pit 3 is also a 5 m × 5 m square and the section exposed in this unit is as follows:

From the surface of the mound slope (c. 6 mbd) down to 6.60–6.70 mbd, a compacted clay layer formed by water runoff from the top of the mound was encountered. The material culture remains from the layer are mixed, but they date almost exclusively to the Halaf period and Ubaid sherd material is infrequent. Thus, for all practical purposes, the Halaf settlement in this area begins at the surface.

In the level between 6.60 and 7.30 mbd, structure walls can be observed in the southern and western sections of the square. These structures were probably circular in ground plan, representing one of the Halaf building levels. They were built atop earlier buildings of similar plan that were found down to 7.80 mbd, presumably part of two additional building levels.

A more complete and complex picture emerges from the northeastern part of the test pit, where the remnants of other structures from at least two levels have been exposed; the upper belonging to the lower of the two levels mentioned above and the underlying one to an even earlier level. The total depth range of these last two is impressive, extending from 7.00 to 7.40 mbd down to 9.40 to 9.60 mbd with the walls of tholos 3/1 surviving to a height of 2.5 m. The diameter of this structure is not less than 6 m with the bulk of it extending beyond the limits of the test pit. Its walls were 0.50–1.00 m thick, and their careful dismantling has allowed us to determine the original construction and subsequent remodeling episodes. Initially, a foundation of red clay that reached a level of 9.20 mbd was erected on a thick surface of compacted clay found at 9.60–9.45 mbd. On this foundation platform, walls made from the same clay were built that were thickened at their base. They were supported both by the platform and by external buttresses that were 0.40–0.50 m thick. The surviving portions of this earlier structure are as follows: in the eastern section is the foundation platform and the lower part of the walls and external buttresses at a depth ranging from 8.40 to 9.45 mbd; in the northern section is the western portion of the wall, external buttresses, and platform; and finally, portions of the outer wall were exposed during the excavation of the test pit. The platform was overlain by the original tholos floor on which a thin fill layer had accumulated at 9.05–9.20 mbd.

Later in time, the tholos was substantially remodeled. The western arc of the outer wall was reinforced from within by an adjoining inner wall. The latter is made of yellowish brown mud bricks mixed with red clay mud brick debris scavenged from the original building. The lower part of the structure was reinforced with a similar layer 0.40–0.50 m thick up to 8.60–8.55 mbd. Also built of similar material was a partition wall, running northeast to southwest, that was found in the northern section. Overlying the above supporting fill layer was a second floor followed by another fill layer 20 cm thick that was, in turn, overlain by a third floor.

The wall remnants in the eastern section also show evidence of remodeling. A new structure can be discerned at the level above 8.40 mbd up to 7.80 mbd, although the mud bricks of the new wall are still made from the same red clay as the original building. The supporting layer and overlying fill are different from that on the other side of the structure, separated by the abovementioned partition. The former is composed of clay/loam accumulations mixed with building collapse rather than the uniform yellow layer in the northern section. However, at the floor level, both sections are relatively homogeneous.

The third floor is overlain by a grayish brown layer of refuse that is found in both the northern and eastern sections up to 7.9–8.0 mbd. To the west of the partition in the northern section, at a depth ranging from 7.4 to 7.6 mbd, there is a uniform layer of yellowish green clay. In the eastern section, a 50-centimeter-deep ash accumulation was encountered. The lower limit of this accumulation is somewhat below the remnants of the rebuilt walls, including the partition, at a depth ranging from 7.65 to 7.80 mbd. Above this level and up to 7.20 mbd, the partition consists of 50-centimeter-long mud bricks made from red clay, separated by gray mortar lines rather than the yellowish brown bricks noted above. At the same level, the rebuilt portion of the outer wall also begins. Above the 7.00 mbd level, the tholos walls have been destroyed, with the debris of these walls lying on the surface found in both the northern and eastern sections between 7.20 and 7.50 mbd. The maximum surviving height of the walls of tholos 3/1 from the base of the original platform is 2.70 m. Rather than dome-shaped, the tholos was apparently cylindrical with a flat roof, since the upper walls do not converge.

Traces of a second tholos of similar size and design (number 3/2) were encountered in the southwestern corner of the test pit. The walls and foundation of the building survived up to a height of 1.20 m. They were found at a depth ranging from 7.50 to 8.70 mbd and can be stratigraphically linked to the second construction phase of tholos 3/1. Thus, tholos 3/2 was probably built simultaneously with the rebuilding of tholos 3/1 and the two structures undoubtedly coexisted for some time.

In the open area between the two tholoi a series of compacted and gypsum-plastered floors were exposed, interlayered with high-density accumulations of sherds and faunal remains. At a depth ranging from 8.15 to 8.55 mbd, a round-bottomed storage pit, pear-shaped in section and gypsum plastered, was discovered. The diameter at its mouth is 0.50 m and its maximum diameter is 0.87 m. The two construction episodes of tholos 3/1 comprise at least two definable building levels. Several additional levels were encountered below this level.

From the northern section, it was determined that the foundation platform of tholos 3/1 overlies and partially incorporates the remnants of an earlier wall made from red clay that survives at a depth ranging from 9.05 to 9.50 mbd. The wall is up to 0.50 m wide and it, in turn, overlies a supporting layer of grayish brown clay that begins at 9.80 mbd. On the western side of this wall five separate floors were found running up to it. Since remains of this structure could only be found in a localized area, the test pit apparently cut through only a small part of the building.

Traces of the next building level were uncovered immediately underneath the supporting layer of the above structure. In the northeastern part of the test pit, a linear wall oriented northeast-southwest was traced for 1 m until it reached the eastern section of the unit at a depth ranging from 9.80 to 10.15 mbd. At the same level, a badly preserved structure was also exposed in the southeastern part of the test pit.

In the northeast corner, directly under this level, the ruins of an earlier rectangular structure were discovered. It was nearly 2.5 m wide, and its axis ran southwest to northeast. Only the southern part of this building extended into the test pit area. A lens of building debris found on the northern side of the unit probably came from the upper walls of this structure. The maximum height of the surviving walls and debris lens is 35 cm, and they are found at a depth ranging from 10.10 to 10.45 mbd.

These remains lie on a level, gypsum-plastered surface under which an ash/charcoal layer was encountered at a depth ranging from 10.45 to 10.50 mbd. Below the ash layer, within an otherwise uniform layer comprised of grayish clay, two surfaces were exposed at 10.60 and 10.70 mbd, also with traces of gypsum plaster. At the level of 10.80–10.90 mbd a clay band interlayered with fill and gypsum lenses was found directly above a layer of red clay that is probably debris from at least one structure. Under this debris layer, at a depth of more than 10.90 mbd, another gypsum-plastered surface was exposed that overlies a uniform brown fill layer. Below this brown layer, reddish brown sterile soil was encountered throughout the test pit at a depth of 11.00 mbd.

Thus, on the average, in test pit 3, sterile soil was reached roughly one meter higher than in the previous test pits. This can be explained by examining the topographic relief of the area where the settlement emerged. Below the eastern, northeastern, and southeastern areas of Yarim Tepe III, the pre-settlement terrain was relatively lower in the direction of the spring. Thus, the natural elevation would have been highest to the west in the vicinity of test pit 3. All in all, within test pit 3 at least six Halaf building levels can be defined between the levels of 6.00 and 11.00 mbd. It should also be noted that at least three more Halaf levels were encountered in the central excavation area above the 6.00 mbd level.

The excavation of test pit 3 has shed light on the western periphery of the settlement at Yarim Tepe III. This area was undoubtedly within the main habitation zone of the site. However, the structures encountered are apparently nonresidential in character. At least thirty plastered surfaces were associated with these structures. Their succession becomes particularly regular below the level of 8.00 mbd and in only one case mentioned above was a storage pit discovered associated with a floor (8.15 mbd). In the majority of cases there were no domestic remains such as ovens, hearths, or accumulations of ash and charcoal.

Down to a depth of 8.50 mbd, the artifact density from the test pit was quite high, comparable to the density encountered in the central excavation area. Further down, it decreased sharply, and was very low below 9.00 mbd, and from 10.50 mbd to sterile only isolated finds were recorded. It should be mentioned that cooking ware and other domestic items predominated over fine wares and luxury items. Additionally, considerable accumulations of faunal remains were encountered. Particularly numerous were bones from cattle, pigs, and onagers. The latter two species are distinctive from the finds in the central excavation area and the other two test pits on the southern and northern slopes of the mound. It should be stressed that over the entire area of the test pit at a depth ranging from 10.00 to 10.50 mbd, only ninety sherds, including ten decorated ones, were found, whereas at a depth ranging from 10.50 to 11.00 mbd, eighty-three sherds (only four decorated) were recovered.

The vessel forms and the decorative motifs of the pottery recovered from the excavation of test pit 3 are typical of the Halaf levels in the central excavation area. In addition to the sherd material, an oval-shaped stone vessel, a wedge-shaped, stone axe, a digging implement made from antler that had a rectangular end with traces of bitumen for hafting, a fragment of a painted female figurine, and a ring-shaped pendant of pink stone were found in the unit. Especially interesting is a fragment of copper ore that was found at a depth of 7.30 mbd, the first evidence for the use of metal at this site.

In the southwestern corner of the test pit and beyond its limits to the west, a burial (number 30) was discovered at 7.30 mbd, approximately 0.70 m below the surface. This

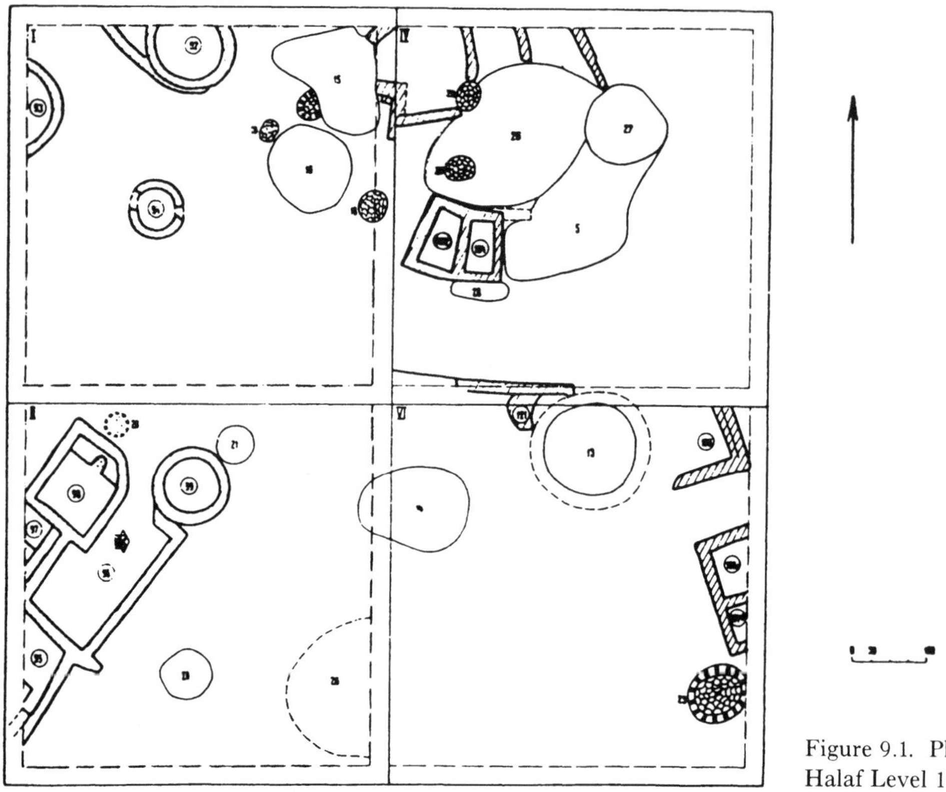

Figure 9.1. Plan of Halaf Level 1

burial is intrusive into and disturbs the previously mentioned upper building phase. The badly preserved remains of a twelve-year old were found in the burial pit. Based on the position of the cranial fragments, left ulna, right radius, fibula fragments, and several finger phalanges, we conclude that the body had been interred in a contracted position with its head toward the west. Four vessels were placed at its feet: three of them painted (two deep bowls and one pitcher) and one plain. From these vessels it can be determined that the burial dates to the end of the Halaf period or to the early Ubaid.

In the central excavation area, the exposed area was reduced below the lowest Ubaid level (c. 3 mbd). The transitional level and the uppermost Halaf level were only exposed over an area of 400 square meters. A summary of the results of this excavation follows.

In spite of the lack of any stratigraphic hiatus between the Halaf and the Ubaid strata, the ground plan of the first Ubaid settlement was entirely different from the last Halaf settlement it overlies. In preparation for the first Ubaid settlement, the earlier Halaf structures were dismantled, pits were filled, and the surface was leveled. At this transition level, Ubaid and Halaf sherd material were intermixed (cf. Mallowan and Cruikshank Rose 1935, 21). In square II, at a depth ranging from 3.30 to 3.55 mbd, large trash accumulations, possibly middens, containing ceramics typical of the late Halaf were encountered, possibly the result of surface cleaning in preparation of the first Ubaid settlement. These accumulations separate complex 7 of the first Ubaid level from underlying complex 9 of the last Halaf level.

The latter complex was exposed in the western half of square II at a depth ranging from 3.25 to 3.80 mbd (Fig. 9.1). Its construc-

Figure 9.2:1. Grain storage structures 92 and 94

Figure 9.2:2. View of complex 9

Figure 9.2:3. Rooms 102 and 104

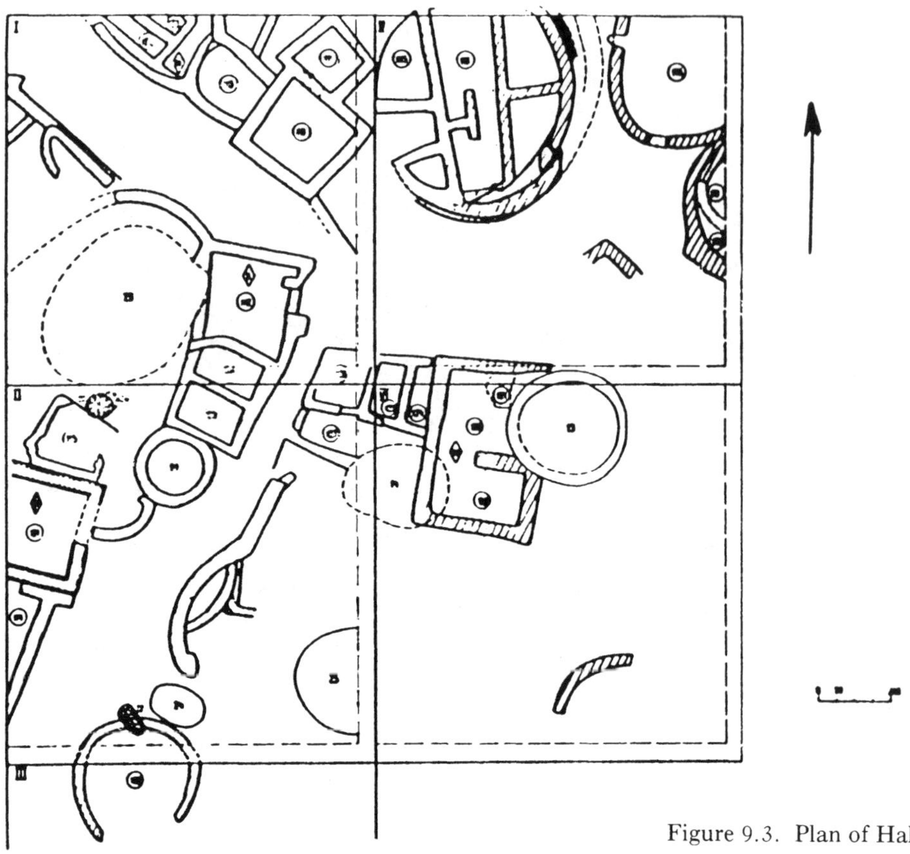

Figure 9.3. Plan of Halaf Level 2

tion marks the beginning of the last Halaf settlement. However, several Ubaid elements are already present. In its ground plan, both rectangular and circular structures (Fig. 9.2) can be found, and the former are already pervasive, forming the core of the complex, whereas the latter obviously have an auxiliary function (Mallowan and Cruikshank Rose 1935). Complex 9 (Fig. 9.2:2) is comprised of at least four rectangular rooms of varying sizes with the largest measuring 4 m × 2 m. Ovens are found both within the rooms and immediately to the outside. The complex is oriented southwest to northeast and differs markedly in plan from the Ubaid complex that is immediately above it. However, it exhibits continuity with the structures below it from earlier Halaf building phases. Rectangular structures dating from the last Halaf phase have also been found in other parts of the excavation area, including the corner of a massive structure unearthed between squares I and IV, two small adjoining rooms (102 and 104) in square IV (Fig. 9.2:3), and isolated linear walls in the eastern half of square VI. The above are not well preserved due to the clearing episode that preceded the construction of the first Ubaid settlement. However, they share in common small tholoi (Fig. 9.2:1) with diameters not exceeding 2 m (numbers 92 and 94) found in square I, and a similar tholos (99) in square II which had been built earlier but overlapped in time with complex 9. The above tholoi were auxiliary in character and were probably used for storage, as is suggested by the finds of carbonized grain within them. Although larger tholoi may certainly have existed in this last Halaf building horizon, the only evidence for them is a poorly preserved wall remnant (number

121) that has been otherwise destroyed by an intrusive pit (number 13) between squares IV and VI.

In the material culture from this level, Halaf ceramics predominate. However, Ubaid sherds are still encountered, and in some cases, otherwise Halaf forms and decorative motifs show Ubaid influences. The last Halaf building level at Yarim Tepe III has interesting parallels with the terminal Halaf layers (5 and 6) at Arpachiyah. In the latter, large tholoi are also absent and the material culture exhibits both Halaf and Ubaid elements. At Yarim Tepe III, the terminal Halaf phase was short-lived, since it is poorly represented, and in a number of sections, especially to the east, it is entirely absent.

The underlying Halaf strata present a considerably more complex and detailed picture of the development of the settlement. One also observes in these levels the development of rectangular architecture, with one structure built immediately atop its predecessor, presenting a series of successive rebuilding levels.

In the level immediately beneath the last Halaf layer (Fig. 9.3), three multiroom dwellings (10–12) were exposed in squares I, II, VI. Associated with these dwellings were five large tholoi. Two of the tholoi occupy the northern half of square IV, a third is located in the eastern part of square II, a fourth extends beyond the southern section of square II, and finally, traces of a fifth were exposed in the southeastern corner of square VI. Perhaps the most interesting is tholos 118 (Fig. 9.4:1,2) found in square IV at a depth ranging from 3.70 to 4.80 mbd. It is a complex structure that probably had a long use-life, since it underwent a number of significant remodelings. The tholos has a diameter of 5 m, and its internal space is partitioned into several compartments. Two Halaf child burials from upper levels are intrusive into the tholos.

Complex 9 in square II overlies a structure of similar plan. This structure existed over a long interval and underwent a number of significant remodelings. From these remodelings, we can delineate a number of successive building episodes within the latter part of the Halaf period at the site. Complex 9-C was found in Level 3 of the Halaf deposit. Remnants of this group of structures lie between 4.30 and 4.65 mbd. Two rectangular rooms within this complex were excavated. The northern room (130) was 4.30 m in length along the southwest-northeast axis and 2.00 m wide, and the southern one (129) was 2.00 m on each side. Both rooms have two floors, one at 4.50 mbd and the other at 4.60 mbd. The fill in both rooms contained the usual domestic refuse, including a large number of plain ware sherds and numerous faunal remains. Additionally, a painted zoomorphic vessel supported by legs, a clay spindle whorl, and a stone seal-pendant were found.

Figure 9.4:1. View of complex 10 and tholos 118

Beneath complex 9-C, a large ash pit (30)

Figure 9.4:2. Tholos 118

Figure 9.4:3. Tholos 133

was exposed which occupied the central part of square II. The pit originated from a surface at 4.65 mbd and extended down to 5.00 mbd. Besides ash, the pit fill included charred grain and straw remnants. After the pit had been filled and its surface compacted, tholos 133 (Fig. 9.4:3) was built beside complex 9-C. The remains of this tholos were encountered at a depth ranging from 4.10 to 4.70 mbd. Its diameter exceeded 5 m, covering the entire eastern half of square II. Although a considerable portion of the structure has been destroyed, remnants of an arc-shaped partition have survived in the interior.

The remains of tholos 131 were discovered in the southwestern corner of square II. This structure had a diameter of 4 m and was found at a depth ranging from 4.27 to 5.20 mbd. This tholos was constructed at an earlier date than the structures described above, though it continued to exist after they had been built.

The above structures can be grouped together into the earliest phase of the second-latest Halaf level. With the exception of tholos 131, they were constructed after pit 30 had been filled.

The underlying building level has been partially destroyed by the one above it. The former constitutes the third latest Halaf level within which several phases can be differentiated. The upper level is represented by room 132 (Fig. 9.5:1), which was exposed at a depth ranging from 4.50 to 5.15 mbd. The

Figure 9.5:1. Halaf Level 3 structures: room 132

walls in the northeastern corner of the room are cut more than 50 cm by pit 30. The room is located near the southern side of complex 9-C. The two are separated only by a narrow space of 25–30 cm, though they are not oriented perfectly parallel to each other. Room 132 is rectangular in shape with the dimensions of 3.20 m × 1.80 m, giving it a total area of 5.76 square meters. Its long axis is oriented southwest-northeast, and its walls are 30 cm thick. Two floors were encountered in the structure at depths of 4.85 and 5.10 mbd. On the lower floor, in the northwestern corner of the room, a hearth with a diameter of 40 cm was exposed. A 45-centimeter-wide entryway was found in the eastern wall whereas the western wall could be traced south under tholos 131. Quite probably, room 132 is associated with the structures found in the southern sections of the excavation area that precede tholos 131.

Two small circular ovens (26 and 29), which lie partly beneath tholos 133, belong to the same phase as room 132. Additionally, a number of other walls that were exposed in

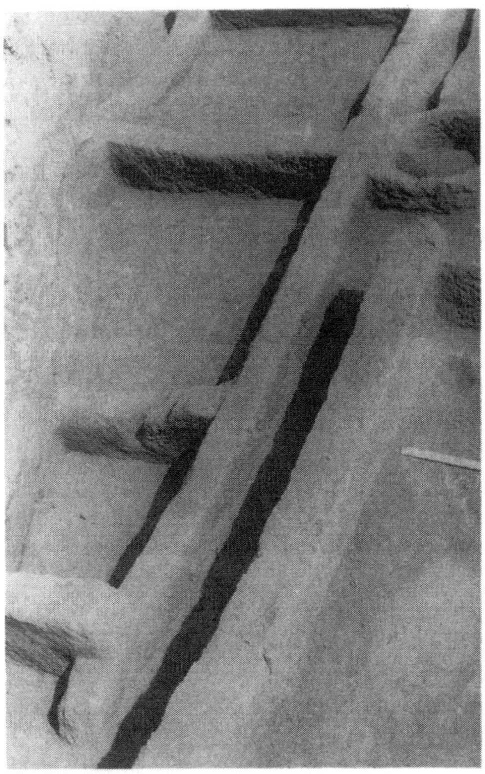

Figure 9.5:2. Halaf Level 3 structures: complex 9-D

the northern (wall 83), central (wall 83), and eastern parts (wall 86) of square II are also in this construction phase. After the excavation of pit 30, which destroyed part of room 132, another wall was constructed adjacent to the remnants of the western wall of room 132. At an even later date, this wall was connected to the eastern wall of complex 9-C. During the construction of this complex, the first two walls were cut to approximately 4.50 mbd and used as foundations for its southern walls.

The underlying phase is perhaps the best represented within this building level. It includes a series of structures that lie beneath complex 9-C that are for the most part similar in plan and orientation to the latter, thus they are grouped together as complex 9-D. However, they are separated from the wall foundations of complex 9-C by a 15 cm fill-layer, and though a number of walls of the overlying complex have similar orientations to those comprising the lower complex, they do not directly overlie the latter. Thus, the temporal gap between the abandonment of complex 9-D and the construction of complex 9-C is quite evident.

Three rectangular rooms make up the core of complex 9-D (Fig. 9.5:2): 134 (1.80 m × 2.00 m), 135 (2.00 m × 2.00 m), and 136 (1.25 m × 2.00 m). Based on the remnant of the eastern wall that continues south, the complex extended further in this direction. The upper floors of the above rooms are all at 5.08 mbd. At the same level, a circular oven (number 27) with a 70 cm diameter was exposed that abutted the eastern wall of the complex opposite the partition wall between rooms 134 and 135. The fill of the three rooms consisted of ash and charcoal accumulations down to the level of the upper floor and, in general, the same can be said for the fill directly above and within the entire complex. Subsequent floors were discovered at 5.25 and 5.60 mbd. On the former, in room 136, an accumulation of carbonized grain was found, and on the latter, in rooms 134 and 135, a high density of sherd material was recorded. The walls of the complex are roughly 30 cm thick, and they survive up to the 4.70–4.90 mbd level. They have been preserved to well over a meter in height since at the end of the 1979 season, the 5.60 mbd floor level was exposed and the base of these walls still had not been reached, requiring further excavation.

To the north of the three rooms forming the core of complex 9-D, another tholos (number 138) was discovered. This structure is one of the most interesting buildings exposed to date at Yarim Tepe III. Since there is a connecting wall between tholos 138 and the outer wall of the above three rooms, the former has been included as part of complex 9-D. Tholos 138 occupies a large portion of the southern half of square I and the eastern portion of its wall is beneath part of complex 10, which is found at a depth of 4.55 mbd. The southern part of the tholos wall extends into square II. There are two outer walls of the tholos, forming two roughly concentric circles that intermittently come into contact or become increasingly separated, though no more than 10 cm apart. The total thickness of these walls reaches 1 m. The outer diameter of the tholos is roughly 7 m, and the inner diameter, approximately 5 m. The tholos walls have survived to an unusual height, from an upper level at a depth range of 4.43 to 5.00 mbd to a lower depth range of 6.60 to 6.90 mbd. Thus, the preserved height of the wall exceeds 2 m. It should be noted that the outer wall encircling was made from mud bricks using a red clay and the inner from bricks using a gray clay. Two parallel inner walls oriented north-northwest–south-south-east form the central corridor through the tholos which varies from 0.90-1.15 m wide. The corridor divides the tholos into eastern and western halves which, in turn, are each divided in half by partition walls oriented west-southwest–east-northeast. Entryways are found on both the northern and the southern ends of the central corridor, beginning at a depth of 5.75 mbd and continuing downward until the tholos floor is reached at 6.70 mbd. Both have been plastered and are 0.45 m wide and roughly 1.00 m high. The area around the tholos has not yet been excavated, and thus, it is difficult to determine its relationship to other structures in the area besides complex 9-D. In the case of the latter, the southern doorway of tholos 138 leads into the complex. It should also be noted that the walls of tholos 138, similar to those of tholos 3/1 in the western test pit described above, are vertical. Therefore, it was cylindrical in shape and probably had a flat ceiling. Since there are no entryways in the walls leading to the side compartments, access to them was probably from above.

On the floor of the northwestern compartment of the tholos at 6.50 mbd, an unfired clay vessel, 30 cm in diameter, was found adjoining the eastern wall. In general, the fill of the tholos varied from compartment to compartment. In the southeastern compartment, a grayish brown fill was encountered that had a high artifact density, and the northeastern one contained a yellowish green soil with only a small number of artifacts. The western ones contained a grayish green fill with a large amount of ash and humus within it. Additionally, both these compartments and the central corridor had high artifact densities, especially on the lower floors. In fact, it would be accurate to state that at a depth ranging from 6.50 to 6.70 mbd, the central corridor was completely blocked by refuse. The sherd material was almost exclusively from domestic wares, including large fragments of pitchers, large cooking pots, and various cups, many of which have carbonized residues on them. In this assemblage, painted sherds are quite rare. Other finds include stone, obsidian, and bone implements, ornamental items, and faunal remains, including cattle.

Equally interesting is a second large tholos (number 137) that was exposed in adjacent areas of squares II and III. The upper portion of its outer wall was excavated at a depth ranging from 4.93 to 5.33 mbd. This wall was built with mud bricks made with red clay. It has a maximum thickness of 50 cm and is plastered on both sides. In contrast to tholos 138, tholos 137 had a single encircling wall and a unique design with four L-shaped internal buttresses attached to the encircling wall in a symmetrical arrangement, forming a cross-shaped interior pattern. The partitions are 1.50 m long on each side, and they have survived to a height of 80 cm. On the southern side the tholos was connected to a rectangular structure that has not yet been fully excavated.

The inner diameter of the tholos is 6.0 m, and the outer wall has survived to a height of 1.25 m. Floors have been exposed at three levels: 5.15, 5.35, and 5.72 mbd. On the uppermost floor, a hearth utilizing the walls of a large storage vessel was found. A similar large vessel, 70 cm in diameter, was discovered lying on the middle floor. The original floor of the tholos consisted of a layer of compacted red clay. The artifact density within the tholos fill was particularly high, especially on the floors. However, unlike tholos 138, tholos 137 contained no refuse-blocked passages. In general, the amount of domestic refuse was much lower, and a much higher percentage of decorated sherd material was recovered in tholos 137.

Undoubtedly, tholos 137 coexisted with complex 9-D and tholos 138, although its construction was at a slightly later date, since its foundation is higher in elevation than the other two. Adjacent to tholos 137 and above it are walls 83 and 86 and oven 26, mentioned above, which are contemporaneous with room 132. These architectural remains, in turn, are overlain by tholos 133 and complex 12. To complete the report on the architecture exposed during the 1978–79 field season, the upper portions of several tholoi which belong to the third Halaf building level were exposed when the surfaces of squares IV and VI were cleaned.

In 1979, two Halaf burials were encountered in the central excavation area. Burial

Figure 9.6. Coarse ware ceramic vessel

Figure 9.7:1–3. Ceramic vessels from the upper Halaf levels

29, exposed at a depth of 4.48 mbd, was intrusive into the northeastern corner of room 130. Within it, the remains of a two-year old lay in a contracted position on its right side with its head oriented south. No grave goods were associated with the burial. Similarly, no mortuary offerings were encountered in the second burial (31). This also contained the skeleton of a two-year old juvenile that was interred in a contracted position on its left side with its head oriented north. The burial lay in an ashy area directly underneath tholos 138, which may have been an earlier hearth or even a small oven.

Among the artifacts recovered from the excavation of the Halaf stratum at Yarim Tepe III, of course, sherds from ceramic vessels predominate (Figs 9.6–9.12). All of the forms characteristic of the late Halaf period are represented in the ceramic assemblage. Of particular interest are sherds from vessels used in domestic contexts. Although a detailed analysis has not yet been undertaken, several general observations are possible. Numerous fragments of very large storage vessels for grain or water have been recovered from the site. These vessels are usually more than 80 cm in height and up to 70 cm in diameter. They are often spherical in shape with a relatively high neck, a simple rim usually 40–45 cm in diameter, and vessel walls roughly 2 cm thick. Their exterior surfaces are usually covered with a slip, and their interiors have comblike incisions. There are also a number of thick-walled coarse ware vessels and basins, including cooking pots without necks, spherically shaped and made from black clay, and gray and red pots with a short or high neck, sometimes bell-shaped or bulging in the central region. Additionally, platters with straight walls and simple rims were recovered. Finally, of particular interest in the coarse ware category are large thick-walled vessels, with both a height and diameter of roughly 40 cm, that have slightly rounded sides and are decorated with painted horizontal lines, both monochrome and bichrome.

Other domestic vessels include several

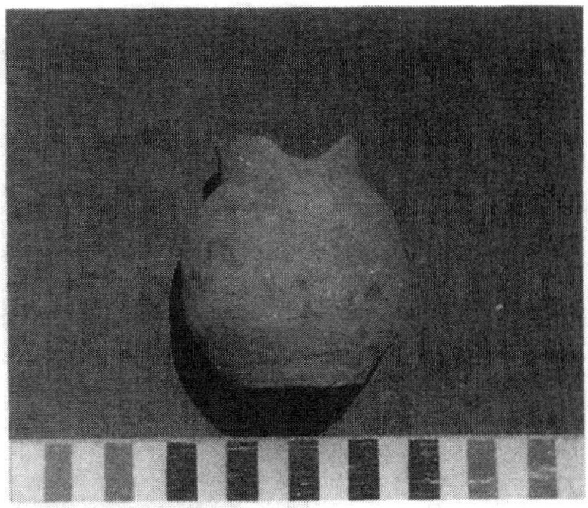

Figure 9.8:1–4. Ceramic vessels from the upper Halaf levels

types of pitchers, including examples with flattened bases, short necks, straight rims, and ear-handles for suspension. Also common are wide-necked cups, basins, round- and flat-bottomed bowls, and small, flat platters. Both large and small spouted vessels, painted ceramic pedestals with cylindrical bases, and miniature and zoomorphic vessels were also found at the site.

Highly stylized painted motifs are typical of the Late Halaf and are represented at Yarim Tepe III by sherds with geometrical designs, including bands, wavy lines, chevrons, angles, triangles, rhombi, lattice motifs, zigzag lines, circles, rosettes, ovals, dots, dotted lines, simple and Maltese crosses, and checkerboard patterns, among others. Additionally, various combinations of the above yield more than three hundred different compositions (Figs. 9.10 and 9.11). Naturalistic representations are rare (Fig. 9.12), with only a marsh bird found on one sherd (though, based on the pattern, the complete vessel probably had a line of similar birds); and a fish covering a large portion of the interior of a bowl found during the 1979 season. Over the course of two field seasons at Yarim Tepe III, the Halaf levels have produced sherds with naturalistic images of a snake, bull, gazelle, cheetah (smaller in size but otherwise identical to an image from Yarim Tepe II), seven sherds with marsh birds, two sherds with

Figure 9.9:1,2. Ceramic vessels from burials 26/1 and 27/2

fish, and finally seven sherds with highly stylized bucrania motifs.

Apart from monochrome designs, usually in red and less frequently in black against a cream slip, there are a number of polychrome designs on the Halaf pottery from Yarim Tepe III; both red and black; and red, black, and yellow compositions. In general, both the shapes and painted decorations of the vessels recovered from the three uppermost Halaf levels at the site are analogous to the ceramic assemblage from both Arpachiyah (Levels TT-6 through TT-8) and Tepe Gawra (Levels XVIII–XX).

Turning to the stone tool assemblage, stone vessels are represented at Yarim Tepe III both by coarsely ground limestone and sandstone bowls and finely made alabaster bowls. Other stone tools include large and small drilled, wedge-shaped axes, drilled mace heads, mortars, grain grinders, and pestles (Fig. 9.13:2–5). The flint and obsidian tools at the site are almost exclusively sickle blades (Fig. 9.13:1). Rimmed stone palettes (Fig.9.14:1), one obsidian rhombus-shaped pendant with two drilled holes, one stone pendant in the shape of a bird (Fig. 9.14:2), and several biconical ring- and barrel-shaped beads (Fig. 9.14:3) made from either marble or carnelian have also been recovered from the site. Finally, of special interest are three stone seal-pendants: one in the shape of an oval, another rounded, and the third conical in shape (Fig. 9.14:2). Bone and horn implements include a digging stick, several awls, a massive point, and a number of polishers (Fig. 9.15:1,2). Clay items besides vessels include beads (Fig. 9.16:1), biconical spindle whorls, "sling projectiles," painted rings (Fig. 9.15:2), and figurines.

More than thirty zoomorphic figurines (Fig. 9.16:3), including cattle, rams, dogs, and sheep, made of poorly fired clay, have been recovered from the site to date. Ten were found this season and, though they are difficult to identify, are probably depictions of sheep or dogs. A considerable number of the above were found in ash-filled pits and were apparently connected with some ritual. Of special interest is one such pit in which several figurines were found along with a clay model of a hoe (Fig. 9.16:2). Fourteen anthropomorphic figurines have been also been recovered, including five this season (Fig 9.17: 1–4). Ten of these are highly stylized varying representations of females (three this season),

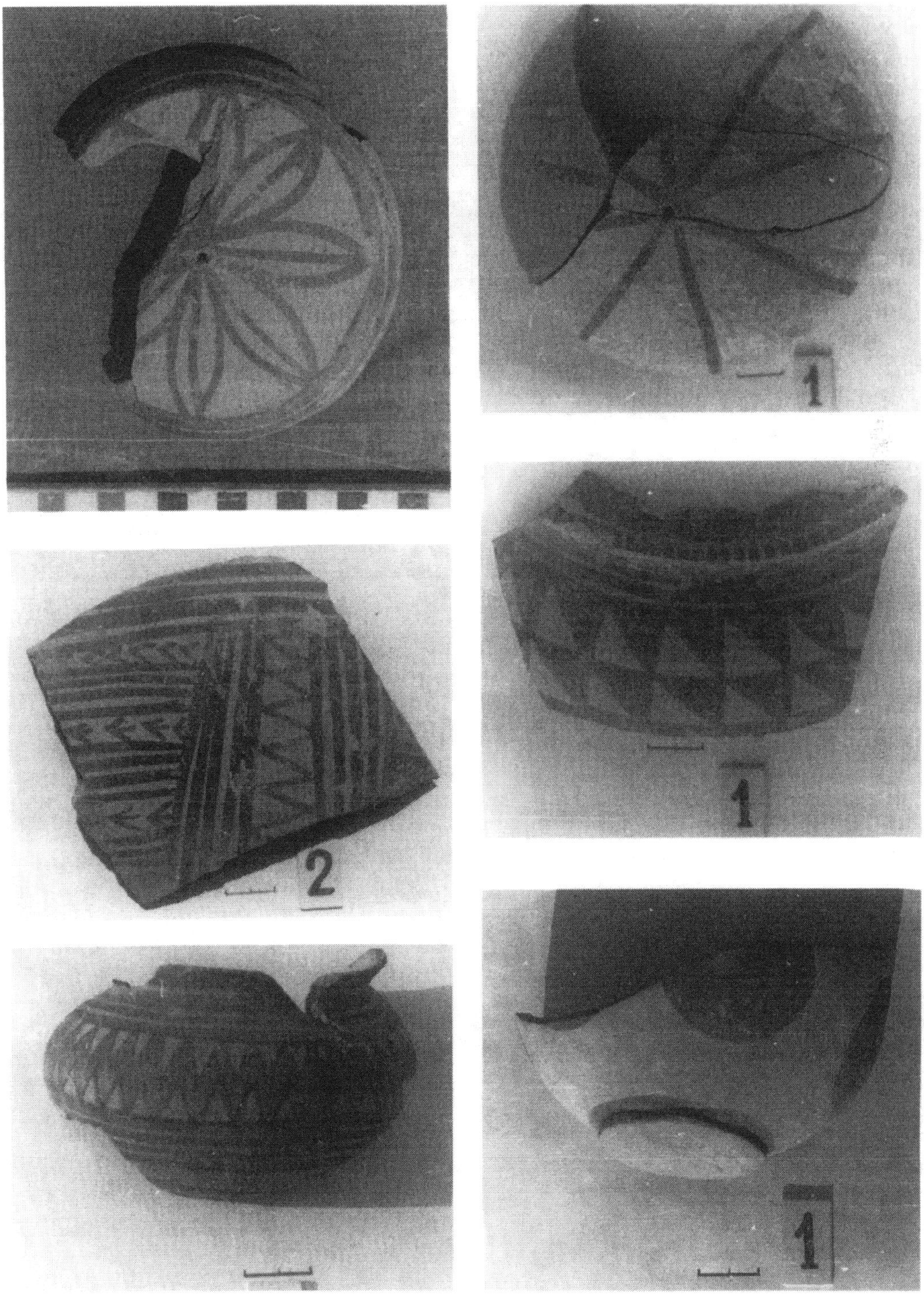

Figure 9.10:1–6. Painted ceramics from the upper Halaf levels

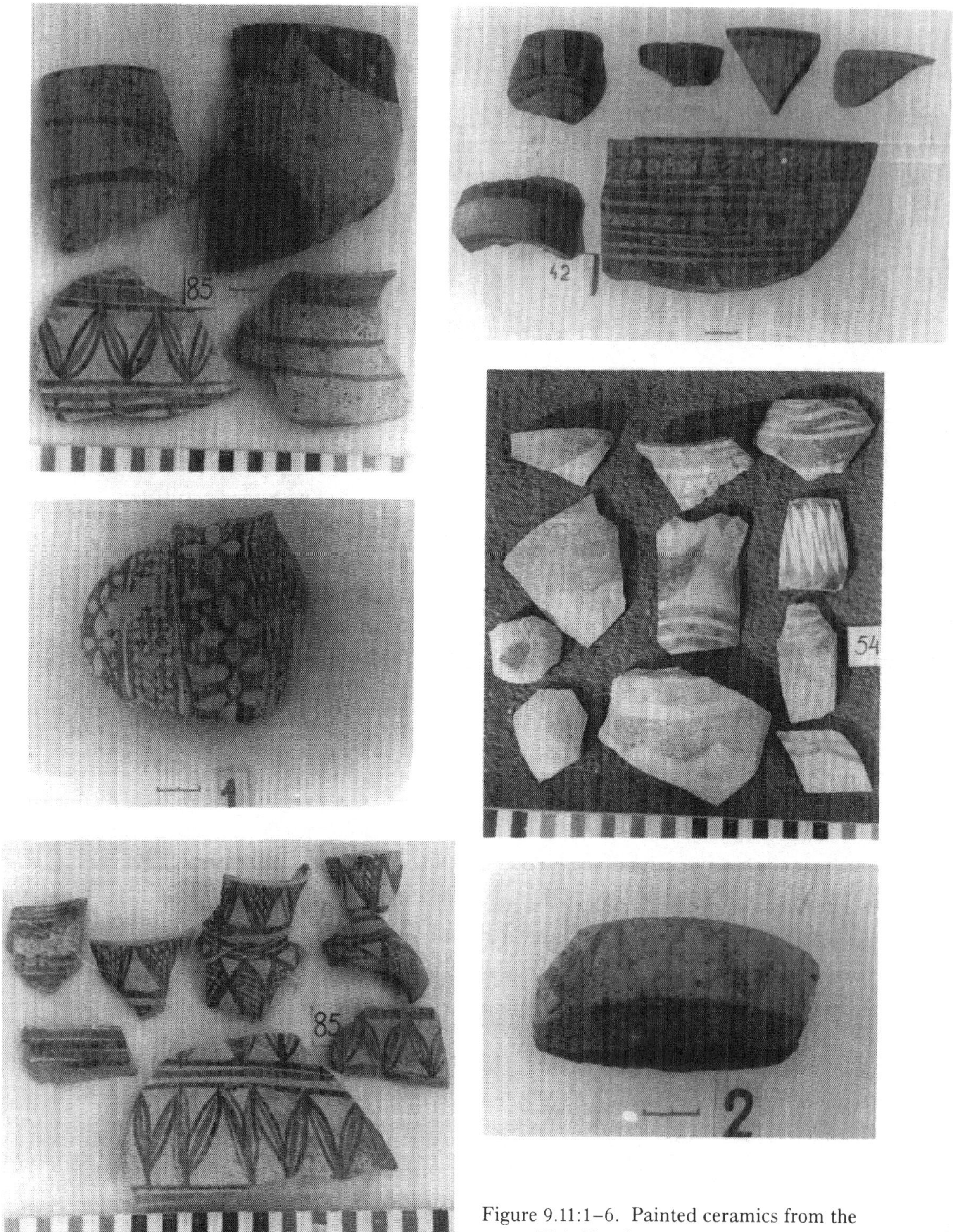

Figure 9.11:1–6. Painted ceramics from the upper Halaf levels

Figure 9.12:1–7. Zoomorphic decorations on Halaf pottery

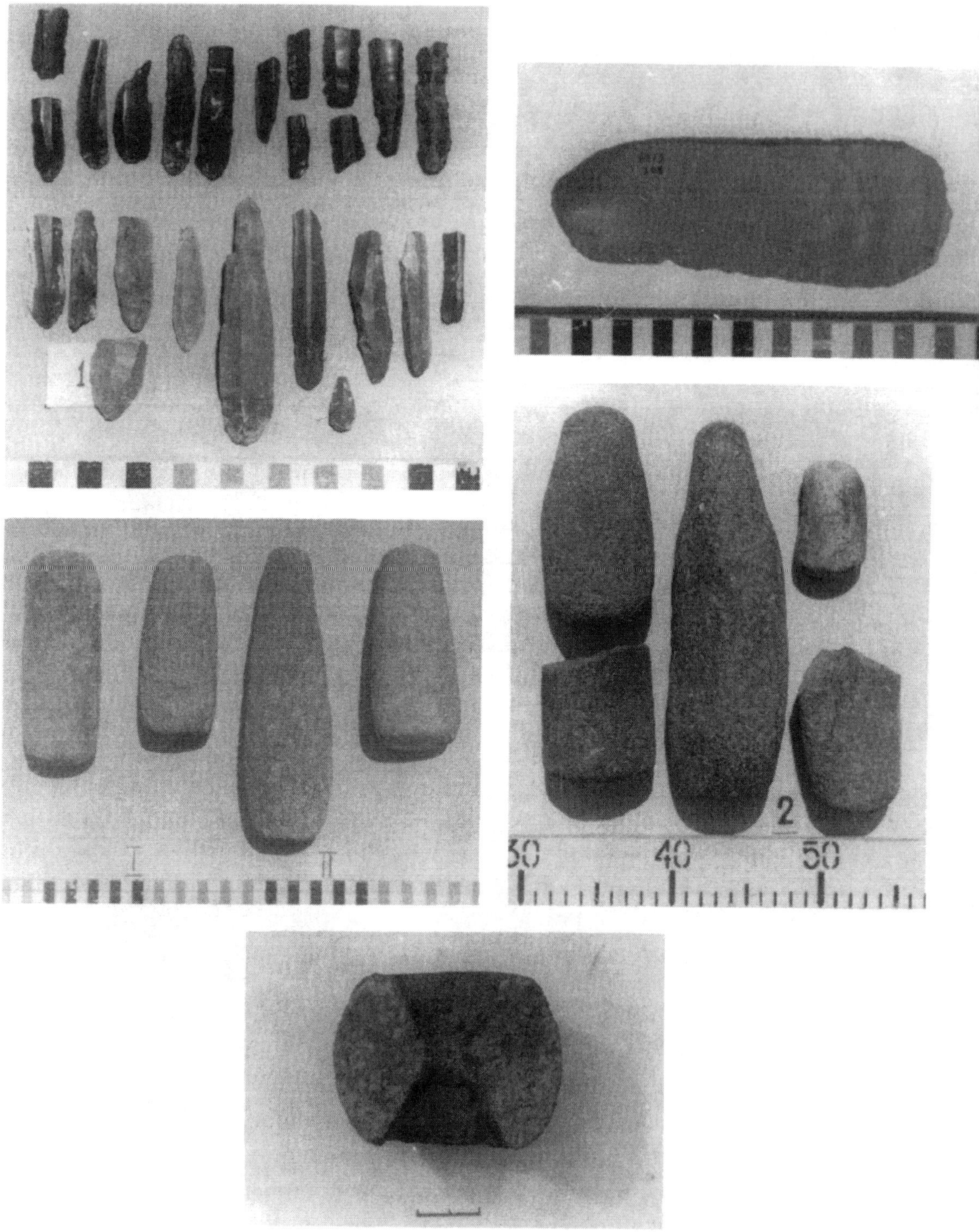

Figure 9.13:1–5. Stone tools from the upper Halaf levels

Figure 9.14:1–3. Stone palette, tools, seals, pendants, and beads from the upper Halaf levels.

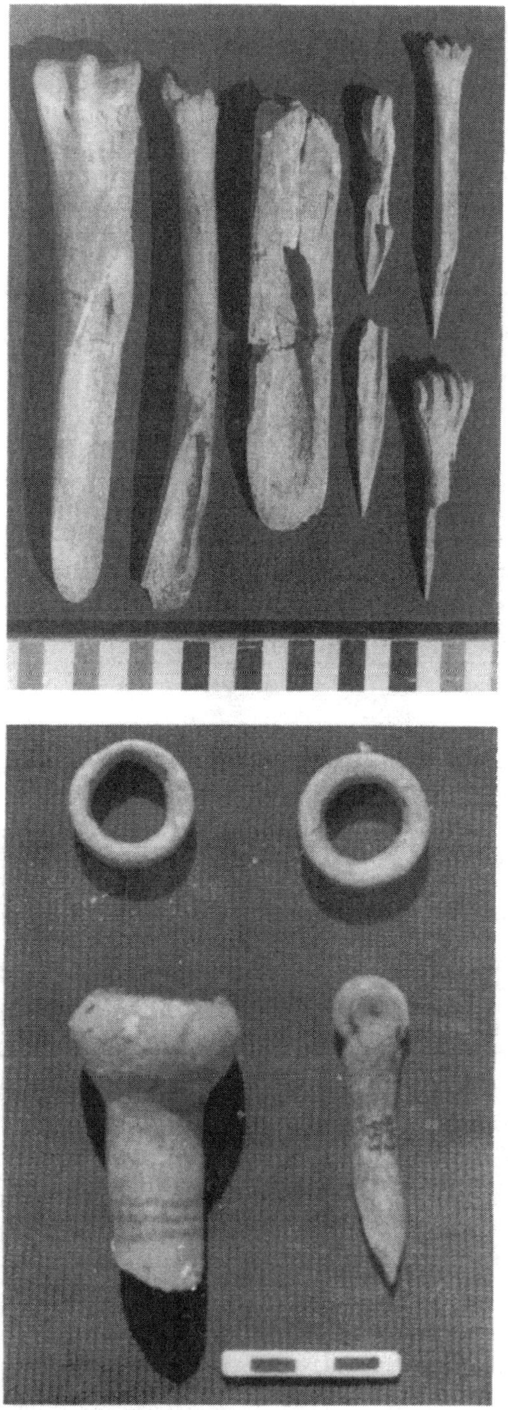

Figure 9.15:1,2. Bone implements and clay rings from the upper Halaf levels.

whereas the other four are realistic depictions of standing or sitting females (two this season). In the majority of cases these finds also have parallels in the abovementioned levels of Arpachiyah and Tepe Gawra and additionally, in the upper layers of the nearby site of Yarim Tepe II.

b. Results of the 1980 Field Season

In 1980 the Soviet archaeological expedition of the USSR Academy of Sciences continued its excavation of the Halaf/Ubaid settlement of Yarim Tepe III.

The 1977–78 excavations demonstrated that the upper 3.5 meters of the cultural deposit belong entirely to the Ubaid period whereas the 8-meter thick underlying stratum is associated with the Halaf period (Bader, Merpert, and Munchaev 1981; Merpert and Munchaev 1982b; chapter 11 this volume). A Hassuna settlement preceded the Halaf occupation at the site, since several pits in the basal level of Yarim Tepe III yielded Hassuna sherd material. Therefore, Yarim Tepe III is probably a three-component site. However, to date only two of its main components have been convincingly demonstrated, the Ubaid and the Halaf.

No evidence for a break in the occupation is observed between the Ubaid and the underlying Halaf layers at Yarim Tepe III, as is also the case in a number of other Halaf-Ubaid multicomponent sites in northern Mesopotamia. The site contains a rather thin layer in which both late Halaf and Ubaid material culture is present. Based on evidence from Tell Arpachiyah, Tepe Gawra and Yarim Tepe III, we conclude that Ubaid settlers in the Sinjar plain and other areas of northern Mesopotamia preferred to build new settlements atop the mounds formed by previous occupations. The stratigraphic sequence supports the idea of a smooth transition from the Halaf to the Ubaid at the abovementioned sites with no significant break between the

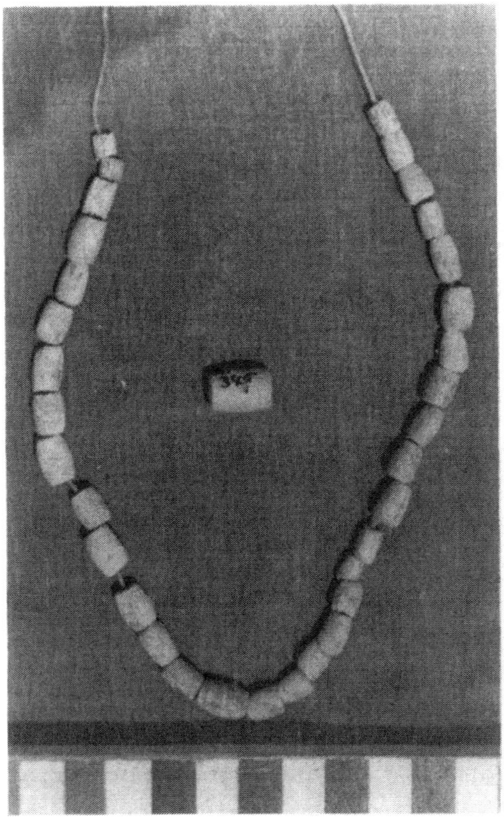

Figure 9.16:1. Clay beads

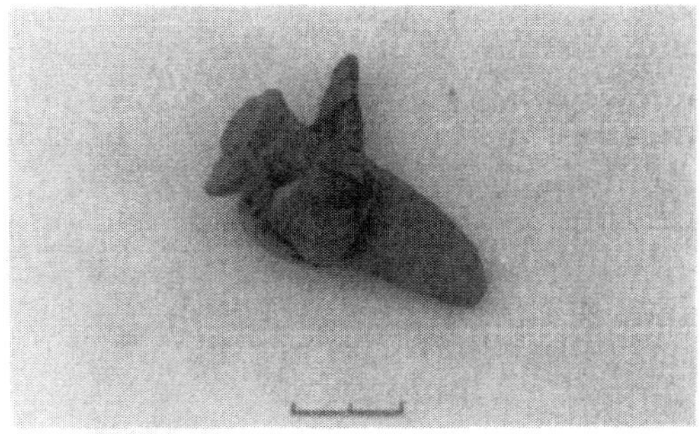

Figure 9:16:2. Clay model of a hoe

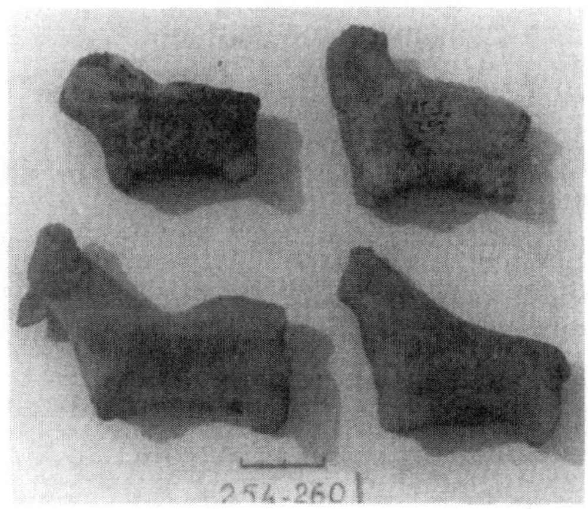

Figure 9.16:3. Clay zoomorphic figurines

two phases, since in each case a transitional layer has been found containing both Halaf and Ubaid sherd material.

In 1979, the excavations in different sectors of the main excavation area progressed to a depth ranging from 4.70 to 6.00 mbd. Among the structures exposed during this season were two large tholos-shaped buildings (structures 137 and 138). The latter was a multichambered circular structure with massive walls that survived in places to a height of 2 meters. The excavation of this structure was completed in 1979, while structure 137 had not yet been exposed down to its foundation.

Thus, the primary task of the 1980 excavation season at the site was to complete the exposure of structure 137 and to study the cultural deposits in the immediate vicinity of the the two tholoi. As a result, the 1980 season focused on the following areas of the main excavation: I-d, II-b, II-d, IV-a, VI-a, and VI-c. In these sectors approximately 1 m of the cultural deposit was excavated with a maximum depth of 7.05 mbd reached in squares II-b, II-d, IV-c, VI-a, and VI-c.

The soil matrix of the cultural deposit in the excavated sectors of the site was the typical humus-rich loam with ash and charcoal lenses. Material cultural remains included plain and painted Halaf ceramics, various stone and bone implements, faunal remains, and other cultural debris. However, before turning to the artifacts, the architectural remains exposed during the 1980 season will first be discussed.

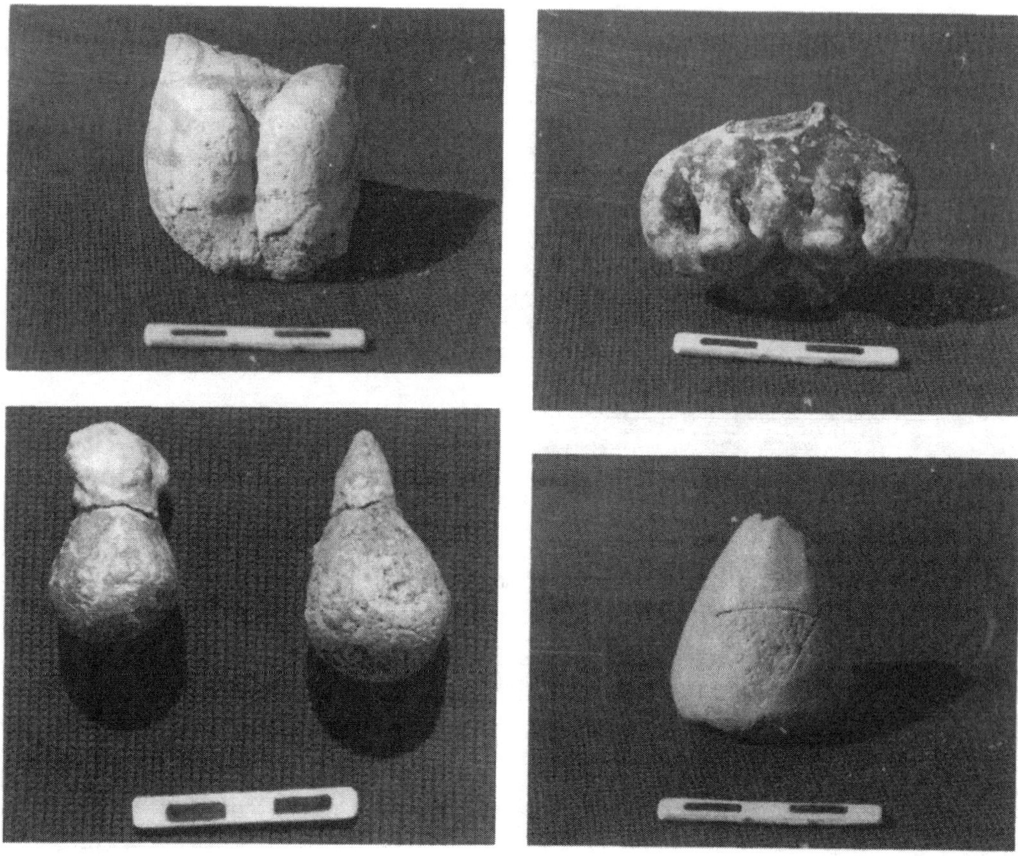

Figure 9.17:1–4. Anthropomorphic figurine fragments

Architecture

Remains of tholos structure 137 were uncovered in squares II-b, II-d, VI-a, and VI-c (Fig. 9.18) at a depth ranging from 4.93 to 5.33 mbd for the top level and 6.84 to 7.00 mbd for the floor level. The tholos is the largest circular structure yet encountered at Yarim Tepe III. Its walls have survived to a height ranging from 1.55 to 2.02 m (Fig. 9.19). However, they are not nearly as massive as the walls of the other tholos (structure 138) located several meters to the west of it in sectors I and II. The maximum thickness of the wall of structure 137 is only 0.5 m and it leans to the southwest up to 0.5 m, which suggests some flaw in the original construction of the building. Unlike the majority of circular dwellings discovered at Yarim Tepe II, this structure was not domed but cylindrical in shape and had, quite probably, a flat roof. The structure had a floor diameter of 5.50 m and a diameter of 5.80–5.85 m along the upper edge of the interior of the structure. Its height was at least 2.05–2.10 m and its walls were made of unbaked clay mixed with cut straw. Both the outside and the inside of the walls were carefully mud-plastered to a thickness of 2 cm. The plaster was made from dark brown clay with very finely cut straw. A grayish green clay was also used in places, though infrequently.

Within the structure were four hollow L-shaped buttresses, almost identical in size, and evenly spaced along the wall of the tholos. These "reinforcing corners," as they were called, survived to the height of the tholos wall or roughly 2 m. Their average length is nearly 1.5 m, and the angle of the walls of each reinforcing corner approaches 90 degrees. These four L-shaped buttresses give the interior of the tholos a cross-shaped plan. The hollow interiors of the reinforcing corners were probably used for domestic purposes, perhaps to store grain and other foods.

The floor of the tholos was a hard packed surface of yellowish brown clay mixed with grayish green clay. It sloped 13 cm upward to the north. Structure 137 had a narrow and nearly rectangular doorway, 0.81 m in height, 0.41 m wide at the bottom, and 0.48 m at the top. It was placed in the southern wall of the tholos. The gypsum-plastered door sill was 0.45 m higher than the floor of the tholos.[1]

Oven 33 was found near the doorway, abutting both the wall of the tholos and the southwestern reinforcing corner. It was a small oven of the mangal type, and its foundation rested on the floor of the tholos. Roughly triangular in shape, the oven was at least 35 cm high. Unfortunately, its upper portion has been completely destroyed and, thus, it is difficult to reconstruct the entire design of the feature. Its wall consisted of clay mixed with gypsum. It is only 4–6 cm thick and had been subjected to extreme heat.

The interior of the tholos, from its floor to the upper edge of its surviving walls, was filled with a humus-loam layer rich in cultural debris. For example, found within the central "cross" of the tholos at a depth range of 6.10–6.35 mbd were six bone and three stone beads, sixty-one "sling projectiles" made of unbaked clay, three bone awls, a bone spatula, two elaborate though fragmentary anthropomorphic figurines, an animal figurine fashioned from unbaked clay, three pieces of tabular flint, fourteen flint flakes, forty-two sheep/goat bones, and finally, over 150 sherds, including fifty-six painted ones.

The tholos fill at a depth ranging from 6.35 to 6.55 mbd had a similar density of material culture. However, the southern part of this level was particularly rich in finds and yielded at least 110 clay "sling projectiles," two bone awls, a stone pestle, numerous sherds from plain and painted vessels, remains of various fauna, a zoomorphic figurine made from unbaked clay, and a distinctively shaped clay animal head (probably a ram). Additionally, three relatively large (up to 6.5 cm in length) conical objects fashioned from unbaked clay that are apparently stylized anthropomorphic figurines were recovered from this area, along with a similar figurine much smaller in size, a clay ring 3 cm in diameter, a flat stone bead, and finally, three small fragments of unidentifiable clay objects.

The fill in the southwestern sector of the tholos had the greatest concentration of material culture debris in the depth range of 6.55–6.75 mbd. In addition to sherd material and faunal remains, this area yielded over seventy clay "sling projectiles," a bone awl or needle, a ceramic ring 3.2 cm in diameter, a flat/concave ceramic object, oval in shape with a handle/projection, and eight poorly fired ceramic figurines. The latter group includes five anthropomorphic forms with missing upper portions and various degrees of stylization. Two of these are cone-shaped, and a third is more realistic, depicting a seated individual with parted legs. In the pubic region of the figurines is a triangle filled with pinhole-sized dots. A fourth figurine is larger (over 10 cm in height) and has the unusual shape of a flat and elongated triangle. Its surface is highly burnished, and in the area of its neck and abdomen there are round depressions. The pubic region is depicted by oblique cross-hatching. The highest degree of stylization is observed in the fifth anthropomorphic figurine, which is cylindrical in shape with two horn-shaped projections. Among the zoomorphic figurines, of greatest interest is a rather large representation (over 10 cm in length and 9 cm in height) of an animal with its head and legs missing, which is, in all likelihood, an ox.

Finally, the tholos fill immediately above the floor level at a depth ranging from 6.75 to 7.00 mbd contained numerous sherds, primarily from cooking vessels, and faunal re-

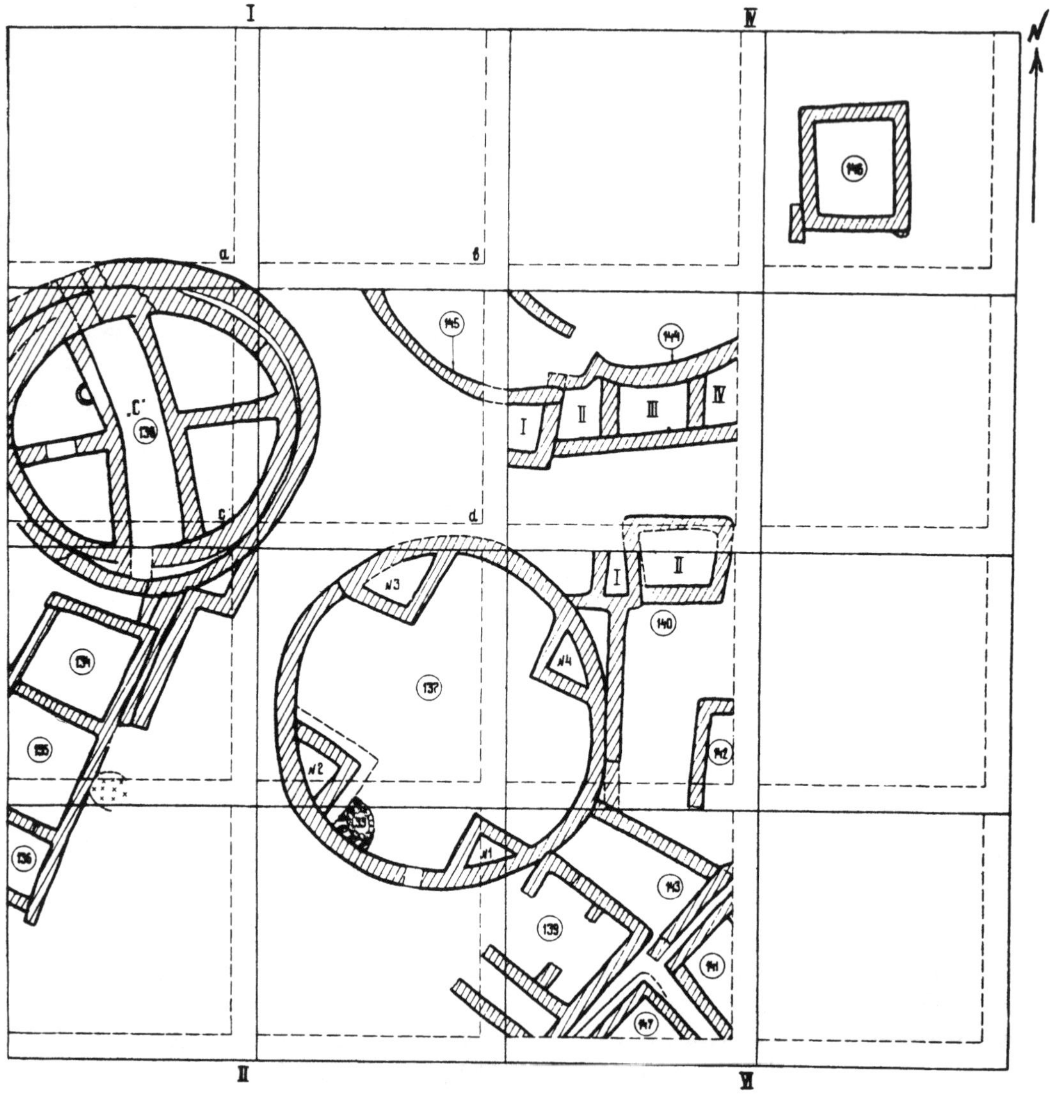

Figure 9.18. The architectural plan of Halaf Level 3

mains, especially of domesticates. Additionally, a flint core, ten large flint flakes, several flint and obsidian palettes, a cylindrical stone pestle, a fragment of grain mortar, three bone awls or needles, three biconical ceramic spindle whorls, a ceramic scraper/polisher, three ceramic rings, several fragments from three clay spoon-shaped objects, two large balls made of unbaked clay (9.5 and 10.0 cm in diameter), and finally, two unbaked clay figurines were recovered from this level. The first of these is highly stylized and is in the shape of a cone, whereas only the upper part of the other has survived and is probably some type of female figurine.

Finds from the fill of the hollow spaces enclosed by the reinforcing corners will now be discussed. In most cases the soil matrix of the fill consisted of loosely compacted humus loam. In the southeastern reinforcing corner

Figure 9.19. View of tholos 137

Figure 9.20. View of structure 139

at a depth ranging from 6.15 to 6.90 mbd no fewer than 700 clay "sling projectiles" were discovered, 616 of them intact. Additionally, two large clay balls, 8.8 and 9.1 cm in diameter, two painted ceramic rings (one rounded and one flat in cross-section, 2.3 and 3.2 cm in diameter, respectively), two biconical ceramic spindle whorls, several fragments of four animal figurines made from unbaked clay, a bone awl or needle, one fragment of a stone cup made from light red and white marble, an obsidian palette, two flint flakes, fifty-nine sherds, and finally four animal bones, including three sheep or goat mandibles, were discovered in this area.

The fill of the southwestern reinforcing corner contained five painted ceramic rings, one round-flattened and two biconical ceramic spindle whorls, a small, crudely fashioned clay cup 2 cm in height, seven bone awls or needles ranging from 7.6 to 11.6 cm in length, and two unbaked clay balls, 5.6 and 9.2 cm in diameter. Additionally, a conical stone pestle, a grain mortar fragment, an obsidian palette, one obsidian and three flint flakes, nine beads, including seven made from cowrie shells and two of white stone, approximately thirty unbaked clay objects in round, round-flattened, and cone shapes, three anthropomorphic figurines of different styles, fragments of five poorly preserved unbaked clay animal figurines, a set of tiny painted clay cups, and a clay pot were found. Finally, numerous sherds including six from roughly fashioned dark gray pots and thirty-two painted ones were recovered along with thirty-five animal bones, including six sheep or goat astragali, and a sheep horn core and mandible.

Material culture recovered from the fill in the northwestern reinforcing corner included six clay "sling projectiles", a fragment of an anthropomorphic figurine and another from a zoomorphic one, both made from unbaked clay, two bone awls or needles, a flint palette, and a painted ceramic bowl, 7.5 centimeters in height. Other material recovered included thirty-seven painted sherds, eighteen coarse, thick-walled sherds, thirty-six burnished plain ware sherds, and finally, five animal bones.

Unlike the rest of the hollow buttresses, the fill inside the fourth reinforcing corner was compacted clay that yielded few cultural remains. Quite possibly, the hollow space within this buttress was intentionally filled with clay at the time of the tholos's construction and use. Only two plain ware sherds, an animal bone, and a flint flake were found in the fill.

In summary, the fill of structure 137 was literally saturated with cultural debris that is typical of a habitation structure. The lower layer of the fill just above the floor of the structure is associated with its original function and, thus, is the most interesting to study. It suggests that the structure was a dwelling, which is further supported by the presence of the abovementioned oven built on the floor of the tholos. In our view, this structure did not have a long use-life. Problems in its original design and construction resulted in a steadily increasing lean of its main wall in a southwestern direction. Thus, the tholos was in constant danger of collapsing. It is possible that the intentional filling of one of the reinforcing corners with packed clay was an attempt to shore up this unstable structure.

It should be noted that whereas circular structures with internal partitions are not infrequent in Halaf settlements, particularly at Yarim Tepe II and Yarim Tepe III, structure 137 was the first tholos excavated of this design with hollow L-shaped buttresses. The only possible analog to this tholos excavated to date is a circular structure with internal partition buttresses exposed in Level XVII of Tepe Gawra (Tobler 1950, Pl. XLII-b).

In squares VI-a and VI-c, several rectangular structures were exposed adjacent to tholos 137 that were associated with its use some time after its original construction. These related buildings will be described below.

The remains of structure 139 were found in squares VI-c and II-d with the top levels of 5.35–5.71 mbd and the foundation at 6.27–

6.30 mbd. The structure was rectangular, with its interior divided into two compartments by a partition wall. The southern chamber was a small rectangular room measuring 2.0 m × 0.60 m and oriented with the long axis running roughly northwest to southeast (Figs. 9.18 and 9.20). The other compartment was almost square in shape with the dimensions 1.90 m × 2.10 m. The latter was in turn divided into two smaller compartments at the time of its use, but the partition between them is so badly preserved that the access between the two compartments cannot be defined. The walls of structure 139, which are up to 30 cm thick, have survived in places to a height of 0.90–0.95 m and were constructed from grayish green clay. The northwestern corner of the structure abuts the wall of tholos 137. The entire western portion of structure 139 was heavily disturbed from building activities in the upper levels. The floor of the structure is a layer of compacted grayish green clay and the structure fill primarily consisted of the same clay mixed with humus containing sherds and faunal remains. Somewhat later in time, a large oven (number 36) was built adjacent to structure 139. This oven was exposed in the eastern corner of square VI-c at a depth ranging from 5.15 to 6.30 mbd. A large portion of the oven extends beyond the limit of the excavation area. The maximum length of this feature along the outer rim is 2.45 m (Fig. 9.21). The oven chamber was circular in shape with a minimum diameter of 1.30 m. The walls of the oven were massive with a maximum thickness of 0.55–0.60 m and they stood 1.10–1.20 m high. This is the first oven discovered to date at Yarim Tepe II or III with walls of this thickness. Because a considerable part of the oven remains unexcavated, the overall plan of this feature remains undefined. As is the case with other ovens discovered at the site, its walls show indications of exposure to high temperatures. The fill of the oven consisted of burnt organic material, fragments of wall collapse, ash, and pieces of gypsum plaster and charcoal.

Prior to the construction of the oven and structure 139, this area of the settlement was composed of square and rectangular structures serving domestic functions (structures 141, 143, and 147). Remains of structures 141 and 143 were exposed in square VI-c (Fig. 9.18) at a depth ranging from 6.15 to 6.70 mbd and 6.39 to 6.70 mbd respectively. All three were built at approximately the same time utilizing the same construction technique of overlapping mud bricks 5–6 cm thick. Only one complete wall running east-west and a portion of another perpendicular to it survive from structure 143. The other walls were probably destroyed when structure 139 was built. The greater part of structure 141 is located in an unexcavated area of square VI-d. Thus, only the northwestern corner of the structure in adjacent square VI-c has been exposed.

Structure 143 abuts tholos 137 and apparently served a domestic function related to the use of the latter at some late stage in its existence. It is probable that during the construction of structure 143, its builders destroyed an earlier small oven on the same location. On the excavation plan this oven has been designated as feature 40. Remnants of this oven were found in square VI-c at a depth ranging from 6.75 to 6.90 mbd, roughly 1 meter east of tholos 137. The oven was circular and its diameter was nearly 0.50 m.

Roughly contemporaneous with structures 141, 143, and 147 are the remains of several buildings exposed in square VI-a. Structure 140 (Fig. 9.18), which probably also served a domestic function, abutted tholos 137 on the latter's northeastern side and was found at a depth ranging from 5.30 to 6.70 mbd. It was composed of two small adjoining chambers. Chamber I was in the shape of a narrow trapezoid, 0.90 m in length and 0.50–0.60 m wide. Chamber I shares its eastern wall with chamber II, which was nearly square in shape (1.35 m × 1.30 m). The southwestern corner of the structure abuts two walls, one running north-south and the other perpendicular to it, forming a 90-degree angle along the wall of tholos 137. The foundations of these walls are at the same level, have almost the same thickness (30 cm on the average), and are both made with grayish green clay mixed with finely cut straw. Chambers I and II of structure 140 were filled with wall collapse and contained Halaf sherd material and several faunal remains.

Remnants of another rectangular structure (142), built at the same time as the above structures (140, 141, and 143), was excavated in the eastern corner of square VI-a at a depth ranging from 6.05 to 6.72 mbd (Fig. 9.18). Its walls were approximately 30 cm thick and were also made of mud bricks. In square VI-b, a large portion of this structure remains to be excavated.

Other architectural remains have been found north of tholos 137 in sectors I and IV. For example, remnants of a multiroomed domestic complex (structure 144) have been discovered in square IV-c at a depth ranging from 6.28 to 7.05 mbd. The long axis is oriented east-west and its western portion is comprised of a small square chamber (I) which extends into square I-d (Fig. 9.18). Chamber I shares its eastern wall with chamber II, which in turn shares its eastern wall with chamber III and the latter similarly with chamber IV in a linear arrangement. The respective dimensions of chambers II–IV are 0.70 m × 0.85 m, 1.20 m × 0.95 m, and 0.70 m × 1.0 m. These chambers are offset slightly to the north of chamber I and are bordered on their northern sides by a curved wall that is apparently part of another tholos. Thus, structure 144 is, in all likelihood, an auxiliary complex related to the use of this tholos and probably functioned as a storage facility for the latter.

Immediately above the surviving portion of chamber I of structure 144, the remains of a later structure (145) were discovered which extend into the adjacent square (I-d) (Fig. 9.18). Apparently structure 145 is also tholos-shaped, continuing into the squares immediately to the north.

Finally, one other structure (146) was discovered in square IV-b. Only 10 cm of its walls were uncovered in the 1980 season at a depth of 5.65–5.75 mbd. The structure is rectangular in shape with the dimensions of 2.50 m (north-south) × 2.20 m. The walls are made of clay slabs that are roughly 25 cm thick. Structure 146 is probably an addition to the same tholos associated with structure 144.

This completes the inventory of the significant architectural remains exposed during the 1980 field season at Yarim Tepe III. Along with tholos 137, these structures comprise the third Halaf construction level at the site, with the earliest structure being the tholos.

Ovens

In addition to the structures described above, the remains of eight ovens (numbers 33–40 [see Fig. 9.21]) were also exposed. Three of these have been described above: oven 33 located in the interior of tholos 137, oven 40 to the east of it, and a third found next to structure 139. We shall now discuss the remaining ovens in more detail.[2]

Remains of the lower portion of oven 34 were uncovered in square IV-c at a depth ranging from 5.10 to 5.36/5.46 mbd. This oven was of the tannur type, quite large and circular in shape. Its outer diameter at the base is roughly 1.5 m. However, the oven wall has not survived along its entire perimeter. The oven was constructed from grayish green clay mixed with gypsum, and near its foundation it was 18–25 cm thick. The access was most likely located in the eastern wall. At the time the oven was in use the immediate vicinity was devoid of any other structures and the area was cluttered with cultural debris. However, the abovementioned remains of structure 145 and associated oven 39 were found in the underlying level of the same stratum. The latter oven was found almost directly under oven 34 in square IV-c at a depth ranging from 5.60 to 6.02 mbd.

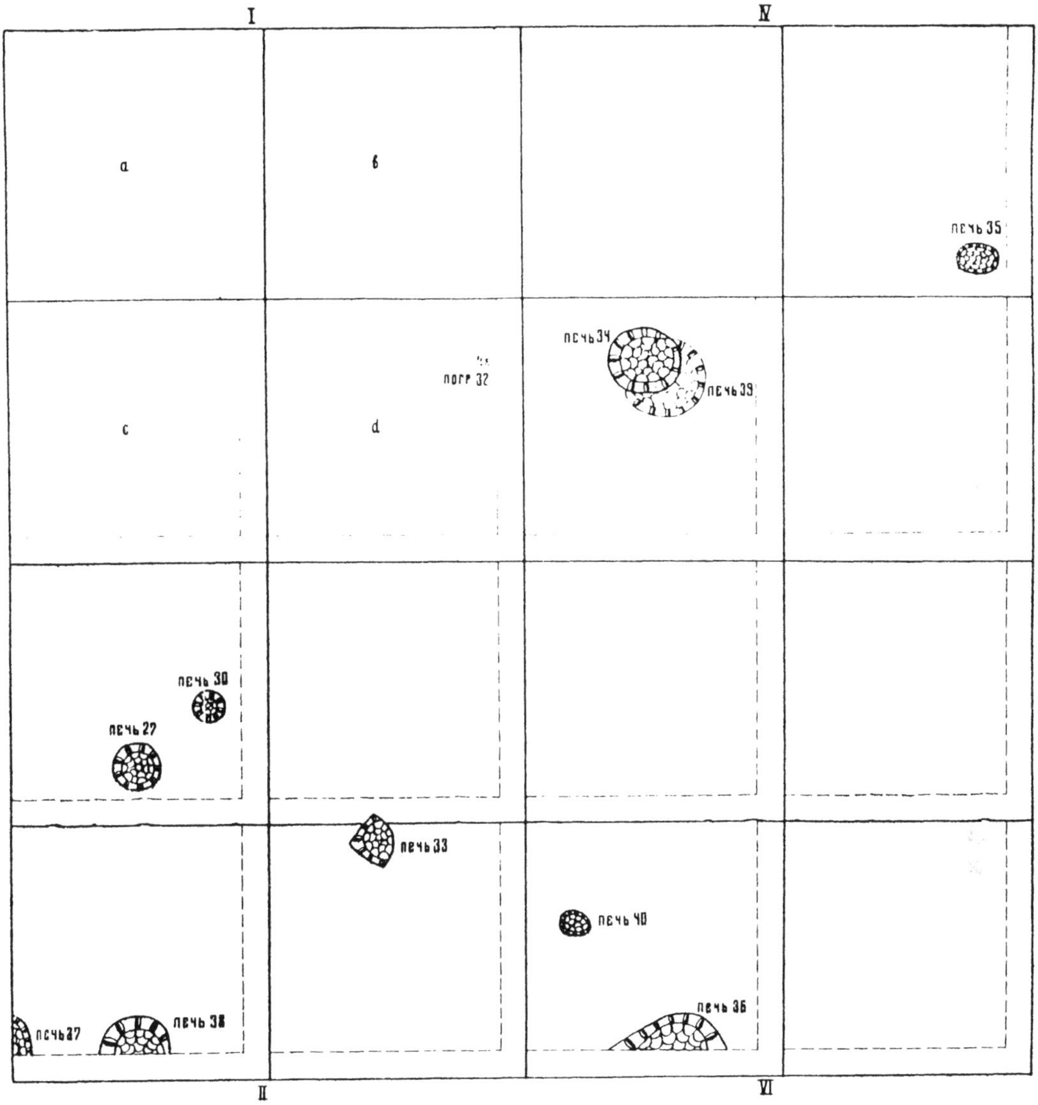

Figure 9.21. Ovens and hearths for Halaf Level 3 (печь = oven).

The state of preservation of oven 39 is somewhat better than that of oven 34. It is also circular in form and its inner dimensions at the base are 1.15 m (north-south) × 1.35 m. Its walls were made of the usual clay-gypsum mixture, 10–15 cm in thickness. The oven was no more than 0.60 m in height, and had a hemispherical shape. Its access was approximately 20 cm wide, located in the western wall.

Remains of two more ovens (numbers 37 and 38) were excavated in square II-c. In the southern sector of the square only a small portion of oven 37 was exposed, whereas the remainder continued beyond the limits of the excavation. It was a large oven of the mangal type and had a rounded square shape. It is probably over 0.70 m in length, nearly 0.50 m wide, and has an average height of 0.35 m. The oven was built on a platform constructed

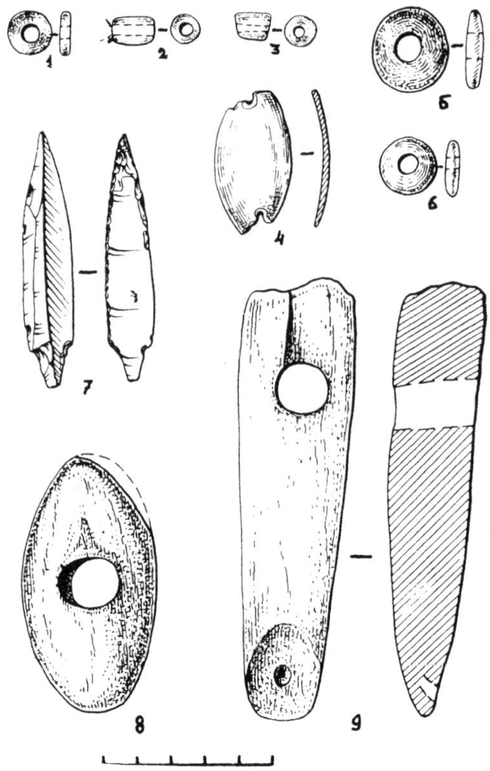

Figure 9.22:1–9. Lithic and bone artifacts from Halaf Level 3

with clay and stone. The walls of the oven are only 5 cm thick and made of clay mixed with gypsum. They show evidence of exposure to high temperatures. The fill of the oven contained large amounts of ash, oven wall collapse, charcoal, and faunal remains.

Oven 38 was constructed earlier than oven 37 since it was found at a lower level (5.35–6.15 mbd) in square II-c. The remains of the oven indicate it was a large structure of the tannur type (Fig. 9.21). It is circular in form, with a diameter of at least 1.90 meters and has relatively massive walls, ranging from 20 to 40 centimeters in thickness.

Remains of oven 35 were uncovered in square IV-b at the depth range of 5.20–5.30 mbd. It was built in an area devoid of other architectural remains. It was of the mangal type, oval in shape, with the dimensions of 0.70 m (east-west) × 0.50 m. The entire plan of the oven cannot be reconstructed since only the base of the feature has survived. The wall of the oven is only 4–5 cm thick and made of the usual clay-gypsum mixture.

It is also interesting to note that in square I-d, at a depth of 5.40–5.45 mbd, the remains of an infant burial were encountered (32 in Fig. 9.21). The skeletal remains are poorly preserved, and many of the bones, including the cranium, have decayed or were destroyed by a small animal burrow that traversed the burial. The burial pit could not be defined, but the remains of the infant rested on a platform of 28 cm (east-west) × 22 cm. Apparently, the body was buried in a flexed position on its left side with its head pointing west.

Small Finds

With the architectural and mortuary data presented, the artifacts will now be discussed. During the course of the 1980 excavation season at Yarim Tepe III, a considerable quantity of diverse artifactual remains were recovered. The large majority of finds were, of course, sherds, and a considerably smaller portion consisted of bone and lithic objects.

Grain mortars make up the majority of the ground stone assemblage. At a depth ranging from 4.65 to 6.90 mbd, three complete grain mortars and three fragments were recovered. In square VI-b, a large boat-shaped grain mortar was found, 51 cm in length, 20 cm wide, and 8–10 cm high. The two other complete grain mortars are flat and oval in shape, one measuring 27.5 cm × 18.0 cm × 6.5 cm and the other 24.0 cm × 16.0 cm × 7.0 cm. For the most part the grain mortars are made from a gray porous stone, possibly of volcanic origin, exotic to the Sinjar region. It should be noted that ground stone implements for grain processing are not common in the Halaf layer at Yarim Tepe III and are not as distinctive in form as those from the overlying Ubaid stratum at the site. In the latter, an elaborate

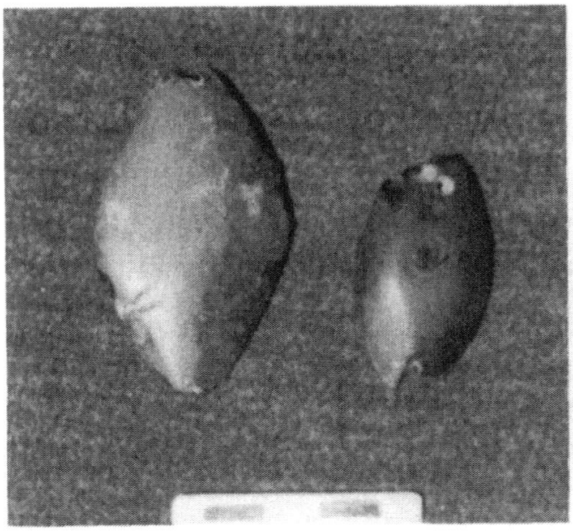

Figure 9.23. Ornaments from Halaf Level 3

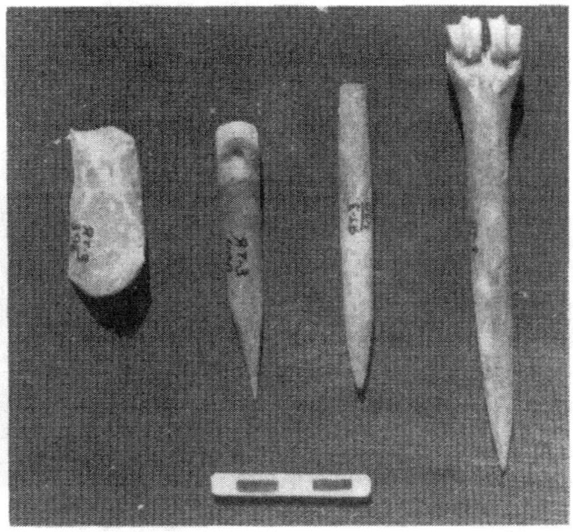

Figure 9.24. Bone and horn artifacts from Halaf Level 3

series of large boat-shaped grain mortars, scrapers, and mortars with pestles have been found (Bader, Merpert, and Munchaev 1981, Pls. XIX–XXI).

In the chipped stone assemblage, flint and obsidian knife-shaped blades for insertion into wooden or bone hafts were only rarely encountered during the course of the last field season. Another interesting stone tool recovered is a flat, wedge-shaped axe, measuring 7.4 cm × 2.3 cm × 1.2 cm. This tool type is commonly found at early agricultural settlements in the region. It was discovered in square IV-c at a depth of 6.50 mbd. In square I-d, at a depth of 5.65–5.90 mbd, a flint projectile point 7 cm in length with a short but pronounced stem was recovered (Fig. 9.22:7). This find is unique for the Halaf Period since, to the best of our knowledge, it represents the first projectile point to date recovered from a Halaf context. Another stone weapon recovered from the site was a mace head made from white alabaster that was discovered in Square IV-a at a depth of 4.85–5.10 mbd. It is 5.5 cm high, with maximum diameter of 4 cm.

A number of stone and obsidian ornaments were also discovered during the 1980 field season. In particular, two obsidian pendants were found, one oval in shape (4 cm × 2.2 cm) and the other a rhombus (5.3 cm × 3.4 cm) with perforated ends (Figs. 9.22:4 and 9.23). The former was discovered in square IV-a at a depth of 4.85–5.10 mbd. Similar stone ornaments are known from earlier Mesopotamian sites, particularly from the Hassuna Period. Oval and rhomboid obsidian pendants are also commonly found at Halaf sites. They have been recovered, for instance, in level TT6 at Arpachiyah (Mallowan and Cruikshank Rose 1935, Pls. VI-b & XI-a) and at other contemporaneous sites.

Of special interest is an engraved seal-pendant made from white stone found in square IV-c at a depth of 6.55 mbd. Seal pendants of this type are characteristic of the Halaf with examples known from nearly every site from the period, including Yarim Tepe II and, especially, Tell Arpachiyah and Tepe Gawra (Munchaev and Merpert 1981, 135, Figures 40 and 41; Mallowan and Cruikshank Rose 1935, Figures 50–51; Tobler 1950).

More than ten beads of different colors made from marble, diorite, steatite, and other types of stone were also recovered during the 1980 field season (Fig. 9.22:1–3, 5–6). Among those found were five cylindrical beads up to 1.5 cm long, several disk-shaped beads with a diameter of up to 1.5 cm, and one barrel-shaped bead. Stone beads of these forms are also typically found at Halaf sites.

Other finds at Yarim Tepe III that are commonly recovered from Halaf sites include stone cosmetic implements in nail form and palettes; examples of each were found at Yarim Tepe III in 1980. The former type was discovered in Square VI-a at a depth of 6.12 mbd and the latter, in square IV-b at a depth of 4.90–4.95 mbd.

Six fragments of small marble bowls were also recovered. They were red, light red, and white in color. It should be mentioned that five of these fragments were found in two adjacent squares: VI-a and VI-c, at a depth ranging from 5.20 to 6.35 mbd. They are all from stone vessels typical of the Halaf Period.

Bone artifacts were not commonly found, and those discovered were primarily awls and needles made of sheep and goat bones (Fig. 9.24) that are known from other early agricultural sites in the region. As was noted above, many of the finds came from the fill of tholos 137. The awls are, on the average, 10–12 cm in length, with the longest being 18 cm. A bone hammer (Fig. 9.22:8) was found in square VI-a near the wall of tholos 137 at a depth of 5.90–6.0 mbd. It is oval in shape measuring 7.0 cm × 4.0 cm and has a central hole to accommodate a handle. Tools of this type are rarely found in Halaf contexts.

Another interesting find from the same square (VI-a) though considerably lower in depth (6.73 mbd) is an implement made from a horn core with a circular hole near its wider edge to accommodate a handle. Its working edge is sharpened and has a circular bore hole (Fig. 9.22:9). This edge is heavily polished through use.

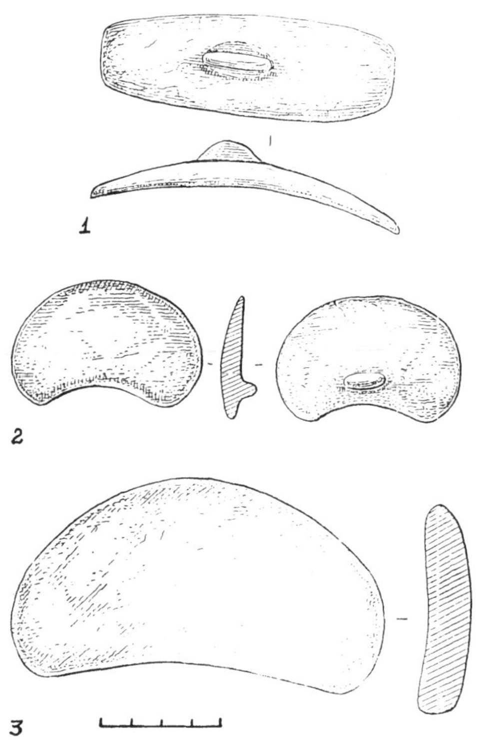

Figure 9.25:1–3. Ceramic artifacts from Halaf Level 3

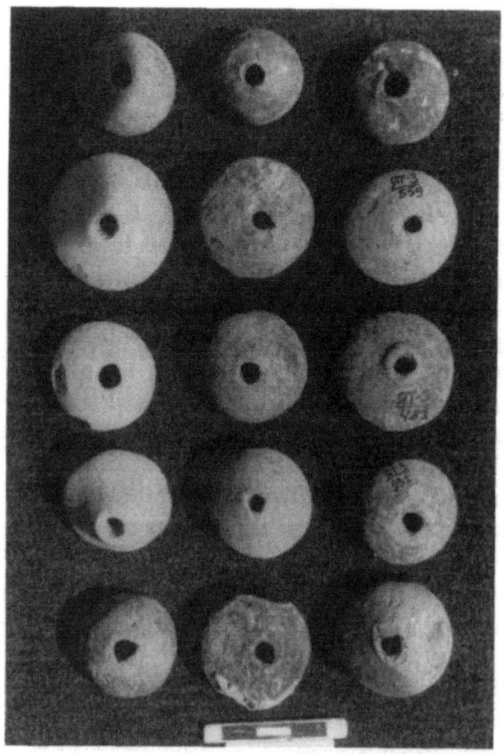

Figure 9.26. Ceramic spindle whorls from Halaf Level 3

Ceramics

The ceramic assemblage is impressive in its size and diversity. It includes vessels, anthropomorphic and zoomorphic figurines, spindle whorls, and scrapers. Scrapers are quite rare at Yarim Tepe III, and the entire group is made from reused sherds (Fig. 9.25:3). Their scraping edges have been worked on the inner side, as a rule. They were used for a variety of purposes, including, quite possibly, the processing of animal hides. This tool type is commonly found throughout the Yarim Tepe II stratigraphic sequence (Munchaev and Merpert 1981, 230).

Nearly twenty ceramic spindle whorls (Fig. 9.26) were recovered from Yarim Tepe III. The large majority are biconical in shape, similar to the forms commonly found at other Halaf sites. Several disk-shaped ceramic objects, perforated in the center, could also have functioned as spindle whorls.

Egg-shaped clay artifacts that have been typically categorized as "sling projectiles" are ubiquitous in early agricultural settlements of northern Mesopotamia. As noted above, large numbers of these objects were recovered from the fill of tholos 137 (Fig. 9.27:1). They were also found in other excavated squares, bringing their total number to over a thousand. All of them are made from unbaked clay, and their exact purpose remains a mystery. However, it is obvious that they could not have served as sling projectiles since much larger spheroid-shaped objects made from unbaked clay (up to 9.2 cm in diameter) were also recovered from the fill of tholos 137 that are better suited as projectiles (Fig. 9.27:2).

Figure 9.27:1,2. Clay objects from Halaf Level 3

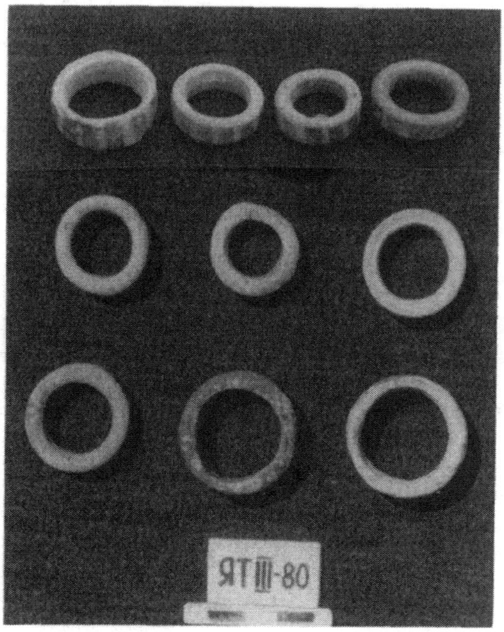

Figure 9.28. Ceramic rings from Halaf Level 3

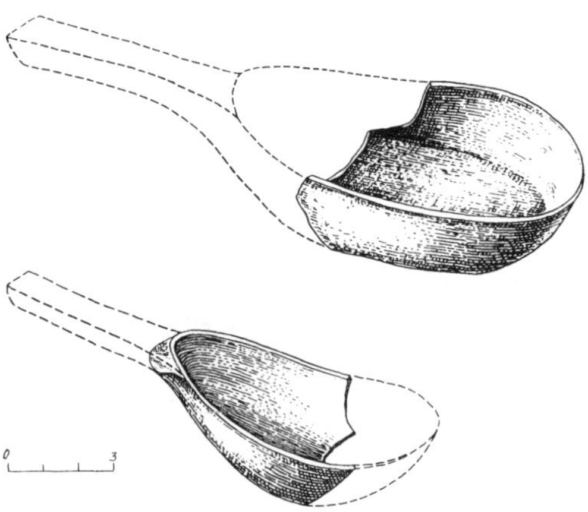

Figure 9.29. Ceramic spoons from Halaf Level 3

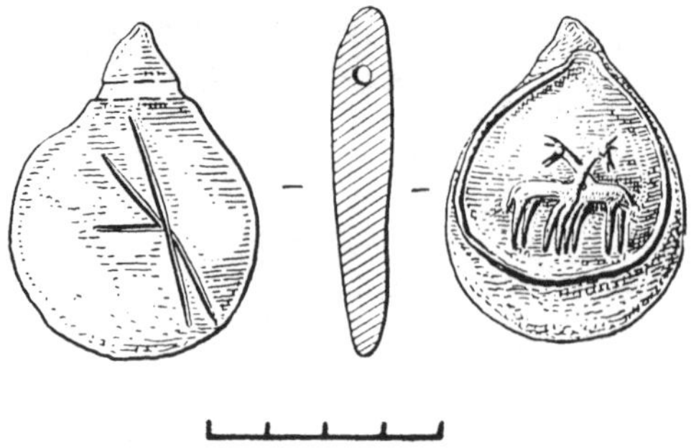

Figure 9.30. Ceramic seal pendant from Halaf Level 3

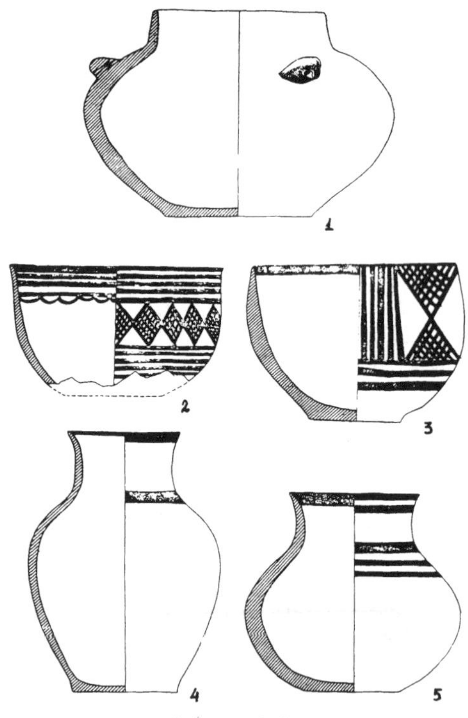

Figure 9.31:1–5. Ceramic vessels from Halaf Level 3

Sixteen ceramic rings were also recovered (Fig. 9.28) during the course of the 1980 season, including eleven found within the fill of tholos 137. Nearly all of them were painted with a reddish brown pigment against a light red background, and the average diameter is approximately 3 cm. Painted ceramic rings are commonly found in Halaf sites and are well represented at Yarim Tepe II.

Of special interest among the ceramic objects found at Yarim Tepe III is a small (6.3 cm × 3.5 cm), flat, crescent-shaped object with a projection/handle on the back side (Fig. 9.25:2). It was found in square VI-a at a depth of 6.25 mbd. A similar object in the form of an elongated concave, perhaps a cover (Fig. 9.25:1), and fragments of two spoons (Fig. 9.29) were found in tholos 137. Finally, a large well-fired ceramic pendant (Fig. 9.30) was discovered in the northern corner of square VI-c at a depth of 6.20 mbd. The pendant is flat and circular in shape except for a pointed projection at the top. It is similar in form to stone pendants commonly found at Halaf sites. One side of the pendant is engraved with intersecting lines, and the other has a relief of two animals and was probably used as a stamp seal. This is the first ceramic seal pendant found at either Yarim Tepe II or III. The other stamp seals are made from different types of stone and in one case from copper (Munchaev and Merpert 1981, Fig. 51–53). This ceramic seal pendant has a rather unique design in comparison with other seals found at either site. Its closest parallels are found in the large and diverse collection of seals recovered from the Ubaid and Halaf levels (XII–XIX) at Tepe Gawra (Tobler 1950, Pls. LXXXVIII, CLXII–CLXXI).

Sherds from vessels, of course, comprise the vast majority of the ceramic material recovered during the 1980 excavations. The ceramic assemblage is typical of the Late Halaf in vessel form, manufacturing technique, and painted designs. A considerable portion of the assemblage consists of cooking vessels and large storage jars up to 80 cm in height for water and food. Among the latter are coarse, thick-walled vessels, both painted and plain. They are similar to the vessels described in detail in the previous report for the site. This also applies to the painted ware assemblage.

A large number of painted bowls were recovered from Yarim Tepe III, from which several types can be distinguished. The most common form is a deep hemispherical bowl with a flat base and ranging from 6 to 14 cm high and from 9 to 24 cm in diameter (Figs. 9.31:2–3; 9.32:1–2). The exterior is completely washed with paint, whereas the in-

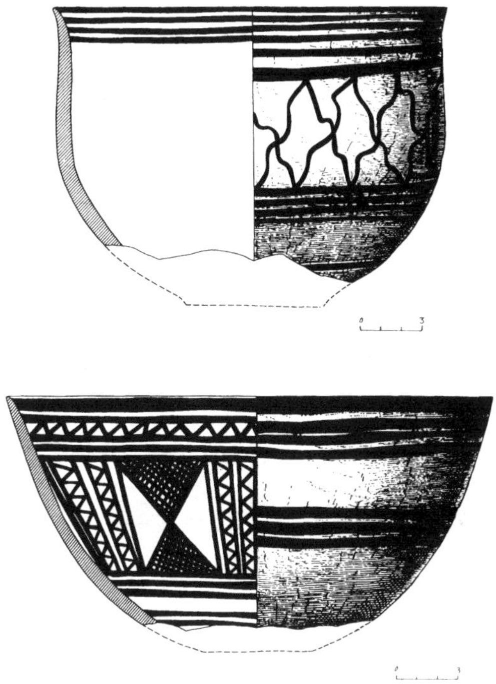

Figure 9.32:1–2. Painted ceramic vessels from Halaf Level 3

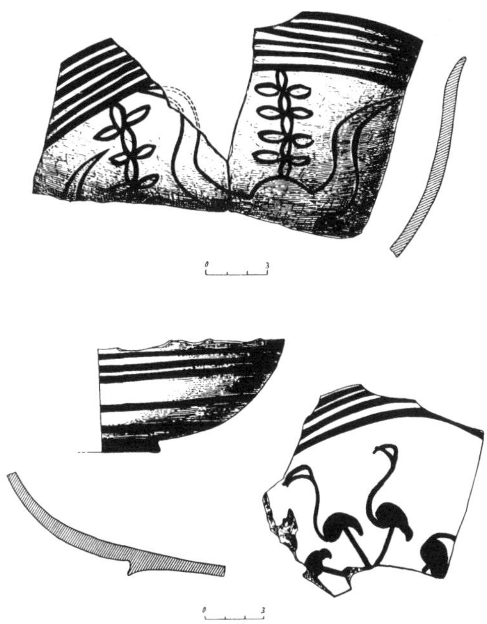

Figure 9.33:1–2. Painted "tree of life" and animal designs on pottery from Halaf Level 3

terior is unpainted except along the rim. The colors range from reddish brown to grayish brown shades against a pale yellow slip. As a general rule, the design motifs are geometric though some are representational. For example, part of a large, round-based bowl, found in square II-b at the depth of 5.8 mbd, had painted circular bands on the exterior and a depiction of the "tree of life" on the interior (Fig. 9.33:1). The designs are brown in color against a light red slip. No less interesting is a fragment of a goblet painted on both the interior and exterior. Inside the goblet there is a painted design of long-legged birds arranged in a circle (Fig. 9.33:2). Once again a brown paint is used for the design and the slip is light red.

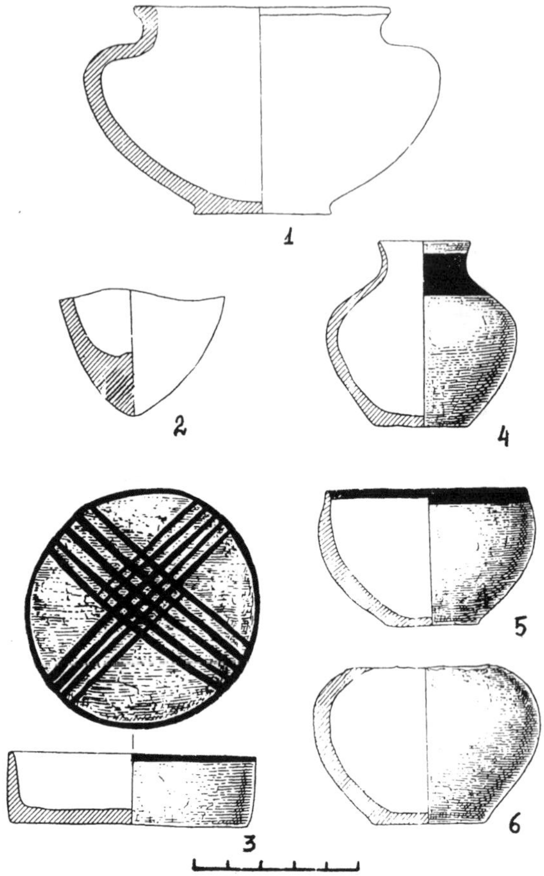

Figure 9.35:1–6. Miscellaneous ceramic vessels from Halaf Level 3

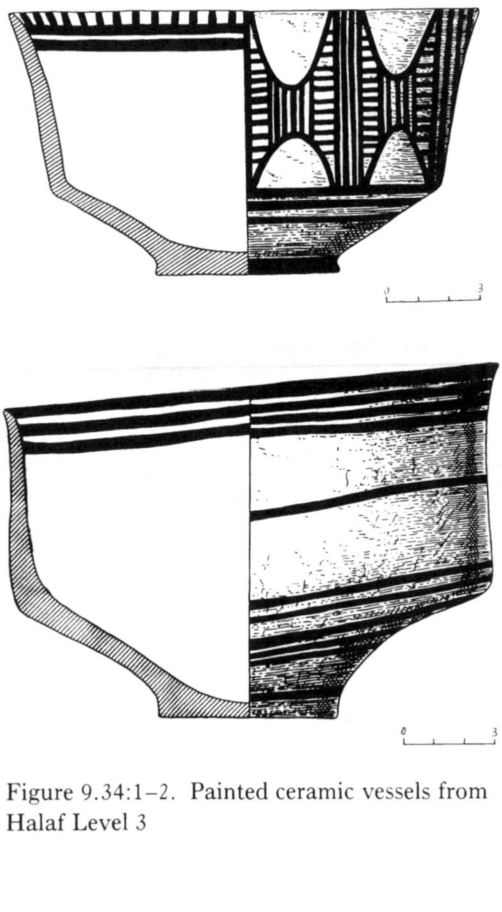

Figure 9.34:1–2. Painted ceramic vessels from Halaf Level 3

Another type of bowl has a sharp carination and a pronounced flat base (Fig. 9.34: 1,2). Designs are painted in brownish gray against a lighter background. Yet another bowl type is more shallow in form, with straight or outward flaring walls. A vessel of this form was found in tholos 137. On the exterior of its base are a series of painted lines forming a cross (Fig. 9.35:3).

Of particular interest are sherds from two bowls found within tholos 137 that have bichrome animal designs. One of them depicts a deer with branched antlers. The contour of the back, antlers, and a portion of the head

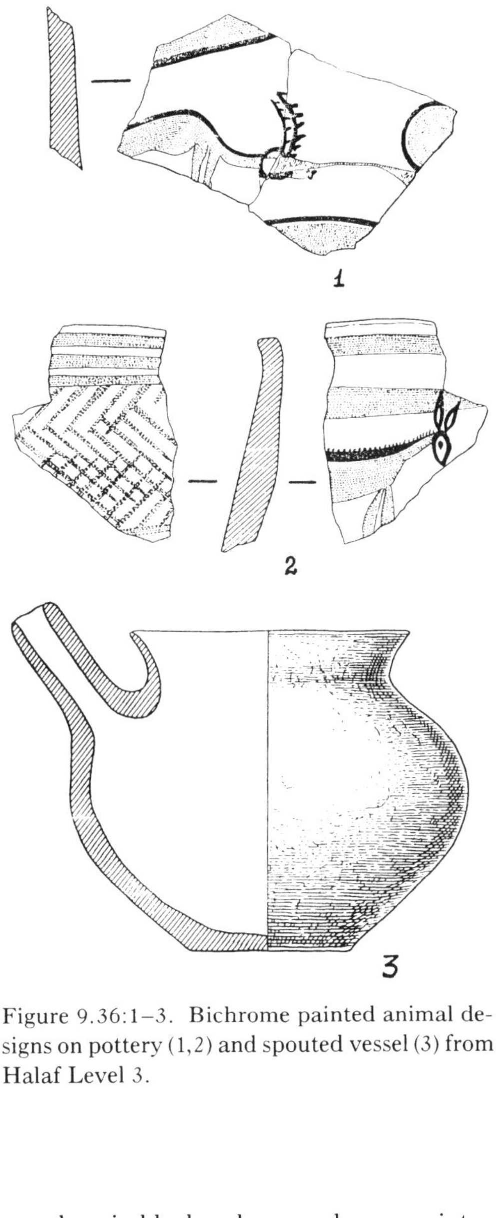

Figure 9.36:1–3. Bichrome painted animal designs on pottery (1,2) and spouted vessel (3) from Halaf Level 3.

was done in black, whereas a brown paint was used for the the remainder of the contour (Fig. 9.36:1). Another fragment with a similar color scheme and style depicts a donkey or fallow deer (Fig. 9.36:2). In both cases the animals are located on the exterior of the vessels and are a part of a larger composition arranged in a band around the vessel.

Another group of vessels consists of various types of small jars (Figs. 9.31:4–5, 9.35:4) including those with constricted necks. One of these has a dark brown paint wash over its entire surface, whereas another is equipped with two pairs of pierced lug handles symmetrically placed on the vessel for suspension (Fig. 9.31:1). Additionally, a spouted pot (Fig. 9.36:3), typical of the late Halaf Period, was found in square IV-c at a depth of 5.65 mbd. A painted pot recovered from square VI-a at a depth of 5.70–5.90 mbd is decorated in a unique style. The designs on the vessel are painted in reddish brown against a light rose background. Apart from simple and cross-hatched bands, the painted designs include two realistic depictions of owls placed symmetrically on the opposite sides of the body (Fig. 9.37). The form of the vessel is not unusual for the Halaf period, with a flat base; wide, rounded body; low, cylindrical neck; and everted rim. Two short spouts were placed on opposites sides of the shoulder of the vessel.

It should be noted that the decorated ceramic assemblage in the Halaf levels at Yarim Tepe II exhibits a wide range of design motifs, though the majority are geometric

Figure 9.37. Painted ceramic vessel depicting owls

compositions. Several vessels are painted over their entire surface, including a very rare Halaf example of a vessel with a pointed base (Fig. 9.35:2). Different decorative techniques were utilized by Halaf potters, including painted designs on a paint-wash or slip background and fine incised lines that often cover the entire surface of the vessel.

Figurines

Although a relatively small area was excavated during the 1980 field season, Yarim Tepe III yielded a substantial collection of anthropomorphic and zoomorphic clay figurines. A total of at least twenty of the former were recovered that can be categorized based on their degree of stylization as either realistic or highly schematic. Only two examples of the former were found at the site (Fig. 9.38). Both of these are similar in design and are typical of a well-known form of Halaf anthropomorphic sculpture. Both figurines are seated females with arms supporting their breasts. Necklaces and, apparently, designs on their garments are painted on the figurines. One was found in square VI-a at a depth of 6.40 mbd (Fig. 9.38:1) and the other was discovered in square VI-c at a depth of 5.35–5.40 mbd (Fig. 9.38:2).

Of the stratified Halaf sites excavated to date, including Yarim Tepe II, Tell Arpachiyah, Tepe Gawra, and, of course, Yarim Tepe III, the more realistically depicted figurines are found in the upper levels, reflecting the latest stage of the culture. Thus, a plausible hypothesis based on this observation is that strata at Tell Halaf, Tell Brak, and Chagar Bazar that have also yielded figurines of this type belong to the Late Halaf phase as well. However, it should be noted that the general

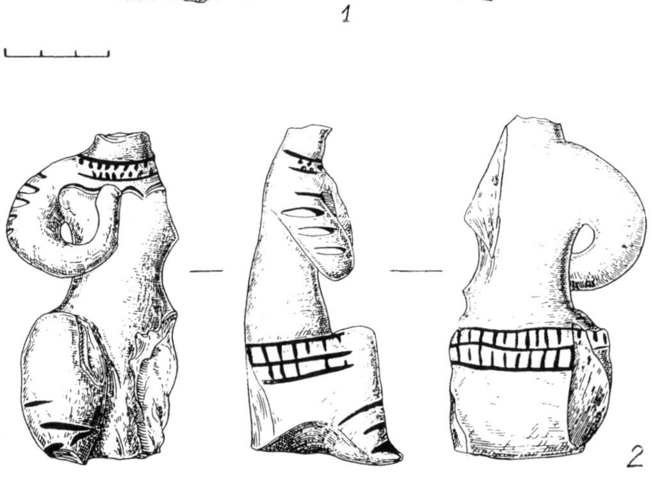

Figure 9.38:1,2. Naturalistic clay anthropomorphic figurines from Halaf Level 3

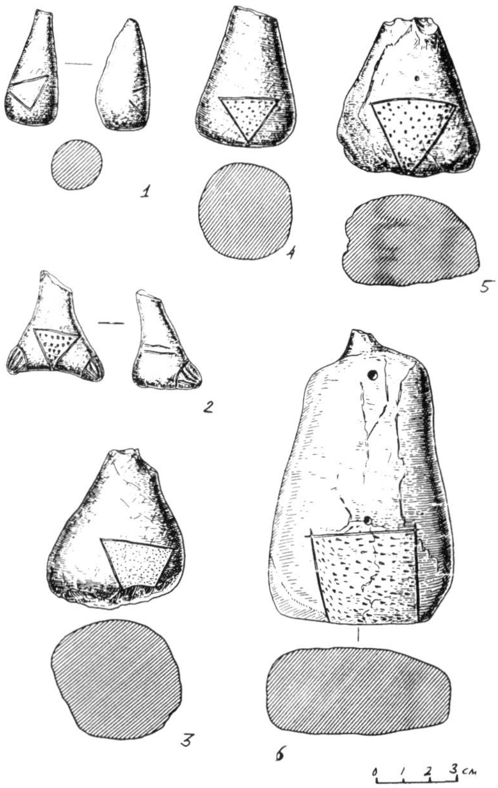

Figure 9.39:1–6. Stylized clay anthropomorphic figurines from Halaf Level 3

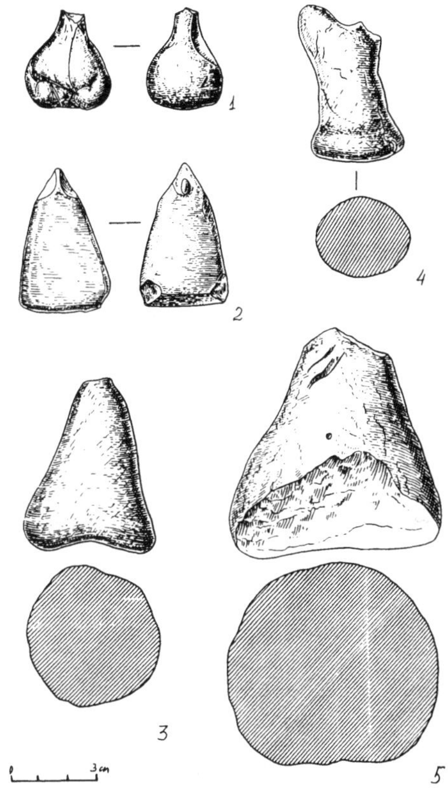

Figure 9.40:1–5. Stylized clay anthropomorphic figurines from Halaf Level 3

form of a female with arms supporting her breasts is ubiquitous throughout the entire Halaf period, beginning with its earliest stages of development (Munchaev and Merpert 1981, 251–52).

The bulk of the anthropomorphic figurines are highly stylized in the form of schematically depicted females (Figs. 9.39, 9.40). Most of these were discovered in the fill of tholos 137. They are poorly fired, if indeed at all, and are usually conical in form, though several are almost rectangular or pyramidal. The majority of them have pronounced female characteristics in the pubic region, which is depicted as an incised triangle, though in one case it is a quadrilateral, dotted with fine pinhole-sized marks. In several instances, the head is depicted in relief or by pinching, and the upper legs are oriented in a manner that suggests that the figurines are seated.

Halaf sites frequently yield considerable numbers of highly stylized anthropomorphic figurines, such as the group recovered from the lower levels of Yarim Tepe II, though it should be noted that differences exist between this latter group and those found at Yarim Tepe III (Munchaev and Merpert 1981, Fig. 64). In general, it can be assumed that the anthropomorphic figurines from Yarim Tepe III are characteristic of the Late Halaf period. It should also be noted that although several other Halaf sites (i.e. Yarim Tepe II and Tell Arpachiyah among others) did yield stylized conical female figurines, only at Yarim Tepe III was a large number recovered of nearly identical design, including the prominently displayed female pubic region.

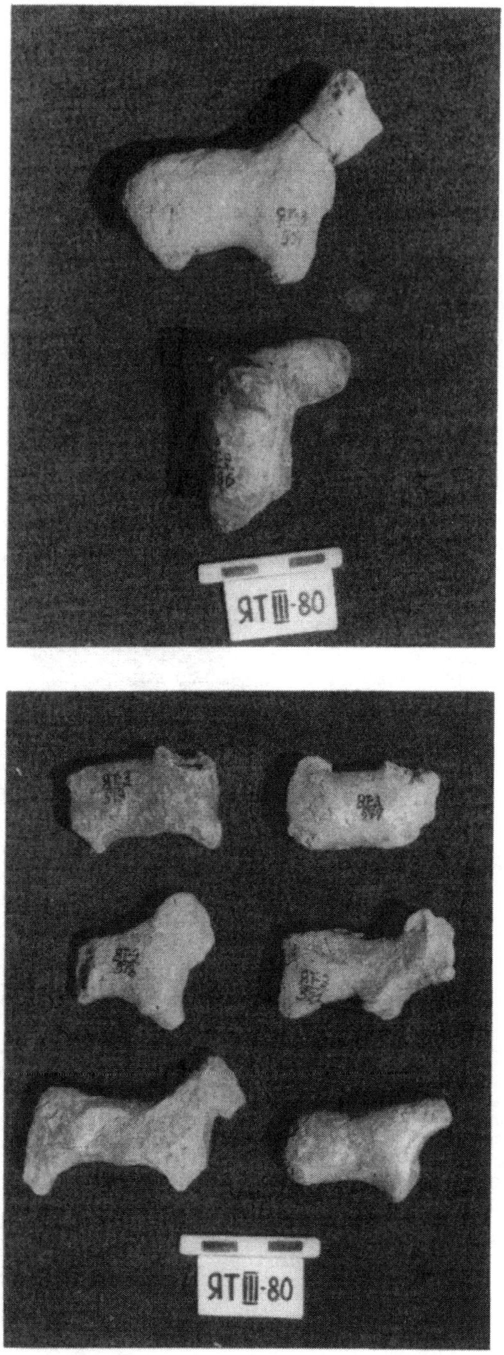

Figure 9.41:1,2. Clay zoomorphic figurines from Halaf Level 3

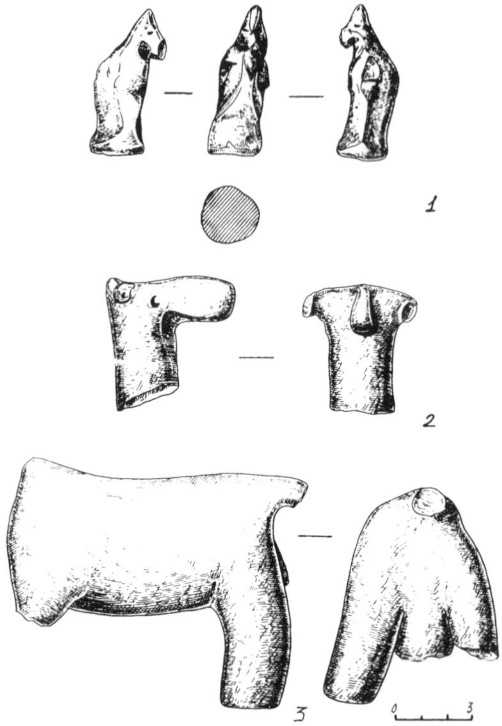

Figure 9.42:1–3. Drawings of clay zoomorphic figurines from Halaf Level 3

The collection of zoomorphic figurines is similar in size to the anthropomorphic figurines (Figs. 9.41 and 9.42). The majority of these figurines are made from unbaked or poorly fired clay. Of this series, five complete figurines and three fragments were found while removing the balk between squares I and IV at a depth of 3.50–3.60 mbd in the Halaf-Ubaid transition level. The greater part of the collection of zoomorphic figurines came from the fill of tholos 137. This group was all quite similar in design. Most of the figurines are medium or small in size and are roughly fashioned representations of standing animals, whose identification is often elusive (Fig. 9.41:1,2). However, in some instances we can confidently identify the figurines as rams and dogs. One figurine worth special mention is that possibly of an ox, which has survived almost intact. Also of interest are the head of a large figurine representing an animal with a long muzzle (Fig. 9.42:2) and a squatting figurine with its head modeled in an atypical fashion (Fig. 9.42:1). The latter two figurines are probably depictions of dogs.

The majority of the above described zoomorphic figurines are quite similar in style to those found at the site during previous field seasons. On the whole, the collection of animal figurines recovered from Yarim Tepe III is typical of Halaf zoomorphic sculpture and has close parallels from the Halaf levels at Tell Arpachiyah and Tepe Gawra (Mallowan and Cruikshank Rose 1935, Fig. 48; Tobler 1950).

In conclusion, the limited excavations during the 1980 field season at Yarim Tepe III were carried out in levels belonging exclusively to the Late Halaf period. An invaluable sherd sample and group of anthropomorphic and zoomorphic figurines were recovered dating to this period. Since more than 5 m of cultural deposit underlie the Late Halaf levels, it is highly probable that a long period of Halaf cultural development is represented at the site, possibly including its earliest phases.

Notes

1. It should be noted that doors are often not identified in the architectural remains since they are situated rather high with their sills often above the level of the preserved walls.
2. The plan of the ovens is given separately from the structural remains.

10

BURIAL PRACTICES OF THE HALAF CULTURE

N. Ya. Merpert and R. M. Munchaev

Evidence for burial rituals in the early food-producing societies of Mesopotamia is thus far extremely limited. The vast quantity of archaeological evidence comes from the settlements themselves. Of the few excavated burials known, most contained juveniles and infants that were for the most part located within the limits of settlements and connected with a special ritual in which burials were placed under walls, thresholds, and floors of houses. This burial practice was widespread among early agriculturalists of both Southwest Asia and Europe. Adult burials occur only in isolated instances, and in a number of cases doubts have arisen as to whether they were deliberate inhumations. Thus, at Tell Hassuna, only juvenile burials are beyond question, especially infants. Adult remains, by contrast, are quite rare, and their character is ambiguous. Those who have excavated this site emphasize that the context of adult burials, in one case in a granary and in the other in a garbage pit, is hardly compatible with any kind of ritual action (Lloyd and Safar 1945). Only juvenile burials have been found at Shimshara (Mortensen 1970, 20), and Umm Dabaghiyah.[1]

Common burial areas of the pre-Ubaid period are known only from Tell es-Sawwan and Tell Arpachiyah. However, the common burial area at Tell es-Sawwan, belonging to the Samarra culture, though it contains several dozen deliberate burials, is also located within the settlement, under the floors of large buildings (most probably shrines). Also, juveniles clearly predominate among those buried, often found in a flexed position and wrapped in mats within narrow pits strewn with ochre. This burial area is of an extraordinary character, confirmed by the striking richness of the mortuary offerings. These include beautiful necklaces and bracelets of beads made of turquoise, alabaster, slate, and other materials; numerous alabaster cups, goblets, pitchers, and zoomorphic vessels; palm leaf ornaments; ritual artifacts; and a unique collection of alabaster and clay anthropomorphic figurines (al-Wailly and Abu es-Soof 1965; al-Adami 1968).

The Halaf burial area at Arpachiyah, excavated by Max Mallowan, is also located within the settlement limits. Although its relationship to the architectural remains is not entirely clear, it may have occupied a section of the western slope of the mound, originally devoid of architecture, to which the settlement spread only at a later date (Mallowan and Cruikshank Rose 1935, 1). However, several of the burials (52, 55, 59) are clearly associated with domestic architecture, since they are situated under walls or thresholds of houses. A total of nine burials were discovered at the site, of which at least three are juveniles. In one case, a secondary burial of a juvenile was noted. In all the others, the inhumations were in a flexed position with variable orientation. Six individuals were accompanied by mortuary offerings, including typical Halaf vessels, obsidian and flint knives, stone adzes and amulets, bone awls, and, in burial 58, grains of barley.

I. Hijara has resumed the excavation of Tell Arpachiyah over the past few years. Over the course of these excavations, he has discovered three burials quite different from the ones described above in that only the crania were found, placed within typical Halaf vessels, whereas the postcranial remains were absent. These unique burials were found within the settlement, although the author emphasizes a certain reduction in settlement size within this particular region of the site, differentiating the burial area from an area containing tholos structures surrounded by a special fence (Hijara 1978, 125–28). The Halaf burials at Tepe Gawra, once again primarily juveniles, several of which were interred in large vessels, are also found within the settlement boundaries (Tobler 1950, 48–49).

Until the 1970s, the mortuary data from Neolithic Mesopotamia was limited to the few finds noted here with the Arpachiyah burials representing the only group relating to the Halaf culture. Soviet research in the Sinjar Valley over a number of years has significantly expanded this database.

As a result of this research, a considerable number of burials dating from pre-Halaf periods have been studied. The earliest of these is a group of burials within the settlement area of Tell Maghzaliyah, located in the Sinjar foothills and dating from the eighth and seventh millennia B.C. (Bader 1979, chapter 2a of this volume; Munchaev and Bader 1979). A number of individual burials were uncovered there. The bodies were tightly flexed, lying on their right sides in shallow, oval pits lined with stones. The heads of the individuals were generally pointing in an easterly direction. The burial pits themselves were capped with stone cairns and mortuary offerings were lacking. These are the earliest deliberate burials so far encountered in this part of Mesopotamia.

A group of burial sites was discovered at Tell Sotto (late seventh and early sixth millennia B.C.). All of these were adolescents and juveniles interred within the settlement limits, sometimes in large vessels. Several of these are accompanied by mortuary offerings, including ceramic vessels and beads made of shell, various kinds of stone, and even copper (Merpert, Munchaev, and Bader 1977,98).

As shown from the above evidence, one can speak of the beginning of the infant-and-juvenile burial tradition within the settlement itself at Tell Maghzaliyah and Tell Sotto. It is interesting to note that this practice is found in the earliest agricultural settlements of the Sinjar Valley. The absence of mass adult burials supports the hypothesis that specialized adult burial areas beyond the limits of the settlement still await discovery.

A similar conclusion can be reached for Hassuna sites in northern Mesopotamia dating from the sixth millennium B.C. Within the settlement area of Yarim Tepe I, several dozen juvenile burials, including several interred within large vessels, were discovered (Munchaev and Merpert 1981). The burials were in a flexed position with variable orientations and sometimes accompanied by ornaments and ceramic vessels. By contrast, adult burials are found only in isolated instances and, as exemplified by the Tell Hassuna material, can by no means always be considered in a primary context. In a number of cases, disarticulated skeletons and large accumulations of human bones in special buildings were

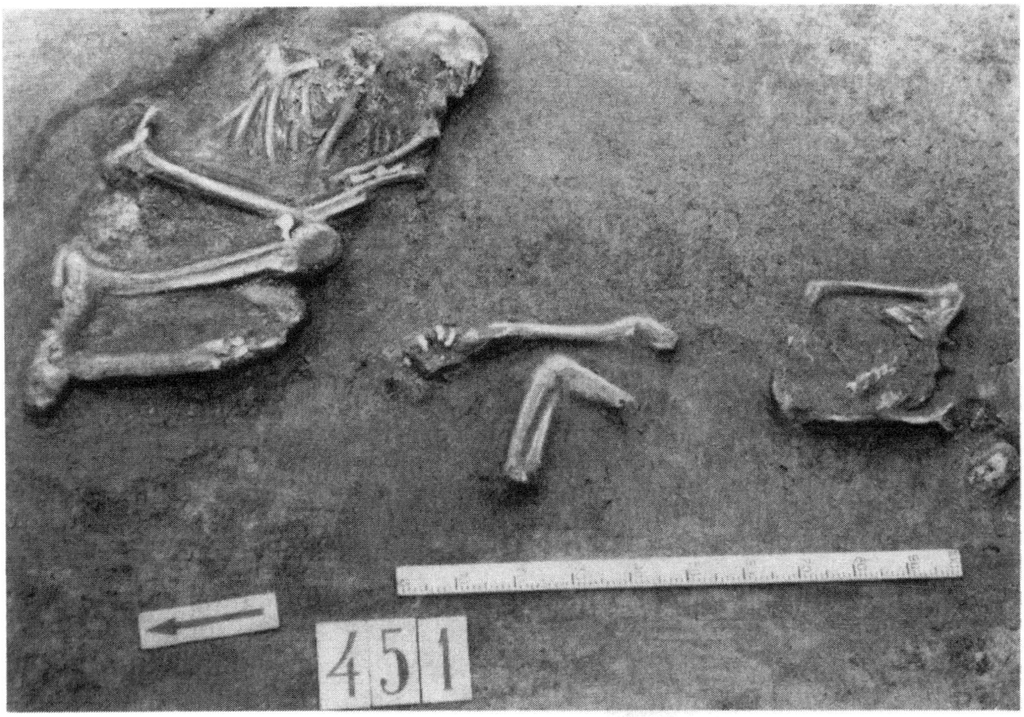

Figure 10.1. Burial 61, Yarim Tepe II: two adults

found. This evidence suggests that special ritual acts were being practiced. On the other hand, the typical adult burial sites of the Hassuna culture, although they certainly must have existed, have not yet been located.

The Halaf Evidence

Due to the Soviet expedition's work in the Sinjar plain, much more evidence concerning Halaf mortuary behavior has been recovered dating from the fifth millennium B.C. At the very outset, it must be emphasized that this material confirms the existence of specialized burial sites beyond the limits of the residential settlement as well as burials within the confines of the settlement.

Beginning with the latter, the best data comes from the settlement of Yarim Tepe II. The excavations at this site, conducted between 1969 and 1976, covered an extensive area (Munchaev and Merpert 1981). To date, this site is the most comprehensively studied settlement of the Halaf culture. The entire cultural deposit at the site reaches a depth of 7 m and belongs entirely to the Halaf period. Nine separate building levels have been distinguished within this deposit, and Halaf burials have been found in almost all of them, yielding a total of no less than fifty.

Beginning with the burials associated with the lower building levels (8 and 9) of Yarim Tepe II, fifteen burials (48–62) were encountered over an area of about 500 square meters (for a more comprehensive description of these burials than presented in this paper, see Munchaev and Merpert 1981). Even this number, representing only the earliest group of burials at Yarim Tepe II, exceeds the previous total number of Halaf burials reported from other sites.

Figure 10.2:1. Burial 44, Yarim Tepe II

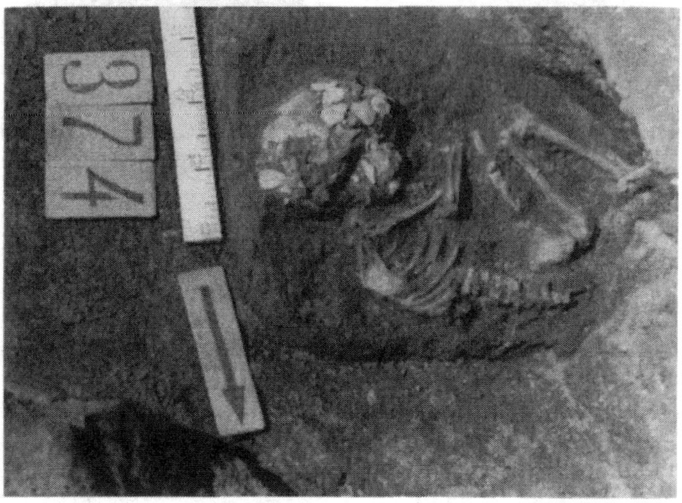

Figure 10.2:2. Burial 46, Yarim Tepe II

It should be emphasized at the outset that both the complexity and variability seen in Halaf burial practices is quite impressive. This is vividly illustrated by the group of fifteen burials mentioned above. Three different types of burial may be distinguished within this group: primary inhumations, cremations, and cranial burials. The presence of secondary burials is also noted, though much rarer.

Six cases of primary inhumation were encountered in the lower building levels (8 and 9) at Yarim Tepe II (burials 57–62). They were located in various parts of the excavation, but primarily in squares 19 and 29, and none were associated with buildings. Four were individual inhumations, whereas two were multiple. Remains of juveniles were found in four of the burials, two in burial 59, and two adults and a child (possibly the parents and their infant) in multiple burial 61 (Fig. 10.1). In all cases where placement could be discerned, the individuals were lying on their sides, usually the left. The various head orientations were south-southwest, southeast, and northwest.

Figure 10.3:1. Burial chamber, Yarim Tepe II

Figure 10.3:2. Outline of burial 45, Yarim Tepe II

Mortuary offerings were found in only two of these burials. In burial 58, a miniature gray ware cup, burnished on both the exterior and interior, was found lying upside down 10 cm in front of the facial portion of the cranium. It was filled with a large number of small ornaments, including pins; round, flat carnelian beads approximately 0.6 cm in diameter; 234 shell beads (0.4 cm in diameter); and 328 beads made of soft stone (0.3 cm in diameter). Also in the area of the cranium, several small charred sheep and goat bones were recovered. In the northern part of burial 59, a shattered, white alabaster, spouted cup was discovered. It was probably deliberately shattered at the time of burial. The above group of burials at Yarim Tepe II have direct analogs at Tell Arpachiyah where, as Max Mallowan noted, nine Halaf burials were excavated (51–59). Eight of these were flexed inhumations.

Inhumations were repeatedly enountered in the subsequent building levels of the site (Fig. 10.2:1,2). In principle they do not differ from those already described, the majority being juveniles in a flexed position. It is interesting to note that some of them were placed in small, lined pits and even in clearly defined burial chambers (Fig. 10.3:1,2). Burial 45, encountered at a depth of 5.15 m below datum (mbd) and associated with Level 6, is an example of the above. Its entryway has been partially destroyed and is traceable only over a distance of 30 cm. However, the chamber itself, which was dug into the previous building level, is well preserved. It is oval in plan, and its dimensions are 40 cm × 35 cm with a height of 30 cm. The floor of the chamber is covered with plant remains on which lay an infant in a flexed position, with its head pointing east.

Traces of other burial chambers are less clearly defined. However, the presence of such funerary structures is by no means incidental, as is shown by the extramural burial site belonging to the Yarim Tepe II settlement that will be discussed shortly. It should be mentioned that some of the primary inhumations noted above contained typical Halaf stone and ceramic vessels (Fig. 10.2:1), and also beads and pendants.

A second group of burials discovered in the lower building levels (8 and 9) of Yarim Tepe II had been cremated, representing the first evidence of this burial practice at a Halaf site. Five such cremations were found at Yarim Tepe II (burials 50–54), all within a single excavated square (number 23). Let us consider this group of burials in more detail. Burial 50 was discovered within the cultural deposit at a depth of 6.0–6.2 mbd. The cremated remains were placed in a pit, and at its northern edge lay fragments of two painted vessels, deliberately shattered and thrown in at the time of the burial (Fig. 10.4:1). The pit fill consisted of ash containing scattered fragments of charred bones from an adult human. Since the walls and bottom of the pit are not charred, it is clear that this cremation burial is secondary and the crematorium is located elsewhere.

In burial 51 (Fig. 10.4:3), at 6.32–6.52 mbd, the cremated remains were placed in a round area with a diameter of 0.9–1.0 m, prepared with mud plaster. The disarticulated (with the possible exception of one leg) charred bones of an adult lay at the northern edge of this space. The mandible was found

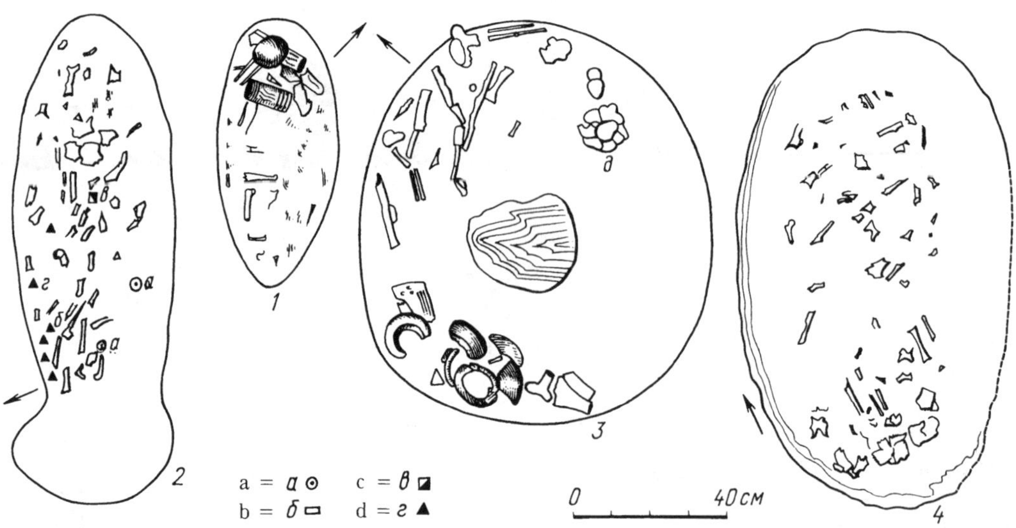

Figure 10.4:1–4. Plans of cremation burials 50, 54, 51, 53 (1–4) at Yarim Tepe II. Key: *a*, Spindle whorl; *b*, Bone awl; *c*, Fragment of ground stone; *d*, Sherds.

Figure 10.5. Cremation burial 40, Yarim Tepe II

between the femora and the radii. Fragments of the cranium were discovered in the eastern part of the prepared area, close to its edge. In the western section lay the fragments of two painted vessels that had been deliberately shattered and bore traces of burning. A fragment of charred wood was found in the center of the burial. Besides this, no other evidence of the cremation, either in the form of ashes or a charred layer, was encountered in the area. Consequently, it must be concluded that the cremation was carried out in a different location, with the cremated remains secondarily deposited along with the vessels deliberately shattered and burned during the cremation event or, more likely, immediately after it took place.

Adjacent to the above cremation, burial 52 (in a small pit measuring 25 cm × 35 cm) was encountered at a depth of 6.35–6.55 mbd. The charred bone fragments of a juvenile were randomly distributed within the pit and the pit fill was devoid of ash or lumps of charcoal. Similarly, in burial 53 (Fig. 10.4:4), located approximately 1 m west of burial 52, charred bone fragments of a juvenile were found in a pit measuring 1.16 m × 0.53 m at a depth of 6.1–6.23 mbd. Thus, though the walls and base of the pit are slightly burned in places, the actual cremation of the body had obviously been carried out elsewhere.

The last cremation burial from Levels 8 and 9 (54) was discovered at a depth of 5.85–6.05 mbd. It was in an oval pit (Fig. 10.4:2), measuring 1.25 m × 0.43 m, that showed evidence of burning on its mud plastered walls and bottom. At the base of the pit, a thick layer of ash 15 cm in depth was encountered. Directly on top of this ash layer, in the eastern and central areas of the pit, lay the severely

Figure 10.6:1–5. Mortuary offerings in cremation burial 40, Yarim Tepe II: *1,* Coarse gray bowl; *2,* Stone vessel; *3,* Stone vessel; *4–5,* Miniature ceramic vessels.

burned and scattered bones of a single individual with not a single fragment of bone found within the ash itself. The bones were overlain by burned pieces of clay, possibly from the walls of the burned pit, and fragments of a small, painted vessel bearing signs of having been in a fire. Fragments of three other vessels were discovered along the northern edge of the pit. In the central part of the pit, in the ash layer, two biconical ceramic spindle whorls, a bone bit, and a fragment of flat, polished red stone were discovered.

Finally, the remains of two more cremations were found in Level 7 at Yarim Tepe II. One of these (burial 43) was found in the same excavated square (number 23) as burials 50–54 associated with the lower levels. A second cremation (burial 40) was found in an adjacent square (28). The latter was a primary cremation in an oval-shaped crematorium pit with the dimensions of 1.10 m × 0.35–0.50 m, which differed from the normative hearth found at the site. After the cremation, the charred bones of an early adolescent, ten to thirteen years of age, together with twenty obsidian beads, were removed from the crematorium pit and placed in a painted ceramic vessel (Fig. 10.5). The vessel was then placed near the right-hand wall of the crematorium pit, over pieces of the burned plaster that had fallen from its walls. The vessel bore no visible traces of having been in a fire, whereas the other vessels found in the crematorium pit did have traces of burning. While the cremation was taking place, several stone and ceramic vessels were shattered and thrown into the crematorium pit. Fragments of one vessel were found in different parts of the pit, having been subjected to varying degrees of scorching.

The following is an inventory of the objects found in the pit. In its western half lay the complete vessel containing the cremated bones and twenty obsidian beads; a roughly fashioned cup broken into pieces; two other ceramic vessels, including a painted one that had been shattered; two shattered stone vessels (Fig. 10.6:2–3); a stone seal-pendant with an incised pattern on one side; a charred plate; and a fragment of a perforated bone pendant. In the eastern half of the crematorium pit were two miniature ceramic vessels (Fig. 10.6:4–5); one shattered stone vessel; one-half of a biconical ceramic spindle whorl; two cowrie shells; twenty-six gypsum beads (Fig 10.7:3); thirteen biconical obsidian beads

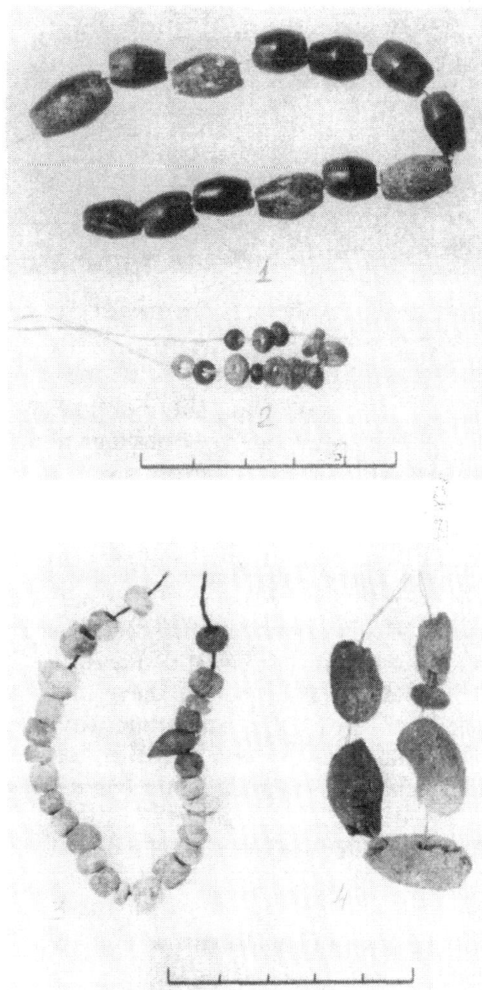

Figure 10.7:1–4. Mortuary offerings in cremation burial 40 at Yarim Tepe II: *1,* Obsidian beads; *2,* Stone beads; *3,* Gypsum beads; *4,* Ceramic beads.

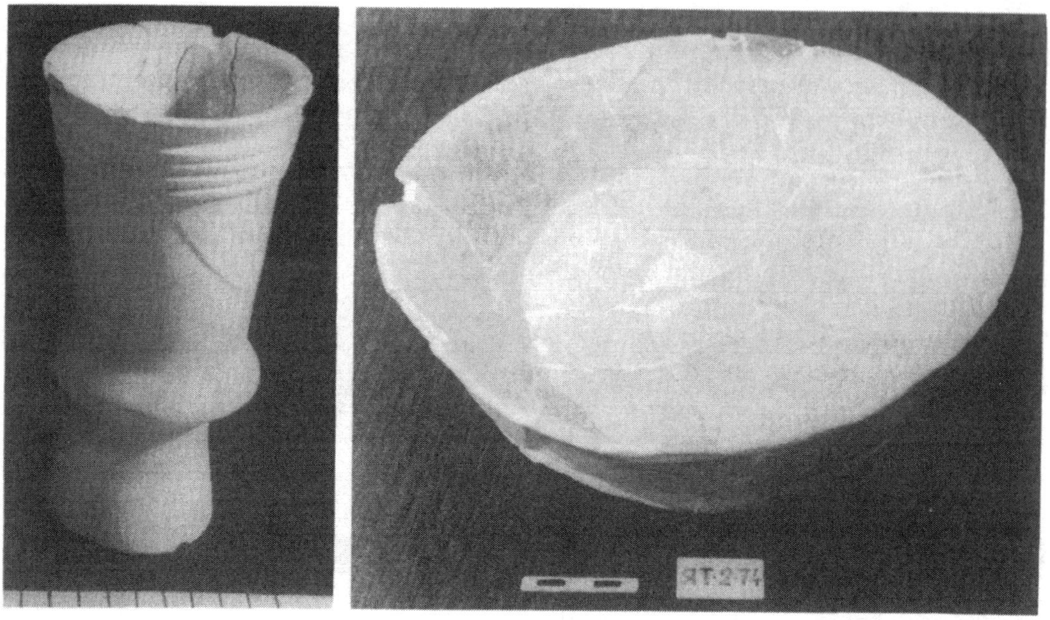

Figure 10.8:1–2. Mortuary offerings in cremation burial 43 at Yarim Tepe II: 1. alabaster goblet. 2. alabaster bowl.

(Fig. 10.7:1); fifteen small, flat, round beads (Fig. 10.7:2), including five of obsidian and eight of rock crystal; and twenty-seven crudely molded and poorly fired ceramic beads (Fig. 10.7:4).

In the other cremation burial in Level 7 (burial 43), the charred bones of a juvenile approximately ten years of age had been placed into a painted ceramic vessel that was then buried under the floor of one of the residential tholos structures. The crematorium was discovered alongside this structure where a thick layer of charcoal and ash from the bonfire was preserved. An alabaster goblet (Fig. 10.8:1), a large alabaster bowl on a circular base (Fig. 10.8:2), and three ceramic vessels, including a painted cup, had been deliberately shattered during the cremation and thrown into the bonfire. Their fragments lay in disarray on a top of the ash layer formed during the bonfire, an area approxiamtely 0.6 m in diameter.

These are the cremations that were discovered at Yarim Tepe II. As indicated above, these burials were all located in the lower half of the cultural deposit, in Levels 7–9. No evidence of other cremation burials was found in the upper levels of the settlement (1–6). Thus, it appears that the custom of cremation was practiced only during earlier occupation of the site. Furthermore, within the Halaf sphere, this burial practice is apparently unique to Yarim Tepe II.

Unfortunately, in view of the limited nature of the data, it is as yet difficult to give a comprehensive characterization of the cremation practices considered here that reflect the complexity of Halaf religious beliefs in the Sinjar plain. However, the authors do wish to emphasize the following generalizations. First, the cremation remains at Yarim Tepe II are limited, as was noted, to one localized area of the settlement, essentially to square 23. Why the cremations were restricted to this small area exclusively remains unclear. Second, of the seven cremation burials described, only two can be confidently classified either as a primary cremation (burial 40) or at least associated with a nearby crematorium (burial 43). In all the other cases there is good reason to conclude that the actual cremation occurred elsewhere. Where these cremations were carried out, either in open courtyards within the settlement or beyond its boundaries, has not yet been established. The only unambiguous statement that can be put forth is that in the majority of these burials, the remains were brought to this particular section of the settlement after the cremation event and buried in special pits or open spaces. During this process, in all probability, some ritual ceremony was performed. The residues of the cremation fires and also the deliberately shattered ceramic and stone vessels were thrown into the burial pits, and thus are somehow connected with this ceremony.

Finally, the third category of burial practice present at Yarim Tepe II is that of the interment of crania separate from the postcranial remains. There are three examples of this practice: burials 49, 55, and 66. It is curious that they are all associated with the earliest building level (9) in the cultural deposit in the same square (23) where the five early cremations were discovered. It is interesting to note that yet another burial (48), in all probability a secondary inhumation, was also found in this square. Burial 48 consisted of a juvenile cranium lying atop a pile of bones that may have been from the same individual. Burials 49 and 55 each consist of a single cranium in a highly organic layer of clay. No burial pit could be discerned in either case. Both crania lay on their left sides with the faces pointing west in one case and east in the other. Burial 56 consisted of three crania found in a shallow round pit approximately 0.6 m in diameter, dug 15 cm into a layer of clay soil. Two of the crania belonged to mature adults and the other, to a juvenile. All were lying on their left sides, with the faces oriented toward the northeast. No mortuary offerings were found in any of the cranial burials.

These three cranial burials at Yarim Tepe II were the first of their kind found within the Halaf region. However, in 1976 two more cranial burials were discovered at Tell Arpachiyah. In one case a cranium was found in a ceramic vessel with one stone and two painted ceramic cups found in association with the burial (Hijara 1978, 125). The other burial, found 0.5 m below the first, consisted of four crania, each in a separate vessel. In addition to these, one stone and six ceramic vessels were discovered in association with the burial (Hijara 1978, 125).

When the cranial burials at Tell Arpachiyah were published, Hijara was not aware of those discovered somewhat earlier at Yarim Tepe II. Thus, he mistakenly claimed that the practice of interring human crania separate from the postcranial remains was unique to Tell Arpachiyah within the Halaf sphere. Hijara attempted to explain this burial practice by suggesting that Tell Arpachiyah had a special status among the other Halaf settlements in the region (possibly as a religious center), and for this reason special ritual activities took place at the site. Among these activities were separate burials of the most important part of the body, the head. Furthermore, Hijara posited that the crania buried were not from ordinary members of the community, but from people occupying special positions in society. He considered it impossible to determine from the evidence at hand whether the individuals interred in these burials were contemporaries, and whether they were inhabitants of Tell Arpachiyah or of surrounding settlements (Hijara 1978, 127).

We do not view the above hypothesis relating to the cranial burials at Tell Arpachiyah as particularly convincing. What is important is the presence of cranial burials in yet another Halaf settlement, indicating that such burials are not extraordinary in character. It is interesting to note that no earlier cranial burials from Mesopotamian sites are known, particularly from Hassuna and Samarra contexts, though they are attested to in a number of very early sites in adjacent regions of the Near East, including Jericho and Çatal Hüyük. Although the latter examples predate the Halaf culture considerably in a chronological sense, we believe that this ritual is based on ideological conceptions that are similar if not identical.

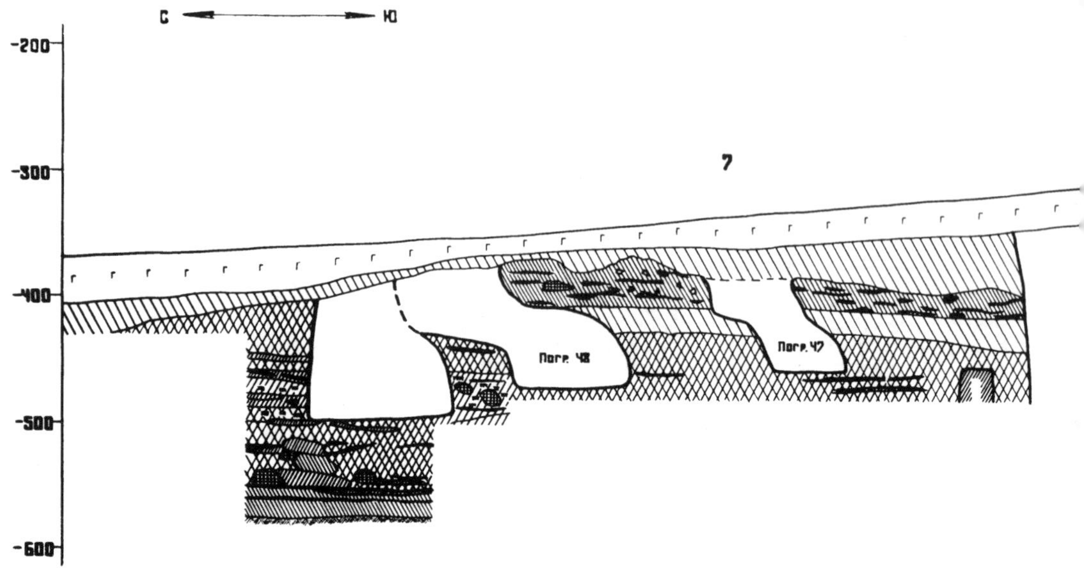

Figure 10.9. Main section at Yarim Tepe I showing burial chambers. Key: *1,* Plow zone; *2,* Brownish soil; *3,* Grayish clay; *4,* Greenish lenses; *5,* Brownish lenses; *6,* Charcoal and ash lenses; *7,* Charcoal inclusions; *8,* Ash inclusions; *9,* Typical clay matrix; *10,* Occupational floors; *11,* Plastered floors; *12,* Stones; *13,* Compacted clay zones; *14,* Wall remnants; *15,* Sterile soil.

The Halaf Cemetery at Yarim Tepe I

A group of Halaf burials were discovered atop Yarim Tepe I (see Fig. 10.9 for the section plan), 250 m to the east of the Halaf settlements of Yarim Tepe II and III. The association of this burial group with one of the latter two sites is beyond question since at Yarim Tepe I itself, no Halaf occupation level is present. Thus, settlement ended at the site with the end of the Hassuna culture, the uppermost level into which the Halaf burials are intrusive. Consequently, at Yarim Tepe I we can claim with confidence the presence of a specialized burial site, established beyond the bounds of the settlement on the slopes of the mound formed by the remains of an earlier Hassuna settlement.

It should be noted that the practice of using Yarim Tepe I as a cemetery has continued through to the modern period. Thus, one finds ancient Babylonian, Syrian, Hellenistic, and Islamic burials among those dating from the Halaf period. The Halaf burials, the earliest at the site, suffered the greatest damage due to subsequent disturbances from later inhumations. For this reason, reliable data have been recovered from only twenty Halaf burials within the limits of the area excavated to date. However, the recovery of several dozen Halaf finds that were certainly associated with burials that have subsequently been destroyed, suggest that the original cemetery covered the entire surface of the mound.

Let us now consider the data obtained from the Halaf burials that have been preserved. First, it is interesting to note that from the ten cases in which the skeletal remains permitted age determinations, nine were mature adults (burials 21, 45, 47, 48, 53, 60, 62, 63, and 72), and only one was a juvenile (four years old, burial 56). This fact alone distinguishes the cemetery from the burials inside the settlement, confirming that those buried within the settlement boundaries were primarily infants and juveniles. Evidence for burial practice within the cemetery shows sig-

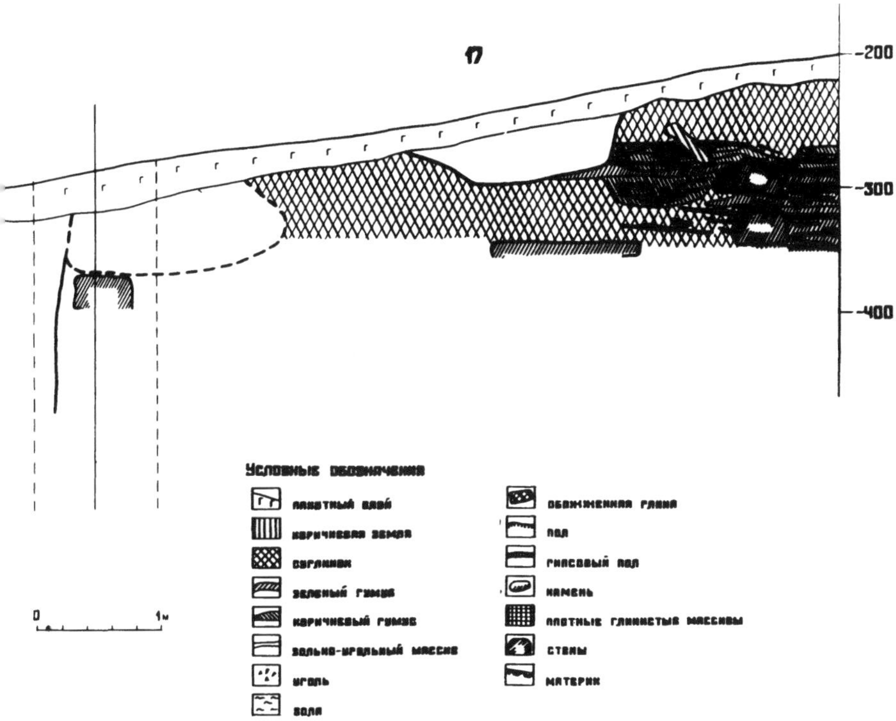

nificantly less variability than inside the settlement. Among the Halaf burials at Yarim Tepe I, one finds no evidence for cremation or cranial burial, and only one secondary burial has been discovered. All other cases were primary inhumations. The bodies were placed in the burial pit either on their right (five cases) or left sides (four cases), in positions so severely contracted that one can assume that the corpses were bound. In general, the wrists lie near the face and the head is usually pointing to the southeast, though a western orientation is found in two cases. In one case, the body was covered with ochre and in another, with charcoal. As a rule, the burials are accompanied by mortuary offerings, including ceramic and alabaster vessels; polished hematite pins and axes; and stone and shell beads and pendants. The vessels have exact analogues from the late Halaf levels at Yarim Tepe II and also at Yarim Tepe III.

From the above evidence, there is good reason to believe that the large majority of Halaf burials are to be found in cemeteries that exist apart from main settlement and that the form of inhumation most commonly practiced was a primary interment in a flexed or tightly contracted position. Other forms of treatment, including cremation, cranial burial, and secondary burial are also represented, but in general are atypical and connected with special ritual actions.

It is important to consider the funerary structures found in the Halaf cemetery at Yarim Tepe I. There is a certain element of doubt concerning the true character of these structures since they are located on the slopes of a mound that has been subjected to thousands of years of erosion and plowing. Also, burials have been discovered at very slight depths, in some cases directly under the topsoil. In only seven cases were the funerary structures adequately preserved to conclusively detect their presence, and the presence of lined burial chambers can be confirmed in at least five of these cases (Fig. 10.9).

Examining these structures in more detail,

Figure 10.10. Burial 45, Yarim Tepe I

Figure 10.11. Burial 47, Yarim Tepe I

we note that in burial 45 only a portion of the oval bottom and remnants of the vault (Fig. 10.10) were preserved, since the rest of the structure had been destroyed by a recent pit. In burial 47 (Fig. 10.11), an oval entryway, more than 1.5 m long and 0.7 m wide, leads into a sloping shaft. At the bottom of this shaft a small lined chamber up to 40 cm wide and 30 cm high was encountered. In this chamber the skeletal remains of an individual with its back facing the entrance was found. In burial 48, an entryway approximately 1.5 m long and 0.9 m wide ended in a step at the entrance to the burial chamber cut into the southern wall of the entryway. The chamber was about 1.0 m wide and its floor was 0.4 m lower than the step. The height of the chamber was more than 0.6 m. Here, as in burial 45, the body lay with its back to the entrance. The adjoining chamber of burial 51 had similar dimensions. Finally, burial 52 had been placed in a lined chamber measuring 0.8 m × 0.4 m with a vault height of 0.3 m.

The possibility exists that funerary structures were present in other burials that, due to later disturbances, were not preserved. This pertains especially to burial 21, since the pit in which the body was found contained rich mortuary offerings and indeed may have been the lower part of a burial chamber. One can be confident of the absence of funerary structures only in the case of burial 60, which is unique in character. In this burial, a sloping corridor terminates in a step leading into a narrow, crevicelike pit reaching 2 m in depth. The top of this step is at the approximate midpoint of the height of the pit wall. The dimensions of the pit are 2.0 m × 0.55 m above, and 1.6 m × 0.35 m below the step. At the end of the entry corridor, adjacent to the step leading into the pit lay the skull of a very large bull (up to 70 cm long) on a platform reinforced with stones. On the step itself lay a pile of fragmented human bones, pieces of charcoal, and fragments of one alabaster and several ceramic vessels. Fragments of two more vessels, one ceramic and one alabaster, a hematite pin, and approximately two hundred gazelle astragali were also found in the pit. The special character of the burial is shown by the construction of the pit, the random piling up of the fragmented bones in this secondary burial, and the considerable amount of gazelle astragali. It is also important to note that this burial is located at some distance from the rest of the cemetery.

Ceramic (Fig. 10.12:1–3) and stone (Fig. 10.13:1–2) vessels from the Halaf cemetery atop Yarim Tepe I, in addition to other finds (Fig. 10.13:3), suggest that this specialized burial site is contemporaneous with the upper levels of the neighboring Halaf occupation sites. However, the discovery of burial chambers within the settlement limits of Yarim Tepe II itself, including those found in the middle levels, indicates that a rather stable burial tradition appeared within the Halaf sphere no later than the middle of the fifth millennium B.C. This appears to be the earliest evidence for prepared burial chambers known to date. The Yarim Tepe structures are more than fifteen hundred years earlier than the well-known tombs of the Levant and Egypt.

Thus far, it is difficult to judge the extent to which this elaborate mortuary tradition was practiced during the Halaf period. However, the presence of funerary structures cannot be denied. Therefore, the possibility cannot be discounted that these structures are somehow related to the round floor plan of the traditional tholos residential structures found within Halaf settlements that sometimes have a cupola-shaped roofs and *dromos* entries.

Conclusions

At the present time, burial vaults and chambers dating from the Neolithic and the Bronze Age are known over a broad expanse of Eurasia, including examples from Jordan, Palestine, Syria, Mesopotamia, Iran, the Southern Caucasus, the Balkans, the islands of the Mediterranean, Eastern Europe, and so

Figure 10.12:1. Ceramic vessel from Yarim Tepe I burial 56.

Figure 10.13:1. Stone finds from Yarim Tepe I burials: Vessel from burial 63.

Figure 10.12:2. Ceramic vessel from Yarim Tepe I burial 52.

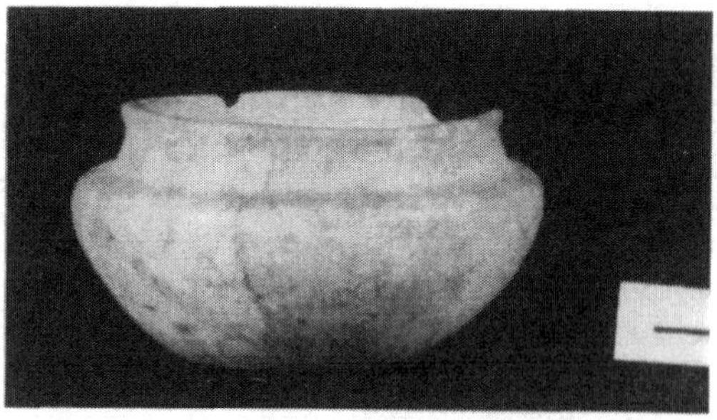

Figure 10.13:2. Stone finds from Yarim Tepe I burials: Vessel from burial 45.

Figure 10.12:3. Ceramic vessel from Yarim Tepe I burial 45.

Figure 10.13:3. Stone finds from Yarim Tepe I burials: Macehead from burial 56.

forth. There is no doubt that they are the products of numerous independent cultural traditions and not diffusion. But as this reasearch problem is investigated further, the extraordinarily wide geographical area of the Halaf culture and the regions it influenced should be borne in mind.

The burials that we have considered—namely those discovered within the Halaf settlement of Yarim Tepe II and in the cemetery associated with it at Yarim Tepe I—demonstrate the variability both in burial practices and the ritual actions connected with them, thereby illustrating the highly developed nature of the Halaf mortuary complex. One can state with confidence that the ideological realm of northern Mesopotamian agriculturalists during this period was quite complex. In addition to mortuary data recovered to date, the religious statuary, the intricate semantics of the numerous design elements on Halaf pottery, and the direct evidence of various rituals all are indicative of the elaborate ideological world that members of Halaf society had created for themselves (Munchaev and Merpert 1981).

Note

1. The scanty nature and obviously extraordinary character of these on the latter site give rise to a supposition that a cemetery exists that has not yet been discovered.

11

YARIM TEPE III: THE UBAID LEVELS

N. Ya. Merpert and R. M. Munchaev

This chapter is devoted to the excavation results obtained from the upper levels of Yarim Tepe III containing the remains of the Ubaid settlement.[1] The Ubaid culture immediately precedes the beginning of Sumerian civilization proper in Mesopotamia. For this reason the Ubaid is sometimes referred to as the "proto-Sumerian" period. Distinct from all preceding cultures, the Ubaid embraced the entire territory of Mesopotamia. At the same time, several variants of this culture have been noted in the northern regions, which has caused some investigators to speak of a specific Northern Ubaid Culture. The question of its origins, its specific character, and the relationship of the Northern Ubaid to the preceding Halaf constitute several of the important research problems in the cultural and historical development of preliterate Mesopotamia. Since the Ubaid strata of only two sites in northern Mesopotamia, Tepe Gawra and Tell Arpachiyah, have been extensively cleared, we should pay particular attention to the excavation of contemporaneous levels at Yarim Tepe III.

The top of the Ubaid stratum is at the surface of the mound. Unfortunately, the upper part of this stratum has been disturbed by numerous intrusive burials, nineteen to be exact. Some of these were placed in massive stone vaults. A number of burials, judging by the scanty grave goods, apparently date from the Hellenistic period, but the majority date from Islamic times.

Eight large houses with a total of over ninety rooms relating to several successive Ubaid construction phases have been uncovered in the area of the site that has been investigated (Figs. 11.1–11.4). The earliest of these stands directly upon the remains of structures belonging to the final Halaf level. In the eastern half of the excavation, floor levels that range from 3.65 to 4.20 m below datum (mbd) have been provisionally assigned to the fourth level. Building complexes 4 and 8 are associated with this lowest Ubaid level.

The original core of complex 8 is comprised of five rectangular rooms (Fig. 11.1) with dimensions ranging from 1.50 m × 0.70 m up to 3.90 m × 2.25 m. In the southwest corner of the largest room (number 88), a rectangular clay oven was found with the dimensions 1.50 m × 1.70 m. In addition, a series of small storage rooms, 81, 82, 86, and 91 (1.50 m × 1.10 m; 1.25 m × 1.25 m; 1.10 m × 1.10 m respectively [only three sets of dimensions given — Ed.]), have been added to the north of this complex. A number of rooms in

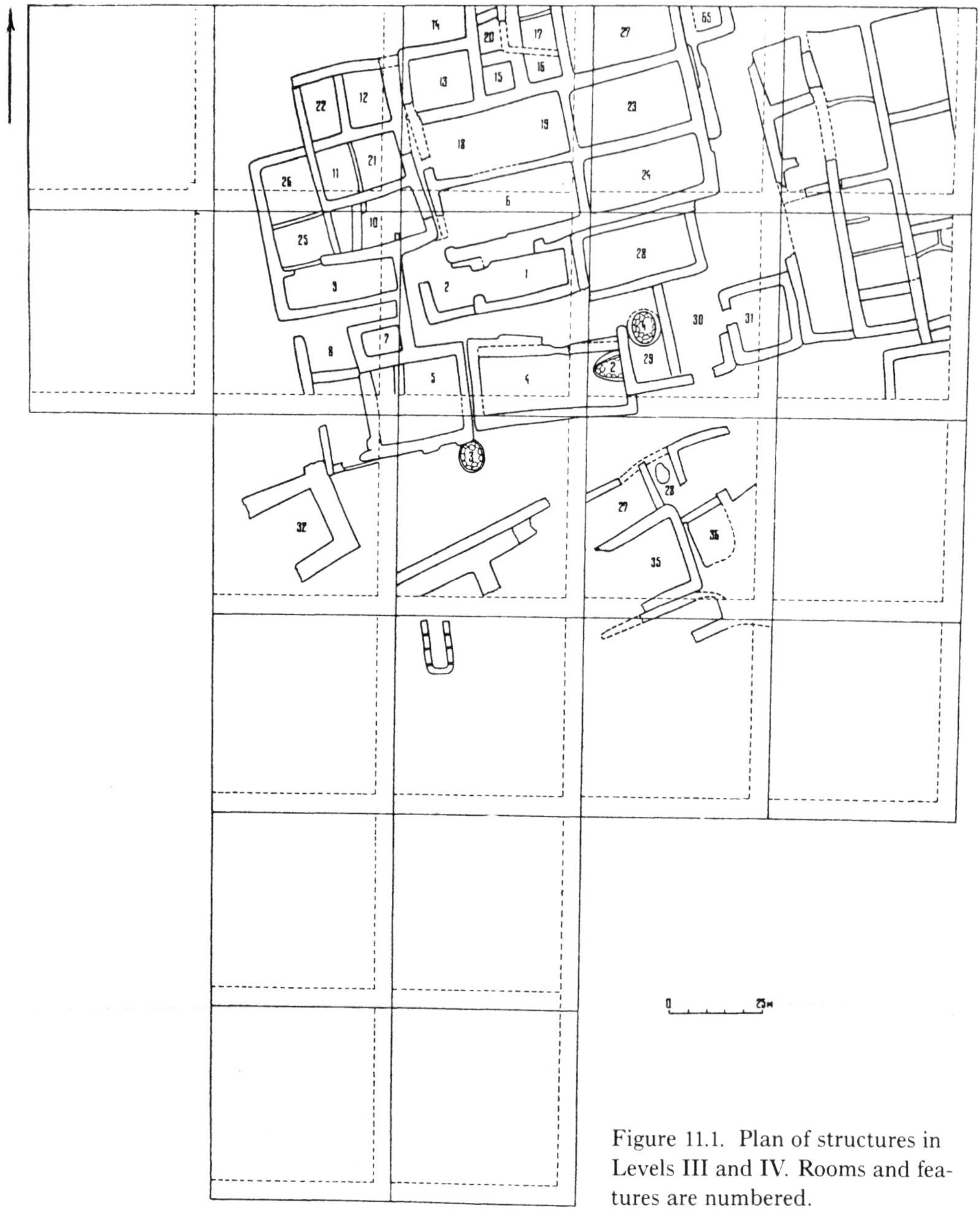

Figure 11.1. Plan of structures in Levels III and IV. Rooms and features are numbered.

the northern and northwestern area of the complex, as well as in its entire southern part, have been disturbed by intrusive pits from a higher level. In general, this building was a large and massive structure with its walls preserved to a height of more than 1 m in places.

In all, complex 8 was composed of more than ten rooms, including several later additions from rebuilding episodes. These episodes are further indicated by the double and even triple walls enclosing a number of rooms. The largest rooms are connected by doorways to

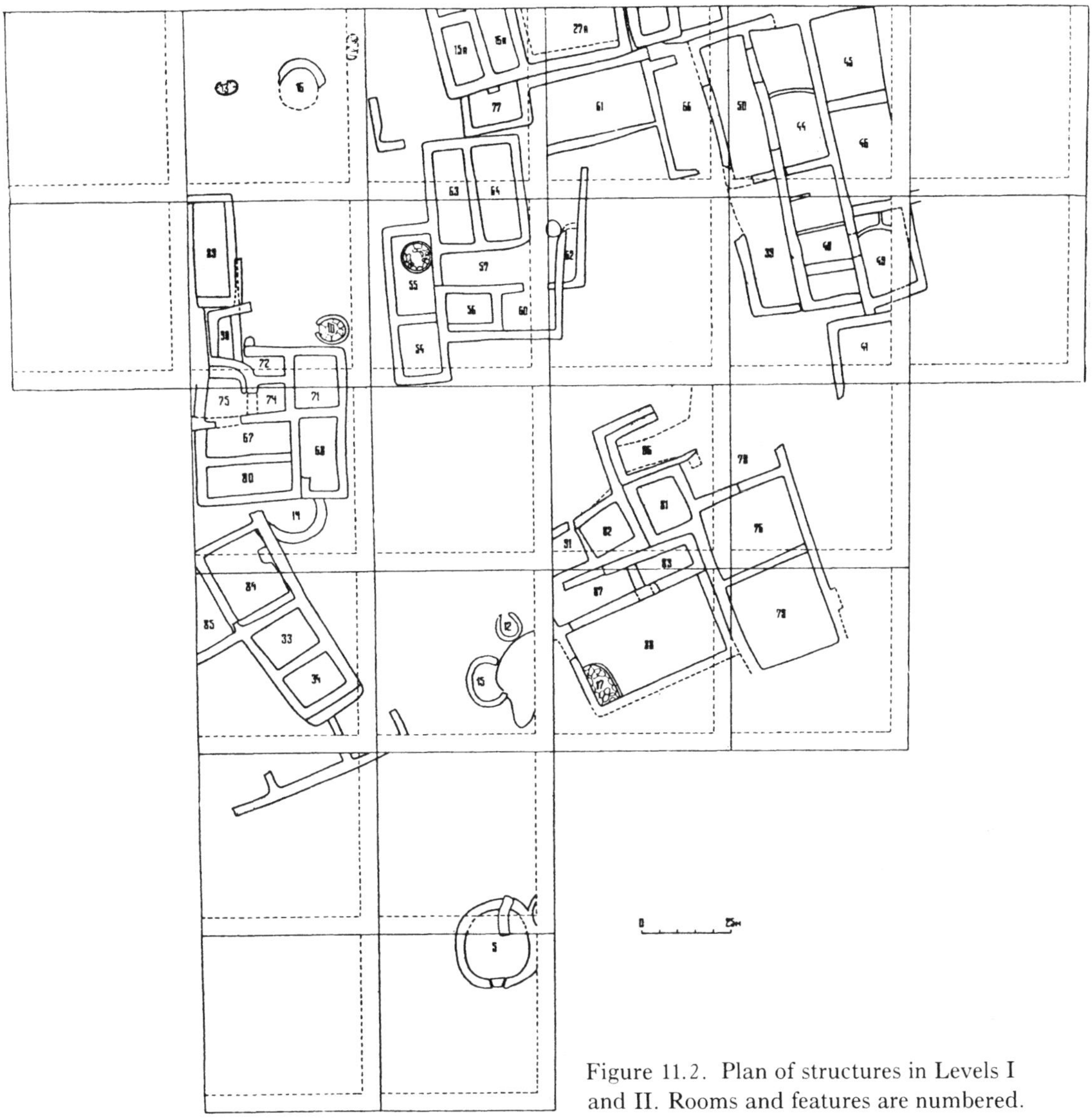

Figure 11.2. Plan of structures in Levels I and II. Rooms and features are numbered.

the adjacent rooms and often contain large grinding stones. The number of finds in the auxiliary rooms is relatively insignificant.

Complex 4 is located to the north of complex 8 (Figs. 11.1 and 11.3:4). The core area consists of five rectangular rooms built at the same time, stretching from north to south with the following dimensions: 3.40 m × 1.25 m, 3.50 m × 1.25 m, 2.25 m × 1.60 m, 2.85 m × 1.25 m [only four sets of dimensions given — Ed.]. Their walls have been preserved to a height of 1.5 m and more. A small square room (1.25 m × 1.25 m) was built as a later addition. It is quite probable that this complex continued both to the south and to the east. Doorways can be traced in three rooms. However, none of the rooms contained ovens or hearths, and in general, they do not appear residential in character. Quite probably the entire complex was an elaborate group of out-

Figure 11.3:1. Structures of the Ubaid levels: complex 1, view from the north.

Figure 11.3:2. Complex 6, view from the east

Figure 11.3:3. Complex 5, view from the south

Figure 11.3:4. Complex 4, view from the south

buildings. A large quantity of ceramics were found in this complex with most of the material coming from floor contexts. In room 45, a grinding stone was found. Two Ubaid burials were discovered in the fill of another.

The few remaining traces of architecture from the next construction level can be provisionally grouped into Level 3. To a considerable extent, these structural remains repeat the plan of the complexes described above. Many of the structures of Level 3 lie between 3.20 and 3.80 mbd and have no underlying architectural precursors. Thus, they are obviously the result of the expansion of the settlement westward. The most significant structure in Level 3 is complex 5 (Figs. 11.1 and 11.3:3). This complex consists of three groups of rooms arranged on the western, northern, and southern sides of a large open courtyard (8 m × 4 m). The building sequence of the individual rooms in each group has been precisely established. Thus, in the western group, room 56 (1.25 m × 1.10 m) was constructed first, room 60 (1.60 m × 1.10 m) second, and then the others. In the northern group, room 15A (1.65 m × 0.80 m) was built first, 16A (1.80 m × 0.86 m) second, and finally 27A (length: 2.65 m). The last group consists of a single structure that we have not yet completely exposed, since one edge of it extends beyond the excavation limit. The walls of this massive structure were repeatedly rebuilt and extended and are traceable from their foundation level to a height of about 3 m, almost to the surface of the mound. When the walls were built, a trench about 20 cm deep was dug around the perimeter of the building. The room built on to complex 4 from the south (77) was part of a later rebuilding episode. This new room connected the northern part of the complex with its western part. The eastern section of the com-

Figure 11.4:1. Complex 1, view from the northeast.

Figure 11.4:2. Passageway between complexes 1 and 2

plex, by contrast, is comprised of rebuilt rooms originally a part of the earlier complex 4.

During later reconstructions of this portion of the complex, the open courtyard was partitioned, and three rooms, 54, 55, and 62, were added to its western part. In room 55 (2.25 m × 1.10 m), a round oven was discovered. In room 62, the carbonized remains of a significant quantity of grain were found, suggesting it functioned as a storage area.

Complex 6 is located to the west, parallel to complex 5 (Figs. 11.1 and 11.3:2) and separated from it by an alley 1.30 m in width. The earliest room (89) in complex 6 is contemporaneous with the core structures of the western and northern groups of complex 5. Later, during a rebuilding episode, the northern and southern walls of this room were taken down. Another complex (7) was excavated to the south of complex 6. The excavated portion of complex 7 consists of four rooms: 33, 34, 84, and 85. Three of these make up one row running from the southeast to northwest. Two rooms (33 and 34) are auxiliary in character, whereas a third (84), with a doorway and the remains of a hearth, had a greater quantity of cultural remains within it and may have been a dwelling.

Complexes 6 and 7 are close together chronologically, and the stratigraphic phase to which they belong overlies the remains of a round oven 1.20 m in diameter. The general absence of fire installations in the rooms of these complexes, with the exception of the hearth in room 84, should be noted. As a rule, the ovens are located in open spaces to the east of the complexes.

Room 47, which is located to the south of the complexes described above, was con-

Figure 11.4:3. Entryway in wall of room 1, view from the west

Figure 11.4:4. Oven in room 4 of complex 3, view from the west

Figure 11.4:5. Wall between structures 63–64 in complex 5, view from the east

structed during one of the rebuilding episodes previously mentioned. Thus, it also belongs to Level 3. During this time, the northwestern corner of the excavation area remained an open space with ovens and hearths the only discernible features. This area remained open during later times as well, with the Level 3 ovens covered first by a 30 cm layer of rubble; this, in turn, was covered by a 50 cm layer of packed clay. The latter also covered the structures of the next construction phase (Level 2). Thus, the density of construction in both Levels 2 and 3 was noticeably sparser here than in the other sectors of the excavated area.

Level 2 is represented primarily by complex 1 (Figs. 11.2, 11.3:1, and 11.4:1). Evidence for both continuity with the preceding phase (Level 3) and new construction is present in Level 2. The former include rooms 15A and 16A, though these rooms were subdivided by partitioning walls such that their floorplans were substantially altered. The same can be said of room 27, which was further subdivided into rooms 15, 16, 17, and 20 in the Level 2 occupation. Thus, part of the northern group of rooms in complex 5 in Level 3 are included with modifications in complex 1 of Level 2. Another part of complex 1, similar to other room groups of the earlier complex (5), also went through several substantial rebuilding episodes. Thus, the walls of the eastern room group of complex 5 were taken down and new walls erected, similar in floor plan but displaced to the south and the west and set off of room 23. To the south, room 24 was built in the place of the open courtyard of complex 5.

Practically simultaneously with the building of the rest of complex 1, rooms 6, 18, and 19 were rebuilt in place of the western group of complex 5, and somewhat later, rooms 13 and 14 were added. All of these rooms made up the core of complex 1, which later was significantly expanded by successive additions. Room 51 represents the westernmost addition to complex 1. On the southern side, rooms 1, 2, and 28, whose wall foundations are still visible on the surface, 2.12–2.15 m high, were built onto the complex at a later date. The entire complex comprises fifteen rooms. All of these rooms are small, with sizes ranging from 3.50 m × 1.40 m to 0.80 m × 0.80 m. The original walls of complex 1 have been preserved to a height of up to 1.30 m.

It must be emphasized once again that the northern room group (rooms 15, 16, 17, 20, etc.) retained the basic ground plan of the earlier group in complex 5. These walls were merely reinforced and raised considerably. Doorways have been noted in the walls of rooms 1 (Fig. 11.4:3), 2, 6, and 18. The preserved wall heights of these rooms range from 0.70 to 0.90 m and their widths, from 0.50 m to 0.80 m. No hearths or ovens were found within the rooms of complex 1. Obviously, the ovens in the open courtyard to the south of room 28 were connected with the use of this complex. This courtyard is connected by a narrow and crooked passage to room 1. The ovens are round and up to 1 m in diameter. They are found at various levels, from the earliest level of complex 1 to higher levels contemporaneous with the later rebuilding episodes.

A second group of ovens are located in the open courtyard to the south of room 51. This courtyard is more than 3 m in width and is defined on the west and the south by walls. On the east it is bounded by rooms 5 and 18. Two round ovens of equal dimensions stood in this courtyard on one of the surfaces (about 2.0 mbd). To the west of the walled courtyard, in another open space, stood a round structure (52) 3 m in diameter, with a wall 26 cm thick and preserved to a height of 40 cm. The walls and the floor showed evidence of fire. The floor was covered with a layer of packed gray clay, which in turn was covered with accumulations of ash and charcoal. The structure stood on the basal level of Level 2 at a depth of 2.0 mbd. Its function is difficult to determine, though the structure's large dimensions preclude us from regarding it as an ordinary oven. However, it must have functioned as some sort of fire installation.

The narrow passage mentioned above

separated complex 1 from complex 3, the latter located to the south. Complex 3 was in a considerably worse state of preservation than complex 1. This complex consisted of a row of rectangular rooms stretching from west to east, including rooms 5 (2.25 m × 2.0 m), 4 (3.75 m × 1.40 m), and 31 (1.45 m × 1.45 m). An oval-shaped oven was preserved in room 4 (Fig. 11.4:4). The debris layer immediately above the floor in this room was clearly residential in character.

To the south, only remnants of the buildings from Level 2 were preserved. These include the remains of room 32 with unusually thick (up to 45 cm) walls, including a massive double wall traceable over a distance of more than 5 m and the remains of structures adjoining it on the south (Fig. 11.2). All of these architectural remains relate to adjacent buildings outside the limits of excavation.

The next major building level within the excavated area consists of the latest group of structures exposed at the site, arbitrarily assigned to Level 1. These structures are traceable only in square I, since this phase has been destroyed by erosion in other sectors of the excavated area. Complex 1 continued to exist at least partially (rooms 1, 6, 13, 15–20) in Level 1. Some of its structures, on the one hand, were completely rebuilt. For instance, room 12 was displaced somewhat to the south and was built in place of room 51. In the space between rooms 18 and 15, a new room (13) was built (1.70 m × 1.30 m). Some structures, on the other hand, were erected in places where no underlying architecture existed. These new structures are found in the western sector of square I and are grouped together as complex 2, and comprised rooms 22, 12, 21, 26, 25, 10 and 9. The dimensions of these rooms are as follows: 1.35 m × 0.80 m; 1.25 m × 0.95 m; 1.45 m × 1.30 m; 1.10 m × 1.40 m; 1.0 m × 2.90 m [Only these dimensions are given — Ed.]. Further to the south, two more small rooms (7 and 8) were located. In square II, this complex is represented only by isolated remnants of walls. The small dimensions of the rooms, the absence of ovens or hearths, and the nature of the cultural debris lead to the supposition that this complex served a storage function rather than being residential.

Thus, the remains of the Ubaid buildings in the excavated area of Yarim Tepe III have now been presented in some detail. We see that this sector of the site was densely built up over the entire history of the Ubaid settlement. The structures shared a common orientation and were grouped around the margins of open plazas. Each of the houses consisted of a considerable number of large and small rooms. All of the houses had rectangular ground plans and were separated from each other by narrow alleys.

The walls of the buildings were constructed from large rectangular, sun-baked mud bricks with the average dimensions of 65 cm × 25 cm × 8 cm. The walls were usually mud-plastered on both sides (Fig. 11.4:5). The floors of the buildings were compacted and also mud-plastered. The dwelling rooms were usually larger and contained grinding stones and hearths (rooms 4, 55, 88). The majority of the rooms, however, may be regarded as storage facilities, which is confirmed by the findings of carbonized grain and the tools connected with grain processing, primarily various types of grinding stones, within them.

As we know, the best-studied sites of the Northern Ubaid are the settlements of Tepe Gawra (strata XVIII–XII) and Tell Arpachiyah (Levels TT1–4), located in the Mosul region. The remains of buildings at these sites are very similar to the architecture exposed at Yarim Tepe III. At Tepe Gawra multi-roomed rectangular buildings predominate in all phases of the Ubaid stratum. Their floor plans, construction technique, dimensions, and orientation of the individual rooms are analogous to the corresponding phases at Yarim Tepe III (Tobler 1950, 25–44), especially the early Ubaid phases at Tepe Gawra (Tobler 1950, Pls. XVI–XIX).

The 2–5 m thick Ubaid stratum at Tell Arpachiyah yields a similar picture. The ex-

Figure 11.5:1–12. Ceramic vessels I

cavators emphasize the poverty of the settlement, which consisted of small rectangular houses. The walls of these houses were built with sun-baked mud bricks one row in width and mud-plastered. The floors were covered with packed clay, and the roofs were made of reeds and thatch (Mallowan and Cruikshank Rose 1935, 11–13). The massive character and complex design of a number of buildings at Yarim Tepe III more closely resembles the structures at Tepe Gawra than those at Tell Arpachiyah.

The sharp differences between the architectural forms of the Ubaid period and the building traditions of the preceding Halaf culture should be especially emphasized. In all three multicomponent sites (Tepe Gawra, Tell Arpachiyah, and Yarim Tepe III), the Halaf strata that underlie the Ubaid strata are characterized by round dwellings with a flat or cupola-shaped roof (Munchaev and Merpert 1981, 186–97).

Before discussing the artifacts from Yarim Tepe III, the six Ubaid burials from the excavated area should be noted. Four of them were found in rooms, either on the floor, under it, or in the room fill. Two other burials were discovered in courtyards. There were four juvenile burials and two adults. The bodies were buried in a flexed position, in one case on the back, and in the others on the right sides, with the heads primarily pointing toward the north and the east. Two of the juvenile skeletons were covered with large fragments of pottery vessels and in one case, traces of matting were found under the skeleton. The mortuary offerings discovered with the burials include three typical Ubaid vessels in one case, and in the other, two cylindrical shell beads found near the right wrist and eight flat annular beads made from black stone near the right ankle.

The artifact assemblage at Yarim Tepe III consists primarily of numerous examples of Ubaid pottery. The ceramic corpus is essentially homogeneous on all horizons throughout all the Ubaid phases and identical in technological, morphological, and decorative characteristics to the pottery from the Ubaid strata of Tepe Gawra (primarily strata XVIII–XVII) and Tell Arpachiyah. In her description of the Ubaid pottery from Tepe Gawra, Perkins emphasized that this monochrome ware is noticeably inferior to the Halaf pottery in all characteristics. The clay is coarser with grit temper, and its color is usually dark brown, light reddish brown, green, or greenish gray. It is well fired but less resistant to breakage than Halaf vessels. The exterior color of Ubaid vessels is either black, reddish brown, or red and easily crumbles. The decorative patterns are crudely fashioned and small in number. Common motifs are chevrons, wavy lines, rows of triangles, ladders, vertical zig-zags, ribbons, cross-hatching, etc. Certain Halaf motifs, including squares, "fish-scales," Maltese crosses, rosettes, and bulls' heads continue, but they are extremely rare (Perkins 1957, 46–47).

This description applies to the Ubaid pottery from Yarim Tepe III. The vessel forms represented here are few and simple, with the basic ones listed below. These include spherical or turnip-shaped, round-bottomed vessels with straight or flaring necks (Fig. 11.5:1–5). Some rims of the thick-walled large vessels have interior projections. Similar vessels are widely encountered at Tell Arpachiyah and Tepe Gawra (Tobler 1950, Pl. CXXII, 107; Mallowan and Cruikshank Rose 1935, Fig. 35; Perkins 1957, 48). Additionally, a number of pots with two loop handles were found (Fig. 11.5:7–9). In general, the decoration is zonal and geometric, with one example of a highly stylized bull's head. The small, round-bottomed vessels in the shape of flagons are comparable to these vessels.

There is also a significant number of flattened, sharply carinated pots with round or flat bottoms and decorative designs on the upper part of the vessel (Fig. 11.6:1–5). Representational motifs include chains of feline carnivores arranged in a circle (Figs. 11.6:5 and 11.7:1) found alongside geometrical designs. At Tepe Gawra this form is typical of strata XVIII–XVII (Perkins 1957, 47). It is

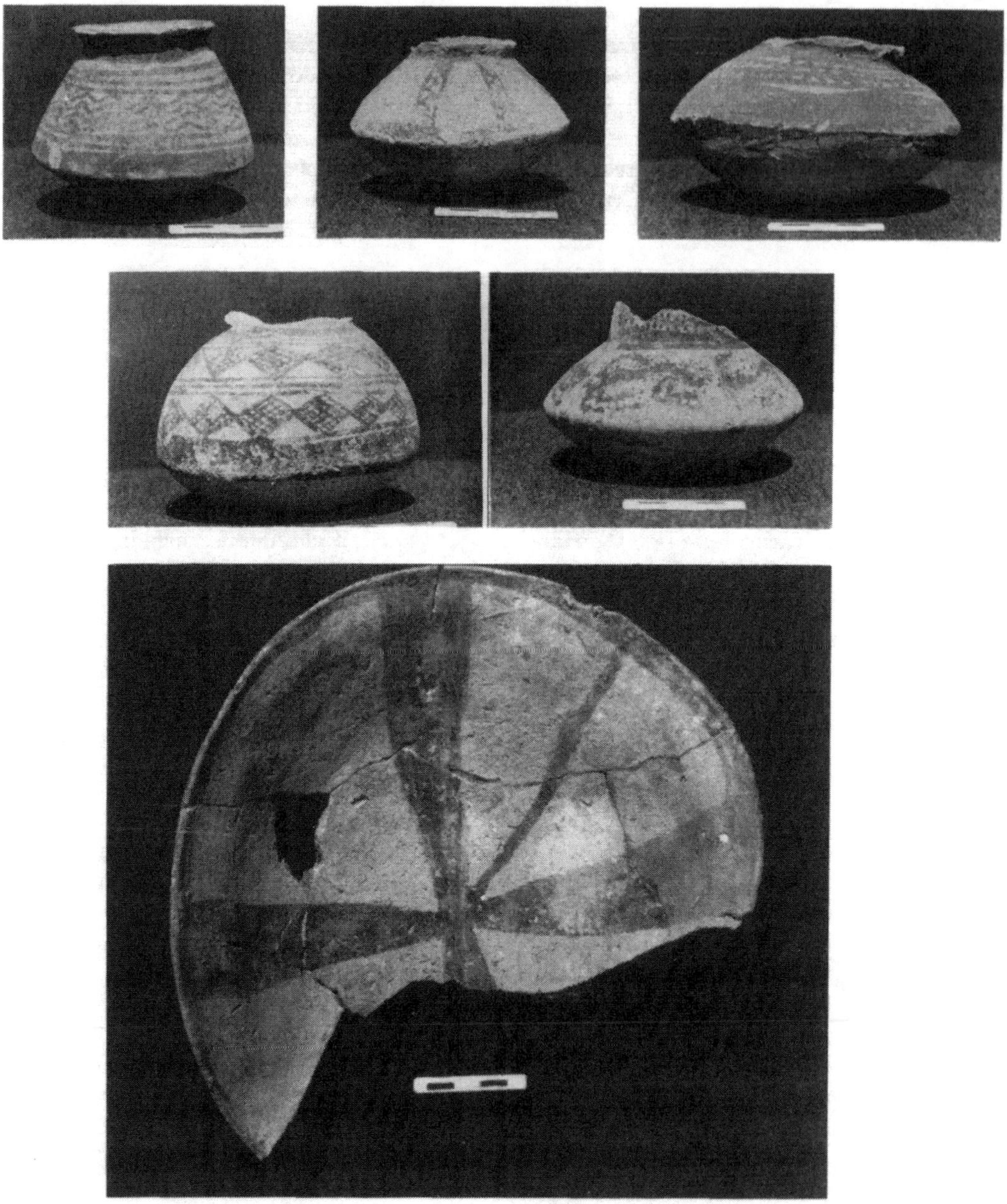

Figure 11.6:1–6. Ceramic vessels II

also well known in the ceramic assemblage from Tell Arpachiyah (Mallowan and Cruikshank Rose 1935, Figs. 34, 37, et al.).

Bell-shaped, round-bottomed vessels with geometrical designs arranged in vertical and horizontal bands are particularly striking (Fig. 11.5:3). These vessels bear a resemblance to decorative arrangements typical of Northern Ubaid sites of Mesopotamia.

Hemispherical cups, usually with a horizontal band composed with simple geometrical designs, are also interesting. The base of

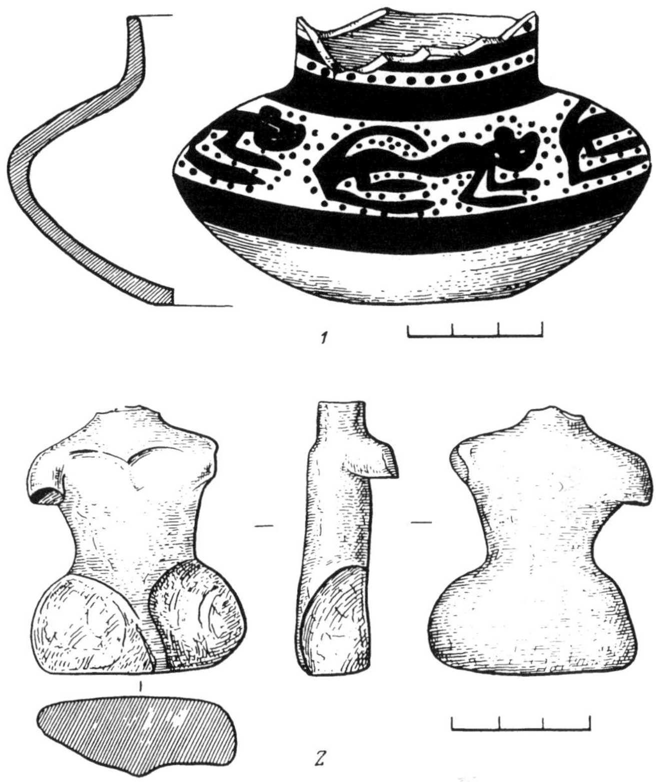

Figure 11.7:1,2. *1,* Vessel with zoomorphic decorations; and *2,* anthropomorphic figurine.

one is marked with a Maltese cross (Fig. 11.6:6), which has parallels at Tepe Gawra (Tobler 1950, Pl. CXXV, 148). Yet another vessel form from the Ubaid stratum of Yarim Tepe III is a flat-bottomed pot with rounded sides, a pronounced neck, and a flaring rim. This form is undecorated except for occasional moldings on the shoulders. At Tepe Gawra, analogous vessels are found below stratum XVII (Tobler 1950, Pl. CXXII, 118).

All of the vessel forms noted above are well represented at Yarim Tepe III. Along with these common forms, rare examples and special forms have also been found. These include "bellows-shaped" vessels with a wide low spout at the side (Fig. 11.5:10), and small double, "binocular" vessels (Fig. 11.5:11). Large spouted vessels were also found at the site (Fig. 11.5:12). We know of no definitive analogues to these unusual forms from other Ubaid sites in northern Mesopotamia. However, general parallels do exist elsewhere, such as crude double-spouted pots found at Tell Arpachiyah (Mallowan and Cruikshank Rose 1935, 18, Fig. 41), and a double vessel made of stone found at Tepe Gawra (Tobler 1950, Pl. CIV). Vessels of these types have also been found in southern Mesopotamia, including at Tell Ubaid itself (Perkins 1957, 71). Alabaster and stone vessels are represented in small numbers in the Yarim Tepe III assemblage. In general, they repeat in miniature the forms of ceramic cups and sharply carinated pots.

The second most prolific class of artifacts at Yarim Tepe III is lithic tools. These include numerous flint and obsidian knifelike flakes and blades from composite sickles. In one case, a whole series of inserted sickle blades were found, bearing traces of bitumen by which they had been attached to a wooden or bone frame. Several adzelike lithic tools of

Figure 11.8:1–31. Stone, bone, and clay tools: *1–9, 13, 14,* Adze-like tools; *10–12,* Axes with drilled holes; *15–17,* Palettes; *18–20,* Bone tools; *21–31,* Clay spindle whorls.

various sizes (Fig. 11.8:1–9, 13, 14), hoes, weighted tips for digging-sticks, marble and hematite fibulae, and fragments of polished axe-hammers drilled to accommodate handles (Fig. 11.8:10–12) have also been found. Many of the lithic artifacts found at Yarim Tepe III have parallels at Tepe Gawra (Tobler 1950, Pl. XCV-c, 2, 3, 5; Pl. XCVII-a, 4, 5, et al.).

In the Ubaid stratum of Yarim Tepe III, one finds an abundance of grinding stones, mortars, and pestles (at least 150 have been found). The grinding stones are characterized by the usual boatlike shape and large dimensions, averaging 65 cm × 20 cm × 6 cm.

Bone tools are rare, represented only by needles made from the long bones of animals and by polishers, some of which are decorated with engravings (Fig. 11.8:18–20). Miscellaneous clay objects include nail-like artifacts (Fig. 11.9:1–5), sometimes painted (Fig. 11.10:5), spheroid "projectiles," biconical and conical spindle whorls (Fig. 11.8:21–31), and rings (Fig. 11.10:1–4). These finds are quite similar to known artifacts from the Halaf and Ubaid complexes of Mesopotamia, particularly at Tell Arpachiyah and Tepe Gawra.

The clay figurines from Yarim Tepe III deserve special attention. Most of them are zoomorphic, possibly depicting small, horned animals (sheep and goats) and dogs (Fig. 11.9: 6–21). Similar clay figurines were found in strata XVIII–XIII of Tepe Gawra (Tobler 1950, Pl. LCIV, 15), and also at Tell Arpachiyah (Mallowan and Cruikshank Rose 1935, Fig. 48). Of all the figurines from the Ubaid stratum of Yarim Tepe III, only one is anthropomorphic (Fig. 11.7:2). The figure is a seated female with a flat torso and separately modeled legs attached with bitumen. The collection of small ceramic sculpture from Northern Ubaid sites is on the whole insignificant and has not been well studied. The figurines from Yarim Tepe III do not stand apart from this assemblage either in form or quantity.

Finally, small, flat, rectangular palettes with molded edges (Fig. 11.8:15–17) and a few ornamental items were also found in the

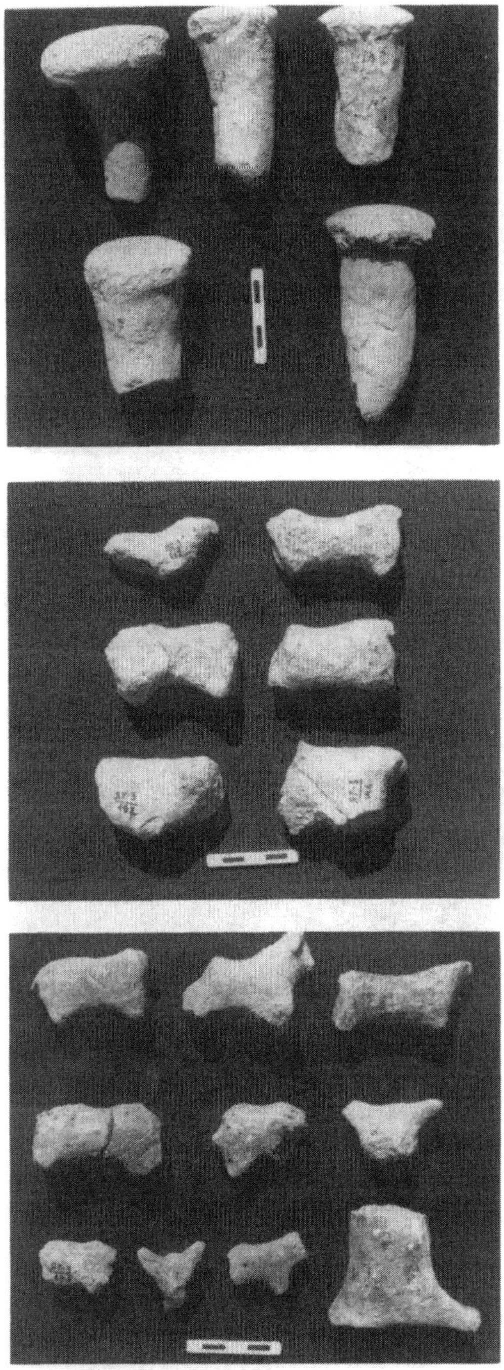

Figure 11.9:1–21. Miscellaneous clay objects: *1–5,* Nail-like objects; *6–21,* Zoomorphic figurines.

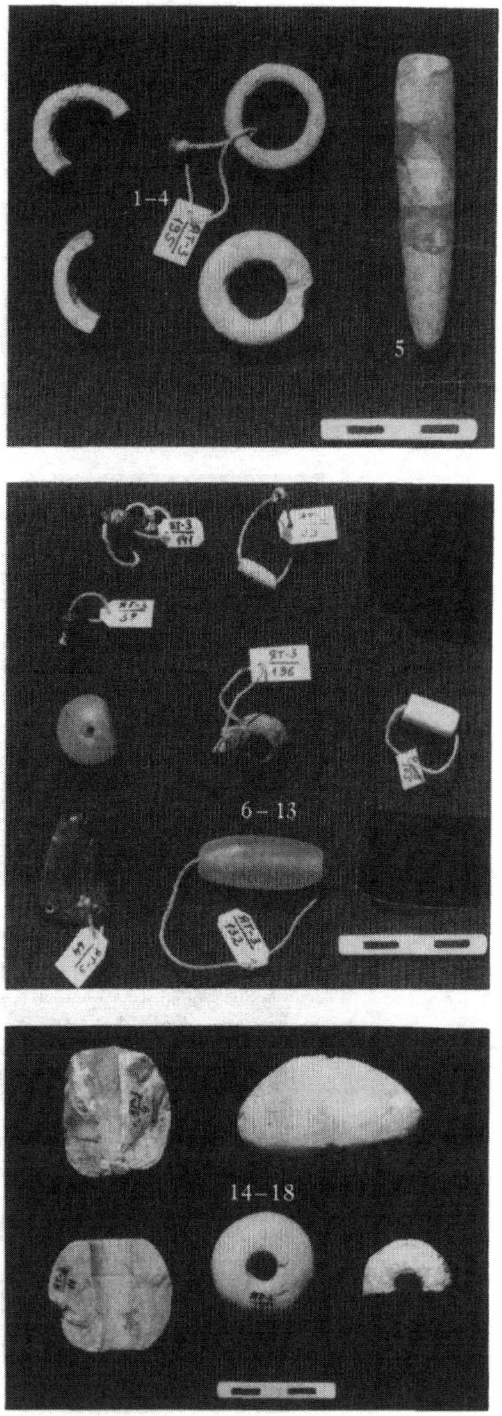

Figure 11.10:1–18. Yarim Tepe III: ornamental objects and small artifacts: *1–4,* Clay rings; *5,* Nail-like object; *6–18,* Beads and pendants.

Ubaid stratum of Yarim Tepe III. The latter include one triangular obsidian pendant (Fig. 11.10:12); biconical, oval, and round beads made from obsidian, marble, chalcedony, and hematite (Fig. 11.10:6, 8–10, 13–18); and cylindrical shell beads (Fig. 11.10:7, 11). All of the objects listed above, including the stone palettes, are also found in Mesopotamian sites dating from the eighth to the sixth millennia B.C.

The artifacts obtained from the Ubaid component of Yarim Tepe III demonstrate quite clearly the cultural and chronological connection of this site with the contemporaneous strata of other well-known Northern Ubaid sites in Mesopotamia, including Tell Arpachiyah and, especially, Tepe Gawra. Within these sites, the Ubaid strata lie directly atop the final levels of Halaf settlement, heralding the advent of a new culture in its northern manifestation. Taking into account a series of radiocarbon dates for the terminal Halaf occupation (Munchaev and Merpert 1981, 266), which lies directly beneath the beginning of the Northern Ubaid, the end of the Halaf can be dated to the last quarter of the fifth millennium B.C. On this basis, the earliest Ubaid level at Yarim Tepe III should date from the end of the fifth or beginning of the fourth millennium B.C. The Ubaid settlement at Yarim Tepe III provides conclusive evidence that the Sinjar Valley was included in the Northern Ubaid culture and was a major route of its westward transmission.

Note

1. Besides the authors, N. O. Bader, O. G. Bol'shakov, V. I. Gulaev, I. G. Narimanov and S. S. Shiriskii took part in the excavations at Yarim Tepe III.

12

THE EARLIEST EVIDENCE FOR METALLURGY IN ANCIENT MESOPOTAMIA

N. Ya. Merpert and R. M. Munchaev

During the past two decades archaeological research in Southwest Asia has pushed back the date of the earliest metal artifacts considerably, forcing us to reconsider the origins of metallurgy as a whole. Not long ago, the finds at Spalka I, dating from the fifth millennium B.C., were regarded as the oldest copper artifacts. In the past, metallurgy was considered possible only after the development of sophisticated ceramic technology. For this reason, copper and lead artifacts found at Çatal Hüyük in contexts dating two millennia earlier (Mellaart 1967, 22, 209, 212, 217–18, Fig. 104; Coghlan 1975, 40–44) were considered a sensational discovery. These finds were only a part of a whole series of "surprises" encountered at this site. Later discoveries indicated that Çatal Hüyük was not unique as a center of early metal use. Metal was also found in sixth millennium B.C. contexts at both Hacilar, in western Anatolia, and Suberde, in the Konya Valley. Additionally, a bead made from rolled sheet copper was found at Ali Kosh in a preceramic context dating from the seventh millennium B.C. (Hole, Flannery, and Neely 1969, 42, 399, 427, 428; Coghlan 1965, 44). Finally, a fragment of copper ore and cold-forged copper artifacts including awls, needles, and beads were found at the preceramic settlement of Çayönü located at the Tigris headwaters in Anatolia (Braidwood 1973, 37; 1970, xi; Braidwood, Çambel, Lawrence, Redman, and Stewart 1974, 570; Coghlan 1975:40). These last finds are significantly earlier than either those at Ali Kosh or Çatal Hüyük, since the Çayönü settlement dates from the end of the eighth through the beginning of the seventh millennium B.C.

The metal finds mentioned here are located in a large area of Southwest Asia, stretching from western Anatolia to Khuzistan, Iran. The almost complete absence of corresponding finds in the central part of this region, which includes northern Mesopotamia, requires further explanation. Several copper beads found in one of the burials from the lowest stratum of Tell es-Sawwan remained the only evidence for the early use of metal in this region until recently. The other metal artifacts from this site lack secure proveniences, with some of the assemblage probably associated with a much later reoccupation at the site.[1] In the subsequent strata that are associated with the later part of the Samarra phase, there are no metal finds from reliable contexts. The sites of the Hassuna culture dating from the sixth millennium B.C.

were also lacking in metal objects until quite recently. Thus, at Tell Hassuna the excavators, Lloyd and Safar, specifically noted the absence of copper at the site (Lloyd and Safar 1945, 263). Though they attached the label "Proto-Chalcolithic" to most of the Tell Hassuna sequence, this term was intended more as a chronological marker tying Tell Hassuna to other sites in the region, particularly Mersin, excavated by Burkitt, than an indicator of technological development. Thus, Lloyd and Safar were "inclined to accept Burkitt's use of the term 'Neolithic' to describe the period immediately preceding the introduction of painted pottery, as opposed to the phase immediately following it which Burkitt labelled as the Proto-Chalcolithic" (Lloyd and Safar 1945, 284). Since Burkitt's "Proto-Chalcolithic" at Mersin had numerous parallels at Tell Hassuna, Lloyd and Safar made use of this term, equating Mersin's "Proto-Chalcolithic" phase with Tell Hassuna stratum 1b and placing only the basal level of the tell (stratum 1a) in the "Neolithic".[2] Thus, the use of the term "Proto-Chalcolithic" had nothing to do with the development of copper metallurgy. During the 1950s V. G. Childe (1956, 173) also emphasized that "we have no direct proof of the use of copper by the inhabitants of Tell Hassuna."[3]

Furthermore, until recently the assemblage of metal artifacts from the succeeding Halaf Period, dating from the fifth millennium B.C., was extremely limited. Only in late Halaf contexts at Tell Arpachiyah, were copper and lead pendants and fragments of pins found (Mallowan and Cruikshank Rose 1935, 103–4; Mellaart 1976, 166). A small copper bead was also discovered in the late Halaf stratum at Chagar Bazar (Mallowan 1936, 26). Beyond these isolated instances, there existed only an unconfirmed hypothesis regarding metallic prototypes for certain Halaf ceramic forms that was viewed with skepticism (Tobler 1950, 211). Even the Early Ubaid copper chisels and awls found in stratum XVII at Tepe Gawra were regarded by the excavator, A. Tobler, as rare exotics from a context that still preceded the "true" Chalcolithic. At Tepe Gawra, Tobler assigns stratum XIII dating from the first half of the fourth millennium B.C. as the first Chalcolithic level, though only one isolated copper awl was found within it. However, in the strata above XIII, copper objects occur in much greater numbers. The metal artifacts found within the Halaf strata at Tell Arpachiyah, by contrast, were considered intrusive by Tobler. Thus, he placed the entire Halaf phase within the Neolithic, and included the greater part of the succeeding Ubaid period within it as well (Tobler 1950, 211–13).

Mellaart, in contrast to Tobler, does not doubt the late-Halaf context of the Arpachiyah metal finds but considers them "a novelty of the period" that "seems strange in light of the considerably earlier use of copper at Çayönü, Çatal Hüyük, Ali Kosh, Mohammed Jaffar, and Tell es-Sawwan" (Mellaart 1975, 166).[4] In fact, more than a thousand years separate the Arpachiyah finds from the metal objects in the lowest stratum of Tell es-Sawwan, and there is more than a twenty-five hundred-year gap between the former and the metal artifacts from Çayönü. It must be reemphasized that Çayönü is unique in its assemblage of metal artifacts dating from the late eighth and early seventh millennia B.C. In the seventh and sixth millennia B.C., metal is known in regions adjacent to Mesopotamia, both to the west and the east, but not in Mesopotamia proper. However, there is no reason to believe that these outlying areas possessed more advanced technologies than the central part of the region. For this reason, the spatial and temporal gap in metal finds, suggesting a lag in metallurgical development within Mesopotamia proper seems contrary to expectation. Obviously, further research in the area was required in order to solve this problem.

The Soviet archaeological expedition in the Sinjar Valley has met with considerable success in providing such a solution. Metal artifacts were located in all three of the primary excavation sites: the pre-Hassuna settlement at Tell Sotto, which undoubtedly precedes Stratum 1a at Tell Hassuna, the Hassuna settlement at Yarim Tepe I, and the Halaf settlement at Yarim Tepe II. The metal finds have been quite numerous, including both artifacts and fragments of copper ore (malachite) of various sizes, from small to more substantial pieces.[5] In view of the special importance of these finds and the need to firmly establish their provenience, each find will be considered in detail below, with specific reference to its stratigraphic association.

Metal Finds at Tell Sotto

The cultural deposit at Tell Sotto reaches a depth of 3.80 m. Four building levels have been differentiated within it (Bader 1975, 104, chapter 3 this volume). Of these, the uppermost (first) level contains standard Hassuna pottery; the next (second) corresponds to the transition from archaic to standard Hassuna; the third level includes both the archaic Hassuna and an earlier period. The fourth and earliest level, which precedes all Hassuna sites known to date, has numerous parallels with the Umm Dabaghiyah material culture assemblage (Kirkbride 1972, 1973a, 1973b, 1974, 1975; Mellaart 1975, 135–41). Thus, the latter represents, in our view, the earliest agricultural settlement of the Jezira region, which we label the Tell Sotto culture. Three metal objects were encountered during the excavations of this site.

In square 10a-4 within Level III at Sotto, at a depth of 1.80 m below datum (mbd), a copper bead rolled from sheet metal and in an advanced stage of oxidation was found in the archaic stratum, overlain by structures of the second level.

In square 17c-1 of the fourth level, at a depth of 2.10–2.30 mbd, two copper beads were found, both heavily oxidized, which made the exact determination of their shape problematic, though they appear to be roughly spherical. They were part of a necklace accompanying infant burial number 22. The body and accompanying goods were placed in a crude ceramic vessel typical of the Tell Sotto assemblage, with supports at the bottom and two molded decorations around the rim. The burial is stratigraphically sealed by the upper floors of the fourth level, and these floors are in turn overlain by ash pits and remains of outbuildings from the third level.

Metal Finds at Yarim Tepe I

Twenty-seven metal objects were discovered over the course of the excavation. Metal objects were found in Levels 5 through 12 (basal layer). In three levels (7, 9, and 11) pieces of malachite were found together with artifacts quite possibly made from this ore. The specific finds were distributed among the levels as follows:

Level 5

Square 48a-4. At a depth of 3.20 mbd, a piece of copper ore (malachite) 3 cm × 2 cm in size, was found inside room 147 of building complex 10.

Level 6

Square 38d-3. At a depth of 4.0 mbd, a piece of copper ore was discovered in an undisturbed context, covered by the collapse of room 118 from Level 5.

Square 38d-1. At a depth of 4.0 mbd, three pieces of copper ore were discovered on the lower floor of a large open courtyard from the Level 6, covered by the remains of Level 5 buildings.

Level 7

Square 47a-3. At a depth of 4.20 mbd, a hollow copper awl, rolled from sheet metal of 0.8 mm thickness, with a length of 15 mm and a diameter of 3–6 mm (the point is flattened on one edge) was located near the wall of room 27 in an undisturbed context sealed by the floors and walls of room 124 in complex 15 from Level 6. Room 124, in turn, is overlain by the remains of buildings in complex 5 within Level 5.

Square 47a-3. At a depth of 4.36 mbd, a fragment of copper ore was found in a stratum extending underneath the foundation of complex 17, sealed by the wall collapse and intact floors of complex 15 from Level 5.

Square 47a-1. At a depth of 4.42 mbd, a piece of copper ore with the dimensions of 3 cm × 1.5 cm was discovered below the southern part of complex 17 in a layer which extended underneath the wall of the complex.

Level 8

Square 37b-1. At a depth of 4.74 mbd, a piece of copper ore was discovered in an undisturbed context overlain by buildings of complex 23 from Level 7.

Square 47a-4. At a depth of 4.60 mbd, a piece of copper ore was found in an undisturbed context covered by the foundation of complex 17 from Level 7.

Square 57d-1. At a depth of 4.52 mbd, a piece of copper ore was discovered in a stratum sealed by the floor of room 252 in complex 14 from Level 7.

Square 57a-1. At a depth of 4.91 mbd, a piece of copper ore was found on the floor of room 296 in complex 32, near the base of the walls, covered by a layer of ash and charcoal underlying complex 20.

Level 9

Square 57a-2. At a depth of 5.0 mbd, a piece of copper ore was located on a surface covered by a Level 9 layer that in turn was overlain by the collapse of Level 8 structures.

Square 47a-1. At a depth of 5.15 mbd, a large fragment of copper ore with the dimensions 4 cm × 3.5 cm was discovered under a series of undisturbed floors within complex 28 from Level 9, which underlies the foundation of complex 17 from level 7.

Square 47b-3. At a depth of 5.0 mbd, a piece of copper ore was located near the base of the wall of room 291 within complex 28 from Level 9 amid a pile of archaic Hassuna sherds. This was sealed by the floor and walls of room 195 within complex 19 from Level 8.

Level 10

Square 27a-2. At a depth of 5.21 mbd, part of a small copper ring made from braided copper wire and semicircular in cross-section (4 mm × 2.5 mm) was discovered. The diameter of the ring is slightly greater than 1 cm. The surface on which it was found was sealed by the undisturbed floors from Level 9 and further by structure 265 of Level 8.

Square 47b-2. At a depth of 5.40 mbd, a piece of copper ore was discovered in the collapse of complex 33 of Level 10, which is covered by complex 28 from Level 9.

Square 47b-2. At a depth of 5.35 mbd, a piece of copper ore was located near the base of a wall of complex 33.

Square 57a-1. At a depth of 5.08 mbd, a copper ore fragment was found at the bottom of the foundation of a granary in complex 36.

Square 27a-2. At a depth of 5.08 mbd, a piece of copper ore found in an undisturbed context within Level 10.

Level 11

Square 27c-1. At a depth of 5.62 mbd, an open-ended pendant of irregular form was discovered. It was made from braided copper wire and had a cross-sectional diameter of 1–1.5 mm and the overall dimensions of 9 mm × 10 mm. It was found inside room 318 from Level 11, sealed by a series of undisturbed floors which were, in turn, covered by the collapse of room 234 within complex 35 from Level 8.

Square 27d-1. At a depth of 5.60 mbd, a piece of copper ore was found under the massive wall of a Level 10 building.

Square 47b-4. At a depth of 5.61 mbd, a copper ore fragment was discovered near the base of a wall inside room 335 of complex 34 from Level 11. This room is overlain by remains of complex 33 from Level 10.

Square 47b-3. At a depth of 5.49 mbd, a piece of copper ore was found in the fill of room 329 within complex 34 from Level 11, under the foundations of a wall in complex 33 from Level 10.

Square 37a-3. At a depth of 5.50 mbd, a copper ore fragment was discovered near the base of a Level 11 wall and was sealed by a series of undisturbed floors.

Level 12

Square 27d-4. At a depth of 5.95 mbd, a perforated piece of copper ore, discoidal in shape, was encountered. The piece had a diameter of 7 mm and a thickness of 3 mm.

Square 47b-3. At a depth of 6.08 mbd, a bracelet made from lead wire and round in cross-section was discovered. It has a cross-sectional diameter of 7–8 mm and an overall diameter of 9.2–10 cm. The bracelet is open-ended, and the fragmentary ends are button-shaped, with the dimensions 12 mm × 10 mm. The bracelet was found at the base of the mound under a wall of room 345 from complex 35 with the impression of the bracelet plainly visible at the base of the wall.

Metal Finds at the Site of Yarim Tepe II

In the Halaf levels, 7–9, thirteen copper artifacts and pieces of copper ore were discovered during the excavation of the site. A number of these finds were overlain by the collapse layers of massive structures, thus their stratigraphic context is beyond question. The finds are distributed among the phases as follows:

Level 7

Square 18b-d. At a depth of 4.70 mbd, a round, copper bead, heavily oxidized, was found near one of the walls of structure 244. This structure was sealed by the floors of a large and multiroomed storage facility.

Square 28b. At a depth of 4.75 mbd, an elongated oval copper pendant, 7 mm × 10 mm, was discovered in a stratum under the floors of Level 6 structures.

Squares 18b and 19a. At a depth of 4.90–5.05 mbd, two pieces of copper ore were found in a fill of tholos 50, sealed by the floors of structure 168.

Square 29b. At a depth of 5.0 mbd, a copper fragment was encountered in an undisturbed context.

Square 24a. At a depth of 5.0 mbd, three pieces of copper ore were found at the bottom of a hearth covered by the wall collapse of Level 6 structures.

Level 8

Square 28b. At a depth of 6.00 mbd, a piece of copper ore was discovered in an undisturbed context overlain by Level 7 structures.

Level 9

Square 24a. At a depth of 6.50 mbd, two fragments of copper ore were encountered in a stratum sealed by the floors of Level 7 structures.

Square 23b. At a depth of 6.40 mbd, another piece of copper ore was found.

Square 28b. At a depth of 6.40 mbd, a small, copper seal-pendant, triangular in shape with the dimensions of 1.5 cm × 1.2 cm was discovered under a series of floors within tholos 67. On the outer surface of the pendant, parallel lines had been incised, and on the reverse side the remainder of a broken projection is visible.

It must be emphasized that Yarim Tepe I and II metal finds were not isolated occurrences, but were present in a number of successive occupation phases, which means that they cannot be considered incidental or intrusive. On the contrary, the evidence from these two sites attests to the regular use of metal throughout the Hassuna and early Halaf periods. It is also important to note that all metal finds connected with the lower half of the Yarim Tepe II cultural deposit date well before the final stage of the Halaf culture represented at Tell Arpachiyah.

Conclusion

On the whole, the facts at hand lead to the conclusion that the early agricultural population of northern Mesopotamia was well-acquainted with metal and certain skills connected with metal production throughout the Tell Sotto, Hassuna, and Halaf periods. The final phase of the Halaf culture found at Tell Arpachiyah spans an extremely long interval, as is indicated by the thickness of the associated strata.[6]

The extremely poor preservation of the organic remains from Tell Sotto and Umm Dabaghiyah have prevented us from obtaining radiocarbon dates from these sites. The only date from this early period was obtained from a sample found in stratum 15 at the site of Telul eth-Thalathat, located 40 km to the east of Tell Sotto, excavated by a Japanese expedition (Fukai, Hariuchi, and Matsutani 1970). This sample (TK-24) dated to 5,570 ± 120 B.C. (J. Sato, T. Sato, Otomori, and Suzuki 1969, 5–13). Kirkbride correctly associates stratum 15 at Telul eth-Thalathat with Umm Dabaghiyah and consequently with Tell Sotto (Kirkbride 1973b, 7). Mellaart (1975, 140) considers the TK-24 date reasonable, but only for the very end of the "Tell Sotto/Umm Dabighiyah culture."

The accuracy of the above radiocarbon date is supported, in our opinion, by the entire group of radiocarbon dates from the sixth millennium B.C. Mesopotamia that have now been systematically analyzed. Based on stratigraphic and typological criteria alone, the final phase of the Tell Sotto culture cannot be chronologically separated to any significant extent from the initial phases of the Hassuna, and consequently, of the Samarra cultures (Bader 1975, 108, chapter 3 this volume). The radiocarbon dates which we have obtained[7] allow us to tentatively date the beginning of these cultures to the middle of the sixth millennium B.C. This, in turn, would suggest that the terminal phase of the Tell Sotto culture dates to the second quarter of the sixth millennium B.C., perhaps closer to its end. However, the burial with copper beads from the basal level of Tell Sotto clearly precedes those phases in a stratigraphic sense and can with confidence be dated to the first half of the sixth millennium B.C., possibly even to its beginning. By contrast, the Hassuna period includes the middle and the second half of this millennium. Evidence for the use of metal is present for the period of development for both cultures thoughout the entire sixth millennium B.C.

A similar statement can be put forth for most of the fifth millennium B.C. The earliest radiocarbon date from this period was obtained from a sample found in the terminal layer of Level 8 at Yarim Tepe II (SOAN-1291: 4710 ± 40 B.C.) At this site, metal was also found in the Level 9 settlement, and the latest discovery of metal was in Level 3 (LE-1015: 4210 ± 130 B.C.).

Thus, the evidence from Tell Sotto and the two Yarim Tepe sites by no means suggests that metal was first introduced into the region during the late Halaf period. Rather, the data obtained by the Soviet expedition in northern Iraq have radically altered the previous scenario of its appearance and use in Mesopotamia proper, which puzzled Mellaart earlier with good reason. We can now speak with confidence of the presence of copper and lead in northern Mesopotamia at least by the beginning of the sixth millennium B.C., and of the uninterrupted use of these metals starting

Table 12.1: Chemical Analysis of Metal Finds at Yarim Tepe I and II

Site	Object	Laboratory Number	Cu	Sn	Pb	Zn
Yarim Tepe I	Ore	2483-p	Basic	–	–	4[a]
Same	Same	2484-p	Same	–	–	4
Same	Bracelet	16623	–	–	Basic	–
Same	Bit	16624	Basic	5	5	?
Yarim Tepe II	Bead	16625	Same	5	–	–
Same	Pendant	16626	Same	4	4	3
Same	Ore	16627	Same	–	4	3
Yarim Tepe I	Same	16628	Same	–	?	4
Same	Same	16629	Same	?	4	4
Same	Same	16630	Same	–	5	4
Same	Same	16631	Same	–	–	5
Same	Same	16632	Same	?	5	4
Same	Same	16633	Same	–	–	4
Same	Same	16634	Same	–	–	4
Same	Same	16635	Same	–	–	3
Tell es-Sawwan	Bead	13616	Same	5	5	5
Same	Same	13617	Same	6	–	4
Same	Ore	2482-p	1	5	–	4

[a]Indices for designation of the ore: *1*, 10%; *2*, 1%–10%; *3*, 0.1%–1%; *4*, 0.01%–0.1%; *5*, 0.001%–0.01%; *6*, ≤0.001%.

at least 1,500 years before the late Halaf period. In light of this evidence, both significant technological progress in metal production and processing in subsequent periods and the early appearance of specific terms dealing with metallurgy present in the earliest Mesopotamian written sources can now be more satisfactorily explained.[8]

The chemical analysis of the metal artifacts and ore samples from Yarim Tepe I and II were conducted by E. N. Chernykh at the analytical chemical laboratory of the Institute of Archaeology, Academy of Sciences, USSR (see Table 12.1). This analysis has led the investigators to make the following conclusions.[9]

Of the ten ore samples from Yarim Tepe I that have been analyzed to date, eight are derived from a copper deposit in which the presence of arsenic ranges from hundredths to tenths of a percent. Two samples were probably from a different ore deposit in which arsenic is absent, although this requires further verification.

The copper oxide in two artifacts, the awl from Level 7 and the ring from Level 11, was also analyzed. Only the metal used to make the ring was found to contain trace amounts of arsenic, measured in hundredths of a percent. The poor condition of the metal does not permit determination as to whether the objects were made from natural or smelted copper. The analysis does not contradict the hypothesis that the metal of the objects came from the same source as the ore found at the site.

The lead bracelet found at Yarim Tepe I weighed 107 grams and was made from an exceptionally pure metal in which only trace amounts of silver measured in thousandths of a percent could be identified.

It is important to note that the elemental analysis of two copper beads and a piece of ore found in the lowest stratum at Tell es-Sawwan yielded results comparable to those from Yarim Tepe I. The Tell es-Sawwan ore sample is chemically similar to the two samples that were apparently derived from a deposit lacking in arsenic.

Of the metal finds at the Halaf site of Yarim Tepe II, a bead from the Level 6 occupation and two pieces of copper ore from Level 7 were analyzed. The bead was made from chemically pure natural or smelted copper. The ores were obtained from several sources, judging by their chemical composition, and apparently from different deposits than those utilized by the inhabitants of Yarim Tepe I.[10]

Notes

1. The association of several of the metal artifacts at Tell es-Sawwan with the later occupation is confirmed by the results of the chemical analysis of these artifacts conducted by E. N. Chernykh in the laboratory of the Institute of Archaeology, Academy of Sciences, USSR, Moscow. This pertains especially to the copper bracelet, the metal of which differs markedly in its trace impurity composition from the early samples.

2. This association has now been called into question, since the excavations by the Soviet expedition at the Hassuna site of Yarim Tepe I indicate that the "earliest" stratum, contemporaneous with Hassuna 1a, is associated with the "painted pottery" period rather than the period immediately preceding its appearance.

3. It must be noted that this is not entirely accurate. An exhibit in the Baghdad Museum includes a drilled piece of copper ore from the Tell Hassuna excavation. However, this find was not mentioned in the published excavation reports.

4. It is unclear which find of the Mohammed Jaffar phase in Khuzistan the author has in mind.

5. The presence of malachite does not necessarily indicate metallurgical activities, since it could be used like other minerals. For example, a malachite fragment worked into a pendant was found at Shanidar in a burial dating from the eleventh millennium B.C. (Selimkhanov 1970, 11). Finds of malachite in the same strata with metal artifacts are more convincing proof of metal processing, particularly given the similar results of the chemical analysis for both ore and metal. In such cases, malachite may, with a certain degree of confidence, be considered as a raw material for the copper artifacts found.

6. If we sum together the strata of the Tell Sotto culture at the eponymous site, the Hassuna stratum of Yarim Tepe I, and the Halaf stratum of Yarim Tepe II, the total thickness of the deposits of the three cultures would be close to 16 m.

7. See chapter 8 of Munchaev and Merpert 1981.

8. Copper is designated by several terms in tablets from the Uruk IVA period (Falkenstein 1936, 14; Limet 1960, 30). In later times, the Sumerian terms both for copper and for lead are well known (Limet 1960, 30, 54, 55).

9. We express our deep gratitude to E. N. Chernykh, who provided us with the results of these studies.

10. The locations of metal sources in the Hassuna and Halaf periods remain an open question. In any event, no evidence exists for the importation of metal from distant deposits in the remote regions of the Arabian peninsula, which are considered primary sources of copper for Mesopotamia in the third millennium B.C. (Limet 1960, 99). The sources exploited on a limited scale during the sixth and fifth millennia B.C. were obviously small ore deposits.

13

SOME PROBLEMS IN THE ARCHAEOLOGY OF MESOPOTAMIA IN LIGHT OF RECENT RESEARCH BY THE SOVIET EXPEDITION TO IRAQ

R. M. Munchaev

In 1969, an expedition of the Institute of Archaeology, Academy of Sciences, USSR began conducting intensive archaeological research in northern Mesopotamia. For twelve seasons the expedition focused on the archaeological remains of the Sinjar Plain near Tell Afar in northern Iraq. Before the Soviet expedition, this part of Mesopotamia was the least-studied region of the Near East from an archaeological perspective, especially the prehistoric period (Munchaev and Merpert 1981, 8–9). Archaeological investigation of the region began about forty years ago (Lloyd 1938, 123–24) and was limited primarily to survey. These surveys revealed a high density of material culture remains in the region (J. Oates 1968, 3–4), from the seventh millennium B.C. onward. Long-term excavations were limited to two sites: the British School of Archaeology in Iraq conducted extensive excavations at Tell Rimah, and in 1967 an expedition from Tokyo University began excavations at Telul eth-Thalathat. This latter site contained strata dating to the Ubaid period and perhaps even earlier (Fukai, Hariuchi, and Matsutani 1970). The excavation of this site continued through the 1970s (Fukai and Matsutani 1981).

In an area of about 60 square kilometers stretching from Tell Afar to Sinjar, over four hundred tells have been located, indicating the significance of this region in the cultural-historical development of Mesopotamia. This density and the fact that the Sinjar Plain had not been previously investigated in any detail were the primary determinants in the selection of this region by the Soviet expedition for long-term and intensive research. A preliminary survey was conducted in the winter of 1968 by N. Ya. Merpert and N. O. Bader, staff members of the Institute of Archaeology, Academy of Sciences, USSR, of a group of six tells situated near the Djubar-Diariasi stream to the west of Tell Afar, known by the general name "Yarim Tepe." From this study it was established that three of the mounds (Yarim Tepe I, II, and III) were settlements of three early agricultural cultures of preliterate Mesopotamia: the Hassuna, the Halaf, and the Ubaid. A year later, in the spring of 1969, the Soviet expedition embarked on what was to be a long-term study of these sites.[1]

In the period from 1969 to 1976, the expedition conducted excavations at Yarim Tepe I (fifth and sixth millennia B.C.) (Merpert and Munchaev 1971a), and Yarim Tepe II (fifth millennium B.C.) (Merpert and Munchaev

1973) and in 1977–80, at Yarim Tepe III (fifth and fourth millennia B.C.). As a result of the systematic survey conducted by the expedition in the foothills and valleys of the Sinjar at least fifty sites from various periods were revealed, including Paleolithic campsites and the earliest sedentary agricultural settlements in the Sinjar Plain (pre-Hassuna). The latter include the sites of Tell Sotto (Bader 1978) and Kültepe (Bader 1982, 50–57), which were excavated 1971 and 1973–1976, and Tell Maghzaliyah (Munchaev and Bader 1979). The latter site is situated on the border between the valley and the foothills of the Sinjar. This site dates from the preceramic Neolithic (Bader 1979), and thus, is of enormous interest. The settlement is the earliest occupation excavated by the Soviet expedition in Mesopotamia.

Tell Maghzaliyah is situated on the right bank of the Abra River, northwest of the Yarim Tepe group. Only the western half of the settlement, with an area of about 4500 square meters, has been preserved. The depth of the cultural deposit at the site exceeds 8 m. It contains fifteen identifiable building levels (Bader 1979, 118–20). Excavations over an area of about 600 square meters have revealed the remains of numerous residential structures and associated outbuildings, part of a defensive wall, and eight burials. During the excavations at this site, a variety of archaeological material was recovered, including an extensive collection of faunal remains from both domestic and wild animals, and paleobotanical finds.

Tell Maghzaliyah is characterized throughout its settlement history by rectangular domestic architecture. House walls were erected on massive stone foundations and constructed with packed clay. The floors of the structures were paved with stone and clay and surfaced with plaster. The passageways between the houses and the outbuildings were also paved with plaster. The roofs of the houses were thatched with reeds and covered with plaster. Distributed around the dwellings, some of which had an area of 100 square meters and consisted of several rooms, were hearths, large rectangular ovens, storage pits, plaster vessels embedded in the ground, and other household-related installations. The site architectural plan was irregular and densely packed. An important feature of the settlement is a thick stone wall that bounds the site on one side, which in places is preserved to a height of two meters. The wall has been excavated over a length of 60 m (Munchaev and Merpert 1981). This is the earliest defensive structure known in Mesopotamia to date.

The archaeological material from Tell Maghzaliyah is varied. However, lithic, especially obsidian, tools predominate. Among them are blades for knives, sickles, and numerous scrapers, cutters, awls, and so forth (Bader 1979, Figs. 5–7, 9; see chapter 2a this volume). There are also a large number of leaf-shaped flint arrowheads made from knife-like flakes (Bader 1979, Fig. 7, 8–13, 20–23, see chapter 2a this volume). The ground stone assemblage includes small grinding stones and pestles. There are also quite a number of bone artifacts in the collection: awls made from fragments of hollow bones, fine needles, including some with eyes, and individual objects with incised ornamentation. The collection of stone vessels, bracelets, pendants, and beads, and also of various small stone, clay, and bone artifacts, is distinctive to the region. Among the latter, the examples of anthropomorphic and zoomorphic sculpture executed in schematic style are of significant interest. The finds of pieces of copper ore and a forged copper awl at Tell Maghzaliyah deserve special attention since they represent the earliest examples of metalworking in Mesopotamia. Finally, the paleobotanical sample includes grains of cultivated cereal plants (single-row, double-row, and dwarf wheat, and also multirow barley). Combined with the osteological remains of domesticated animals, these finds point to the initial phases of development of agriculture and animal husbandry in this region of the Near East.

Tell Maghzaliyah dates, according to preliminary data, from the end of the eighth or the first half of the seventh millennium B.C. This date is supported by similarities between the Maghzahliyah assemblage and assemblages from sites located in adjacent territories, including Abu Hureyra and other sites in the Euphrates basin in Syria; Jarmo, Tell Shimshara and Ali Kosh in Iraqi Kurdistan and Iran; and particularly Çayönü in southeastern Turkey. The sites we have mentioned, including Tell Maghzaliyah, may allow us to solve one of the key problems of the earliest stages of sedentary communities, that of the formation and subsequent development of food-producing economies in Mesopotamia and in the Near East as a whole. In this analysis, the settlement of Tell Maghzaliyah, located at the northern extremity of the Jezira along the major paths by which the settling of the Mesopotamia began, takes on special significance.

The next stage of the cultural-historical development of Mesopotamia, characterized by the wide dissemination of agriculture and animal husbandry, the settling of the Sinjar Plain proper, and the origins there of typical agricultural settlements is represented by the sites of Tell Sotto and Kültepe. These were small settlements, each with an area of less than a hectare, situated at 3 and 9 km respectively from the Yarim Tepe group. Of the two sites, the excavation of Tell Sotto was more extensive. The depth of cultural deposit there reaches 3.80 m. Four building levels are identifiable within it: the three upper ones belong to the Hassuna, and the fourth and lowest, dates from the pre-Hassuna period. In the latter, the remains of a large, multiroomed house made from sun-baked mud bricks, with a long corridor between the rooms, child burials in jars, storage vessels for grain, tub-shaped basins for water, and clay pots dug into the floor for containers were uncovered. Other archaeological finds include grinding stones; flint insets for sickles; small, wedge-shaped, roughly-polished stone axes; stone and ceramic vessels; bracelets; beads; bone awls and perforators; and also copper beads (Munchaev and Merpert 1981, 309–10). The pottery of Tell Sotto deserves special attention. In this assemblage are crude, thick-walled vessels in the form of simple jarlike pots and cups. Some of these are decorated with various applied motifs, including anthropomorphic and zoomorphic designs. In addition, a few vessels are decorated using red ochre. Isolated examples of thin-walled vessels of high quality, which are considered to have been imported, were also found here. Quite interesting are the finds of individual anthropomorphic and zoomorphic clay figurines, executed both realistically and in schematic style (Bader 1975, Figs. 6,7; see chapter 3 this volume).

Other sites in adjacent regions that have comparable assemblages to that of Tell Sotto include Kültepe, Tell Afar, Umm Dabaghiyah in the region of Khatry, which was excavated by a British expedition in the 1970s (Kirkbride 1972), and a number of sites in the Jezira. The assemblages from these sites, which characterize the initial stage of the settlement of the Sinjar region and the development of sedentary agricultural life here, are set apart as a unified cultural and temporal phase from the succeeding Hassuna culture. The distinctive character of the pottery from sites of the Tell Sotto-Kültepe type and its decorative style, which has a narrow geographic range, permits us to regard these sites as representing a new and previously unknown culture within Mesopotamia (Munchaev and Merpert 1973, 31–32; Kirkbride 1972, 15). We have proposed that it should be named the "Tell Sotto culture" (Munchaev and Merpert 1973, 32; 1981, 12, 300). According to preliminary data, this culture should be dated from the second half of the seventh and the beginning of the sixth millennia B.C.

The results of the expedition pertaining to the Hassuna culture, which reflects the stage of the cultural development in northern Mesopotamia following the Tell Sotto-Kültepe stage (the middle and second half of the sixth millennium B.C.) are quite important.

In this period, almost the entire Sinjar Plain and the Jezira were settled, and relatively large settlements appeared alongside small hamlets like Tell Sotto and Kültepe. The expedition studied Hassuna assemblages at three settlements: Yarim Tepe I, Yarim Tepe II, and Tell Sotto; it also located a number of other settlements containing materials of this culture. The most important work, in terms of scale, was done at Yarim Tepe I.

The settlement of Yarim Tepe I covers an area of about 2 hectares, and the cultural deposit exceeds 6 m in depth. The entire occupational sequence has been divided into twelve levels, all of which belong to the Hassuna culture (Munchaev and Merpert 1981, 18–20). The eastern half of the mound, with an area of about 1720 square meters, was the primary object of excavation. No other settlement of the Hassuna culture has been studied to such an extent. As a result, abundant and diverse data characterizing the architecture and building techniques, the development of agriculture and animal husbandry, and also pottery and other crafts were uncovered. No less than a thousand residential structures and outbuildings, fire installations of various types including pottery kilns, and a significant number of burials, have been studied. An impressive amount of archaeological material has been recovered, much of it ceramic. The latter includes characteristic examples of Samarra ware and also the richest collection of Hassuna anthropomorphic figurines known to date (Merpert and Munchaev 1973, Pl. XLIII; see chapter 6 this volume). There are numerous tools connected with agricultural activities, including stone hoes, grinding stones, pestles, and flint insets for sickles. Wedge-shaped, roughly-polished axes, projectile points, and pins were also found. It is important to mention the various small objects of stone, clay, and bone which have been recovered, including a small series of stone seals that are the the earliest examples known to date. Also, pendants and beads made from various kinds of stone, including turquoise, were exposed (Munchaev and Merpert 1981, Figs. 33–42).

The metal finds are of particular interest. They were found within various layers of the mound, beginning with the earliest phase. More than twenty pieces of copper ore, copper awls, a fragment of a pin, a pendant and a massive lead bracelet with a diameter of 9.2–10 cm were found (Merpert and Munchaev 1977). The discovery of copper artifacts, and also of malachite in the sites studied by the expedition, from the aceramic site of Tell Maghzaliyah through the Halaf settlement of Yarim Tepe II, permits us to shed considerable new light on the initial stages of the development of metallurgy in Mesopotamia. It was also established that in this area, as in the adjacent areas of Turkey and Iran, the use of metal and the development of metallurgy proper began approximately at the same time (the eighth to sixth millennia B.C.), which is significantly earlier than once thought.

Excavations at Yarim Tepe I, particularly of its earliest strata, which characterize the archaic Hassuna culture as well as the Tell Sotto-Kültepe culture that preceded it, have notably changed our conception of the development of the early agricultural villages of northern Mesopotamia and their distribution in this region of the Near East.

The expedition devoted no less attention to the study of the subsequent prehistoric stage represented by the well-known Halaf culture. At the same time, it should be emphasized that before the beginning of the Soviet expedition's investigations, only a few isolated sites of this culture (e.g., Tepe Gawra and Tell Arpachiyah) had been systematically studied, and in none of them did the excavations reach the initial occupation layers of the mounds.

The Soviet expedition to Iraq examined two Halaf settlements, Yarim Tepe II and Yarim Tepe III, and also a significant number of burials from the same period. The most extensive work was done at Yarim Tepe II. This mound, which gave its name to the whole Yarim Tepe group, is only half-preserved. At one time this site was apparently a circular mound with a diameter of 100–120 m and a height of 7 m (Munchaev and Merpert 1981, 161–63). The site is located on the right bank of the Djubar-Diariasi stream. The Halaf occupation is directly on top of a small Hassuna settlement. Five hundred square meters of the site have been excavated down to the basal occupation layer. Thus, Yarim Tepe II is the only Halaf site to have been exposed over such a considerable area down to the bottom of the cultural sequence. The stratigraphy of the site has been studied extensively. The mound contains nine building levels, reflecting all stages of development within the Halaf period in the Sinjar Plain, from the time of its appearance until its replacement by the Ubaid culture.

The spread and development of the Halaf culture, covering virtually the entire fifth millennium B.C., marked a new stage in the cultural-historical development of Mesopotamia. Research conducted by the Soviet expedition has established that the Sinjar Plain was the most important region and perhaps the center of origin of this culture, which is notably different in many attributes (e.g., form of architecture, pottery, sculpture) from previous cultures in the region.

The excavations at Yarim Tepe II have shown quite convincingly that a new architectural tradition appears in the Sinjar Plain in the fifth millennium B.C. At the very beginning of the Halaf, the round, one-room house with a domed or, more rarely, flat roof, to which outbuildings were attached on one or several sides, became the basic architectural form. At Yarim Tepe II the remains of about seventy tholos (beehive) structures have been excavated; these dwellings had diameters ranging from 3–5 m, and outbuildings with diameters of 1.5–2 m. The round structures, subdivided into a number of compartments by internal partitions, apparently served as granaries. Structures analogous to these have been uncovered in the Halaf stratum of Yarim Tepe III. Narrow, multichambered, rectangular structures, which probably also served as granaries, are also worth mentioning.

The Halaf culture is known for its distinctive painted pottery, characterized by advanced technological attributes that are illustrated in a variety of forms and decorative motifs. The latter are primarily geometric, though representational designs are also known. The ceramic assemblage recovered at the sites of Yarim Tepe II and Yarim Tepe III represent the largest collection of Halaf pottery to date. This corpus reflects the basic stages of development within the Halaf period. The form of the vessels and the decorative motifs characteristic of both the western (Euphrates) and the eastern (Tigris) groups of Halaf sites are equally represented in the Yarim Tepe ceramic assemblage. This assemblage also includes unique examples of anthropomorphic and zoomorphic pottery (Munchaev and Merpert 1981, 251–53). The discovery, at Yarim Tepe II, of the first pottery kilns of the Halaf culture, which are extremely complex, has permitted us to establish, finally, the type of kiln that Mesopotamian potters of the fifth millennium B.C. used to fire their ceramic vessels, and how such a sophisticated and painstaking heat treatment necessary to manufacture Halaf pottery was achieved (Bibikova 1981, 287–307). The collection of anthropomorphic and zoomorphic figurines found at Yarim Tepe II, and especially Yarim Tepe III, is also impressive. It includes examples of the kind of sculpture known up to this time as well as new forms, executed as a rule in a schematic style.

The remaining material from the Halaf sites excavated by the Soviet expedition is extensive and varied, including numerous paleobotanical and paleozoological finds (Bakhteev and Yanushevich in Merpert and Munchaev 1971b). Special mention should be made of the discovery at Yarim Tepe II of a number of curious cult objects and also of several dozen Halaf burials, which significantly exceeds the total number of burials from this period known before the beginning of Soviet excavations (Munchaev and Merpert 1981, 198–214). The first intensive analysis of the mortuary offerings and burial rituals of the Halaf culture suggests that the ideological realm of sedentary agriculturalists in northern Mesopotamia during this period was extremely complex (Merpert and Munchaev 1982a; chapter 10 this volume). Thus, the research conducted by the Soviet expedition in northern Iraq has shed considerably more light on the cultural-historical development of northern Mesopotamia in the fifth millennium B.C. Additionally, this expedition has made an important contribution to the complex, and to date, unsolved problem of the origin of the Halaf culture.

Finally, another important problem confronting Mesopotamian archaeology that has been and remains an object of study by the Soviet expedition to Iraq is the character of the Ubaid culture. This culture portends the beginning of Sumerian civilization and the origins of the earliest city-states in Mesopotamia. In contrast to the earlier Halaf and Hassuna cultures, the Ubaid culture embraced the entire territory of Mesopotamia, and its influence spread both east and west. At the same time, in the northern regions of Mesopotamia its development took on distinctive forms that have caused some investigators to speak of a separate Northern Ubaid culture. The questions of its development, its distinctive character, the relationship of the Ubaid culture with the preceding Halaf culture, the nature of the transition between them, and so on constitute some of the most interesting problems of the archaeology of preliterate Mesopotamia. These problems are being studied by the Soviet expedition in light of materials obtained from Ubaid strata at the site of Yarim Tepe III.

This site is about 6 hectares in area, almost adjoining Yarim Tepe II on its northeastern side. The depth of the cultural deposit of the mound has reached 12 m in places. The upper strata, to a depth of at least 4 m, date from the Ubaid period. The remains of eight large residential structures have been uncovered in several consecutive building levels. These houses were placed around the sides of open plazas. Each structure was multiroomed (both large and small rooms). The rooms were rectangular and interconnected by narrow passageways. Their walls were built of large mud bricks with an average measurement of 65 cm × 25 cm × 8 cm and mud-plastered on both sides (Merpert and Munchaev 1982b; see chapter 11 this volume). Comparative analysis suggests architectural parallels between Yarim Tepe III and such well-known sites of the Northern Ubaid culture as Tell Arpachiyah and particularly, Tepe Gawra. Other aspects of the archaeological record of these sites are also highly similar, especially the ceramic assemblage, which is identical in technological, morphological, and decorative attributes to the assemblage from Ubaid strata at Tepe Gawra and Tell Arpachiyah.

It is important to note the large quantity of ground stone tools connected with the milling of grain that were found in the Ubaid strata of Yarim Tepe III, including grinding stones, mortars, and pestles (a total of at least 150 of these have been found). Stone hoes, and weighted tips for digging sticks were also discovered, as well as a significant quantity of preserved grain, primarily wheat. The Ubaid strata of Yarim Tepe III, dating from the late fifth through early fourth millennia B.C., suggest that the Sinjar Plain was an integral part of the territory where the Northern Ubaid culture developed and was an extremely important avenue for its westward transmission.

Summarizing the results of the Soviet expedition to Iraq, we should report that, as a result of its work, the total archaeological database of the earliest periods in northern Mesopotamia has been increased substantially and that important new sites have been discovered, including some that characterize a previously unknown early agricultural horizon within Mesopotamia whose occupants probably initiated the settlement and widespread cultivation of the Mesopotamian plain. The Soviet expedition has studied a wide range of sites, embracing a broad chronological interval from the eighth to the fourth millennia B.C., that reflect specific stages of the cultural-historical development of preliterate northern Mesopotamia from the beginning of a food-producing economy to the establishment of Sumerian civilization. For this reason, it has now been established that the study of many of the key problems pertaining to the archaeology, ancient history, and culture of Mesopotamia and the Near East is now impossible without taking into account the results of investigations by the Soviet expedition to Iraq (J. Oates 1973, 153; Mellaart 1975; D. and J. Oates 1976; Lloyd 1976).

Note

1. Along with research workers of the Institute of Archaeology, Academy of Sciences, USSR (R. M. Munchaev, N. Ya. Merpert, N. O. Bader, V. A. Bashilov, V. I. Gulaev, A. I. Kuza, and S. S. Shirinskii), the following participated in the expedition: O. G. Bol'shakov (Leningrad Division of the Institute for Oriental Studies, Academy of Sciences, USSR), M. U. Yunisov (Institute of Ethnography, Academy of Sciences, USSR), O. M. Djaparidze (Tbilisi State University), and I. G. Narimanov (Institute of History, Academy of Sciences, Azerbaijan SSR).

14

MESOPOTAMIAN INTERACTION SPHERES

N. Yoffee

The Concept of *Interaction Sphere* and Its Appropriateness in Mesopotamian History and Prehistory

The term *interaction sphere* entered archaeological literature in 1964 in J. R. Caldwell's description of the Hopewell interaction sphere of eastern North America. For Caldwell, the specific problem was how to describe nominally distinct Middle Woodland assemblages (ca. 250 B.C.–A.D. 250) that also possessed significant similarities in the material residues of mortuary practices. He invented the term *interaction sphere* to denote that there were social, ideological, and trade connections among populations that shared thereby a restricted corpus of material cultural—in the Hopewell case of pipes, figurines, copper axes—namely, those objects associated with the interment of the honored dead. Struever and Houart (1972) observed that such characteristic Hopewell items not only reflected membership in a "burial cult," but also denoted the location of "transaction centers" through which the goods moved intraregionally.

The interaction sphere concept, as formulated by Caldwell, describes the condition in which those otherwise locally "autonomous" societies were also connected on a regional basis—that is, local social systems could be identified by distinctive settlement patterns in specific ecological or geographic circumstances, the practice of appropriate subsistence techniques, and the maintenance and reproduction of historically determined cultural ways and associated material culture. Nevertheless, the circulation of certain goods "bounded" these local systems within a large, regional or super-regional area. In order to perpetuate the flow of these goods, furthermore, a common code of values and beliefs, manifested in a shared corpus of symbols, was invented to facilitate the social interaction needed to exchange the goods. This common code, if not conceived by elites, soon became controlled by them. For Caldwell, thus, the formation of an interaction sphere had evolutionary implications: it connected distinct peoples above the local ties of kinship; it promoted the adoption of innovations and ideas among different people; and it increased the status of local elites and so formed the foundation from which more stratified societies could emerge (see Schortman and Urban 1987 for an assessment of why and in what historical context Caldwell invented the term *interaction sphere*).

The problem of inferring the existence of interaction spheres that connect local societies was also addressed by David Clarke, who strongly asserted that "cultures" in prehistory should not be confused with "anthropological" cultures (by which he meant local ethnolinguistic groups):

> The anthropologist looks at aspects of the social system of cultures [while] archaeologists . . . look at the material system of the same cultures [and find that] . . . the systems are not the same yet neither are they unconnected. Serious dangers await those who transfer observations about the one class of system to the other and yet it is important that the coupling between the different systems and their attributes should be . . . made explicit. . . . The archaeological entities reflect realities as important as those recognized by . . . other disciplines. . . . [They] are equally real . . . and simply different. (1978, 61, 369)

An interaction sphere, therefore, is a useful term in prehistory since it implies that certain material features found over a large area reflect a set of cultural relations that transcend localized nests of institutions and distinct peoples embedded within it (see Willey 1991 for a review of the term *horizonal styles* in Mesoamerican and Andean archaeology that has much of the force of the concept of interaction spheres; and see below in this chapter for a discussion of Willey's essay).

Although this chapter will explore the utility of the 'interaction sphere' concept in the "later Neolithic" of Mesopotamia, especially since the data unearthed by the Soviet expedition allow new interpretations of the Hassuna, Samarra, and Halaf periods, I begin with a brief look at the nature of social interactions in historic Mesopotamia. As historians of early Mesopotamia are aware, there is no Mesopotamia at all in the political sense of the term (Yoffee 1988a). Instead of a single Mesopotamian state, one is wont to speak of Mesopotamian civilization, the set of cultural traditions that overarched the large number of independent political states and the many ethnic and linguistic groups in the land. I shall briefly delineate the traditions that facilitated interactions among these various groups and which, indeed, made Mesopotamia Mesopotamia.

These defining traditions of Mesopotamian history, of course, have their roots in Mesopotamian prehistory. In this chapter I use the concept of interaction sphere in order to delineate some major organizational boundaries in Mesopotamia and to explain their evolution. If the term interaction sphere is to escape a certain inherent fuzziness (imposed by the two common nouns), however, it must denote certain kinds of behavior but not others, and the marked behavior must be palpable, not merely abstractly logical.

In particular, interaction sphere does not denote the sort of interactions P. L. Kohl has discussed as a West Asian "world-system" (1978). Kohl, in emending Wallerstein's idea of *world-systems* (1987a, 1987b), has shown that local developments in third millennium Mesopotamia, such as the production of certain goods for export (especially textiles) and the attendant social stratification associated with such changes in these industries, can only be understood when placed in an interregional perspective of organized trade that goes far beyond any conceivable borders of Mesopotamia. Although, as Kohl argues, the system of long-distance economic exchange can affect important aspects of the internal social and political structure in all societies participating in the long-distance trade, interaction sphere must refer to a more compact geographical area and it must, at the same time, point to a more encompassing set of behaviors than is considered the subject of world-systems theory. World-systems are by definition so geographically enormous that the core institutions that differentiate one society from another cannot be determined by contacts effected through long-distance trade. An interaction sphere, by contrast, refers exactly to the systematic, consistent, and normative set of activities that link people in such a way that a significant part of their identities are defined by such interactions.

At the other end of the scale of things that interaction sphere does not denote is the behavior of ethnic groups (contrast Schortman and Urban 1987; Schortman 1988). In historic

Mesopotamia, for example, ethnic and regional boundaries have been studied by Brinkman (on Kassites and Hurrians 1981), Dosch and Deller (on Kassites 1981), Charpin (on Hurrians 1977), Kamp and Yoffee (on Amorites 1980), Eph'al (on Arabs, Egyptians, and Jews 1978) among many others. Whereas these studies of ethnicity demonstrate that membership in an ethnic group might prescribe certain behaviors, the interaction of the many and varied ethnic and linguistic groups in Mesopotamia can only be explained by a more encompassing theory of social organization.

Interaction Spheres in Mesopotamian History

In this section I explore briefly (see Yoffee 1988a for a lengthier exposition) some salient historic cultural institutions that are "Mesopotamian" and consider certain mechanisms by which Mesopotamian culture is reproduced in historic periods. Although the discussion must be brief, this overview of selected Mesopotamian historic institutions is relevant, for it allows a certain structure to be imposed on the succeeding prehistoric investigation.

This historical lens on the past, however—if it has the virtue of not separating Mesopotamian evolutionary processes on the basis of one new data set (written documents)—also carries with it a certain theoretical liability: there is no necessity that the prehistoric past must constitute an unbroken cord to the historic present. That is, the process of change might result in dramatically new organizational forms from which an extrapolation backwards would represent a Procrustean bed in which the prehistoric past is condemned to resemble a form of the historic present. Nevertheless, this perspective from historic Mesopotamia does provide a correction to the historian's occasional view that prehistory is a long, largely incomprehensible, and immaterial backdrop to the events documented in cuneiform writing. Indeed, Mesopotamian histories often have the urgency of a von Däniken script: after a long period in which little of consequence occurred, vast building schemes and imperial adventures were undertaken by great men who transformed their world. This sort of historian's myopia is, of course, matched by the prehistorian's tendency to fold his or her tent and steal away at the dawn of history (after Kohl 1978, 475), as if prehistoric institutions were utterly different from those in historic periods.

Two significant issues of Mesopotamian historical interaction bear on the prehistoric investigation in this chapter. The first is that Mesopotamia is divided into two distinct regional entities, Assyria and Babylonia. The second is that within each region numerous named social groups coexisted as organizational units.

The political autonomy of Assyria and Babylonia is clearly seen in the distinct trajectories of their collapse in the mid first millennium B.C. As is well known, the Assyrian state had transformed itself from its decentralized structure in the early and middle second millennium to a more highly nucleated government in the late second millennium. Although this transformation occurred initially as a response against the hegemony of its neighbors (especially of the Hurrian state of Mitanni), in the first millennium Assyrian warrior kings came to dominate most of the territory from the Mediterranean to the Zagros Mountains. The policies of imperial expansion included disenfranchising the old-line Assyrian nobility in favor of new military officials and administrative bureaucracy, importing tens of thousands of foreign laborers who were employed on vast Assyrian estates, and founding new capitals that functioned as treasuries for the booty acquired by the Assyrian army (see Yoffee 1988a for references).

With the defeat of the Assyrian army by Medes and then by Babylonians (614–610 B.C.), the supply of foreign goods and services to the new Assyrian elite and Assyrian control of a largely non-Assyrian countryside ceased abruptly. Furthermore, absence of previous Assyrian institutions of urban and rural leadership, which had been systematically supplanted by the neo-Assyrian military kings, prevented any reformulation of an Assyrian state. The collapse of Babylonia was not as precipitous as that of Assyria. Nevertheless, after the conquest by Cyrus the Great of Persia in 539 B.C., the rulers of Babylonia no longer thought of themselves as Babylonian, and no Babylonian state ever again emerged. Differing from the Assyrian case, however, Babylonian culture (literature, customs, and language) did not disappear with the collapse of the Babylonian state. Embedded in Achaemenid, Seleucid, and other political systems, however, Babylonian culture became reduced to one among many competing ideologies and linguistic choices and, no longer providing any selective advantage in political or economic life, eventually atrophied and died. The last document written in the cuneiform script in A.D. 75 may serve, in a suitably Mesopotamian way, to mark the end point of Babylonian civilization.

If Assyriologists are well aware of the differences between Assyria and Babylonia, however, they also know that there are aspects of a profound interregional cultural boundary that spanned them both. On the linguistic level, whereas Assyrian and Babylonian are dialects of Akkadian, a "Standard Babylonian" literary dialect was created in which both Assyrian and Babylonian hymns and epics were written.

Indeed, the model of Mesopotamian cultural unity can best be seen in "the preservation, transmission, and revision of whole bodies of texts — or what Oppenheim has aptly called 'the stream of tradition' — which were assembled from all parts of [Mesopotamia] and over many centuries" (Machinist 1985, 185).

These texts include, signficantly, the corpus of god-lists, in which tutelary city-deities are grouped into a common pantheon and whose assembly is located in a Mesopotamian city — usually Nippur, later Bablyon. P. B. Machinist (whose argument this is) also notes that in the first millennium B.C., Assyrian relations with Babylonia not only dictated military intervention in Babylonian affairs, but also justified the removal of cultural treasures to Assyria. Machinist analyzes these actions as an Assyrian attempt to replace Babylon with Assur as the center of Mesopotamian culture (Machinist 1976).

If Mesopotamian culture is thus separate, both as unit of analysis and as tangible reality, from Mesopotamian political systems, a few words must be devoted to how it was learned and transmitted and why it was of social importance to its members. This subject is of particular significance since, from the earliest historic records, Mesopotamia, both in its regions and within the earliest city-states, consisted of many ethnic groups whose number, if anything, did not diminish throughout the three thousand years of Mesopotamian historical documentation.

By ethnicity I specifically mean (see Kamp and Yoffee 1980; Yoffee 1989) a mode of social incorporation in complex societies in which a number of individuals see themselves as alike by reference to a common or fictitious ancestry and who are so regarded by others. That is, ethnic identity involves the maintenance of a set of ideas and symbols by which claims may be made to solidarity or opposition in intergroup or interpersonal relations. This "interactional approach" (Barth 1969) defines ethnicity, not on the basis of a list of traits, such as language, territorial contiguity, biological descent, ecological adaptation, or political organization, but solely on the basis of self-ascription and ascription by others.

This interactional approach to ethnicity also allows for the existence of more than one type of social and economic status within the single, ethnically bounded group, and this is the case in the Mesopotamian examples. Amorites, to cite one case (Kamp and Yoffee 1980), were nomads, villagers, urbanites, peasants, and mayors. They could be loyal subjects of the Ur III state and also "tribally" organized in the countryside and a threat to the state because they were not under its control. After the collapse of Ur III (ca. 2000 B.C.), Amorites could compete successfully for political power (mainly with other Amorites) because of the variety of ethnic relationships that Amorites could mobilize, especially in the absence of a centralized opposition in city-states that had been subjects of the Ur III kings.

The history of Mesopotamia is replete with such examples of ethnic group behavior (see above cited articles for literature): ethnicity was a useful mechanism through which political interactions within and among city-states could be channeled. Furthermore, ethnic relationships, which transcended city-states and regions could not only accommodate, but also further the maintenance of overarching Mesopotamian cultural ideologies. Amorites and Kassites, having taken political power sequentially in Mesopotamia in the early and middle second millennium, used Mesopotamian cultural institutions (issued law-codes in the time-honored method [Yoffee 1988b], revived ancient languages, and patronized scribal schools to preserve and reinvigorate ancient literary forms [Lambert 1961]) precisely because, in doing so, they could legitimize their political gains. While maintaining aspects of their own ethnic orientations, they also were (or became) Mesopotamians.

To sum up this bird's-eye view of Mesopotamian historical interactions, it is seen that the major regional units of Assyria and Babylonia were participants in an overarching Mesopotamian civilization. Although a political ideology, for example, that depicted in the Sumerian King List (Michalowski 1983), existed to justify imperial aspirations of joining the regions, a unified Mesopotamian state never really existed. Within and beyond the city-states in Babylonia and Assyria, moreover, were the various ethnic groups, each with individual cultural identities, but which shared a common Mesopotamian culture. It is the task of the next section of this chapter to trace the prehistoric roots of this Mesopotamian culture.

Prehistoric Mesopotamian Interaction Spheres

The Later Neolithic in Light of Recent Finds of the Soviet Expedition

By the "later Neolithic" (that is, the Hassuna, Halaf, and Samarra periods) of Mesopotamia, as many studies have demonstrated, the economy of permanent villages was based on morphologically domesticated species of flora and fauna (see recent summaries by Harris 1989; Moore 1985; Oates 1973, 1980; Zohary 1989; Zohary and Hopf 1988). That is, by about 6000 B.C. (Aurenche, Evin, and Hours 1987), human beings had already intensified by existing technology the amount of labor required to maintain residential stability in and out of areas of naturally occurring and abundant cereal, animal, and lacustrine resources (Sherratt 1980a, 1980b; Bar-Yosef 1980; Bronson 1975). The sequence of sites from Nemrik (Kozlowski 1990a, 1990b), Qermez Dere (Watkins and Baird 1987), and Maghzaliyah (Bader, chapters 2 and 5 this volume) spans the late ninth through mid-seventh millennia and so now attests to a process of sedentism and food production occurring much earlier than had been suspected only a few years ago. Furthermore, the sites of Sotto and Kültepe, according to Bader, connect the "earlier Neolithic" material cultures of Maghzaliyah, Jarmo, and Umm Dabaghiyah to the archaic phase of the Hassuna (but see Merpert, chapter 7 this volume and Bader's own distinctions among the material cultures of the "earlier Neolithic" in chapter 5 this volume).

The intensification of labor in these "early Neolithic" villages can be inferred from elaborated storage facilities, permanent architecture, use of prodigious amounts of plaster (see Kingery, Vandiver, and Prickett 1988), and from traces of the plants and animals being exploited. Furthermore, this process of intensification led to new divisions of labor in the production and distribution of resources and to a new concern for the transmission of property (see Yoffee 1988c for this latter inference;

for notions of broadly similar ways of life in these early villages, see Voigt 1985, 1986; Cauvin 1988; Hole 1984; Watkins 1990). These profound social and economic changes are reflected in the large size and diversified architecture, including fortifications (such as at Maghzaliyah), of Neolithic sites. These changes in social organization are here briefly followed through the Hassuna, Samarra, and Halaf phases of the "later Neolithic."

During this period trends towards further social and economic differentiation occur significantly through the medium of regional forms of interaction. Although the sites of the "later Neolithic" are scarcely larger than those of the "earlier Neolithic," it is the formation of new geographic scales of social and economic behavior that have evolutionary significance. These regional "interaction spheres" are then pursued through the subsequent organizational restructuring in the Ubaid period and political and demographic transition of the Uruk. The purpose of these few remarks, of course, is not to provide an in-depth account of evolutionary trends in Mesopotamia; but it is considered that certain aspects of the prehistoric material record constituted formative stages, and not simply in a chronological sense, in the development of Mesopotamian states and civilization.

The Hassuna, Samarra, and Halaf phases (for descriptions and dates see Mellaart 1975; Lloyd 1976; Nissen 1988; Porada 1965; Aurenche, Evin, and Hours 1987) were first known as, and are still basically typed with reference to decorated ceramic assemblages. At Tell Hassuna, the type site for the Hassuna phase, the stratigraphic profile of the ceramics (Lloyd and Safar 1945; Perkins 1949) indicates the following stylistic overlaps: Hassuna wares are the earliest, but are partly contemporary with Samarra wares, which are in turn earlier than, but partly contemporary with Halaf wares. At Yarim Tepe I, the most important Hassuna site in terms of the architectural elaboration and site structure (chapters 6 and 7 of this volume, Munchaev and Merpert 1981), the characteristic Hassuna ceramics of the early levels are mixed with Samarra wares in the upper ones. The Halaf settlement of Yarim Tepe II, was founded upon a small Hassuna village, and the cemetery of Yarim Tepe II was placed on the top of the abandoned village of Yarim Tepe I (chapter 10 this volume).

Samarra wares are found in sites of central Mesopotamia (see J. Oates 1973, 1976, 1978, 1980 for summaries of the sites), the most important of which are Tell es-Sawwan and Choga Mami. Although the stratigraphy and plans of the former are difficult to unravel, the number of large buildings, the rich infant burials under impressive structures, and a possible fortification system all point to a socially and economically stratified society. At Choga Mami were found irrigation ditches prepared by exploiting the difference in altitude between two roughly parallel streams, and at Sawwan botanical remains imply that irrigation was being practiced (Helbaek 1964).

Joan Oates has written (1973) that the Samarra culture may be considered an adaptation to the needs of irrigation agriculture in central Mesopotamia. Hassuna, on the other hand, reflects an adaptation to the northern dry-farming zone. It is, of course, not implied that the decorated pots themselves, which form the basis for these "cultural" distinctions, were ecologically adaptive, but that the people living in each zone interacted more intensively with their neighbors than they did with those living in the other zone.

The distribution of Halaf ceramics, which overlap in time the Samarra wares but continue later than Samarra examples, barely reaches the Mesopotamian alluvial plain (see Fig. 14.1). Rather, Halaf pottery, often described as the finest prehistoric ware in Mesopotamia, is found from Lake Van in the north (and, in fact, Halaf ceramics are found in Transcaucasia according to P. L. Kohl), to the Mediterranean in the west, to beyond the Tigris in the east (Watson 1983). Remarkably, over this wide expanse, "Halaf [has been considered] one of the most homogenous prehistoric cultures anywhere in the world" (Le

Blanc and Watson 1973, 132; but see Watson 1983 for an assessment of local variations within other aspects of the Halaf assemblage, including *tholoi*, figurine styles, and stone "stamp seals").

Prior to the discovery of Yarim Tepe II, the single most important Halaf site was Arpachiyah (Mallowan and Cruikshank Rose 1935; Hijara 1978, 1980). In the center of the site, in level TT-6, stands the largest building exposed at Arpachiyah, the "potter's workshop." This structure was so named because of the amount and quality of Halaf polychrome ceramics that were stacked in several rooms. In connection with these pots were found flint and obsidian tools, red ochre and other pigments, painters' palettes, and bone implements for trimming and burnishing clay. Near the house were several kilns. Mallowan considered that this structure was a specialized activity location for the production of high-quality ceramics for the massive site of Nineveh, which presumably dominated Arpachiyah in the Halaf period but whose Halaf levels have only been sounded.

In addition to the potter's tools and the finished pots themselves, this large structure also contained turquoise and shell jewelry, stone vessels, and amulets that together seem to reflect the high status of the resident(s). These finds along with the house's central location and impressive size do not lead one to suppose that the resident was a simple craftsman. Rather, the house is presumably that of an elite (or elite group), someone involved in the production and distribution of high-quality Halaf ceramics.

The neutron-activation studies of Davidson and McKerrell (1976, 1980; Frankel 1979; Watkins and Campbell 1987) indicate that high-quality Halaf ceramics were produced in a limited number of centers and were circulated to the hinterlands of those centers (although Davidson and McKerrell's statistical significance level for similarity of ceramic composition among sites is not high [50%], thus calling into question this conclusion [Gil Stein, personal communication, from James Blackman's communication to him]). The stylistic unity of the design motifs that, as shown in LeBlanc's investigation (1971), extends over the huge territory regarded as Halafian (Fig. 14.1) is thus the more remarkable.

One plausible interpretation that accounts for these data is that production and distribution of Halaf ceramics was "controlled" by emerging local elites who maintained regular networks of communication and were bound by particular sets of interests. This interpretation, which is borrowed from Flannery's hypothesis of exchange between elites in the Olmec and Oaxaca region of southern Mexico (Flannery 1968), has an economic rationale: in the Halaf, the circulation of Anatolian obsidian, copper, turquoise, and other materials was facilitated by a system of symbols that provided a shared "cultural" bond.

Interaction Spheres

James Mellaart has tended to equate prehistoric Mesopotamian material cultures with distinct peoples, their customs, and even their languages (Mellaart 1975). According to Mellaart, Halafians were northerners, originally at home in central and eastern Anatolia, controllers of the supply of obsidian localized there. The extent of the Halaf is due to their southward migration from their Anatolian homeland. This view is supported by Munchaev and Merpert (1981), who point to the abandonment of Yarim Tepe I (a Hassuna site) and the founding of Yarim Tepe II (a Halaf site) as well as the establishment of a Halaf cemetery atop Yarim Tepe I. Furthermore, it is held in this theory of demographic or ethnic replacement that the rectangular house form of the Hassuna is supplanted by the round (*tholos*) form of the Halaf. An alternative explanation for culture change in the "later Neolithic" is presented here.

Using the concept of interaction spheres, it is proposed that the Hassuna and Samarra may each be recognized as a geographically localized bundle of interactions that "united"

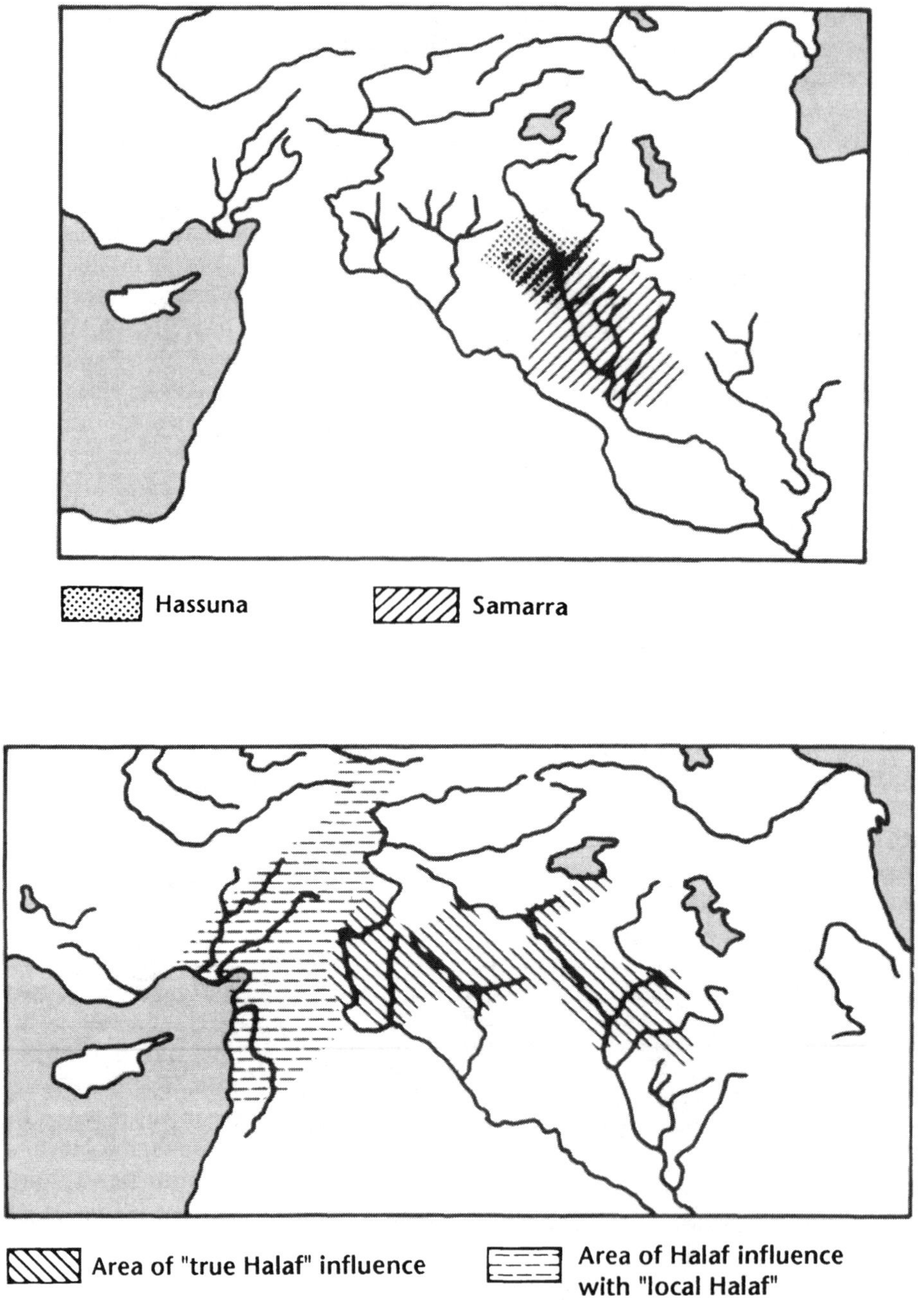

Figure 14.1 Extent of Halaf, Samarra, and Hassuna material culture. Top: Rough indication of Hassuna and Samarra "interaction spheres" (After Mellaart, *The Neolithic of the Near East,* 1975). Bottom: Rough indication of Halaf "interaction sphere" (After Mellaart, *Earliest Civilizations of the Near East,* 1965).

their regions. Each set of interactions, which included exchanges of goods and (presumably) marital partners, is discerned in the shared corpus of pottery motifs in each region; it is inferred that each of these regionally distinctive ceramic repertories represent shared values and a shared history. When a Samarra site is found in Syria (Baghouz) or Samarra ware is found in northern Mesopotamia, the appropriate conclusion is, not that irrigation agriculture (as is identified in Samarra sites in central Mesopotamia) was imported to these regions, but that central Mesopotamians were (most likely) trading in those areas. In any case, "central Mesopotamia" is obviously not a sharply bounded geographic enclave, and movement of peoples between regions is perfectly expectable. "Overlaps" in the ceramic stratigraphy between Hassuna and Samarra simply indicate communication between the regions.

The Halaf, to the contrary, reflects neither an adaptation to a specific environment, an ethnolinguistic group (which seems precluded in that no documented group of people in West Asia sharing a common culture has ever exhibited the uniformity of behavior over the extent of territory in which Halaf is found), or a relatively tight geographic bundle of regional interactions. (Although this argument does not preclude northern influence into Mesopotamia, it rejects the interpretation that the "meaning" of Halaf is that a new group of people invaded and "displaced" an older one, after which the new group became displaced. In excavations at Tell Sabi Abyad, a Halaf site in northern Syria, the excavators report that "earlier Neolithic occupation levels immediately preced[ed] the occurrence of Halaf at the site. No gap in occupation is indicated, and Halaf at Sabi Abyad was apparently the result of a gradual and continuous local process of cultural change" [Akkermans and Le Mière 1992, 3]). Rather, the Halaf may be described as an interregional sphere of interaction. The uniformity of Halaf motifs coupled with the circulation of pottery from specialized production sites implies that local elites were attempting to control the production not only of the pots but also of the important symbols — the metalinguistic code — that facilitated the exchange of goods over long distances and across many local ethnic boundaries.

To summarize the argument thus far, it is posited that two kinds of interactions characterize the later phase of the Neolithic in Mesopotamia. In the first, regional bundles of interaction account for the development of the material cultures we call Hassuna and Samarra, in north and middle Mesopotamia, respectively. In the second, a wide-scale, interregional exchange sphere can be inferred in the Halaf in which local elites sought to control the circulation of goods across vast geographical and social distances.

The Formative Stages of Mesopotamian Historic Civilization

The Ubaid has been described as being the first and last unified material culture assemblage in Mesopotamia (e.g., Porada 1965; see Oates 1983, 1987; Jasim 1985, 1989). Both in the northern region and, for the first time, in the south similar ceramic types and, significantly, temple architecture (at the major sites of Gawra and Eridu) seem to reflect a cultural unity (described in authors from Perkins 1957 to Stein 1991). Not only is Ubaid found in the northern Mesopotamian plain and in southern Mesopotamia, but it is also present in Syria (Simpson 1988), in Iran (Henrickson and Thuesen, eds. 1989), and in a number of sites along the eastern Arabian coast (Masry 1974; Frifelt 1989). Based on petrographic analysis (Oates et al. 1976), it is thought that the Arabian Ubaid ceramics were made in Mesopotamia and brought to this area by Mesopotamian traders.

Although C. Redman (1978) once considered that the southern Ubaid was a creation of migrations from the north into an unoccupied niche, the recent evidence of pre-Ubaid sites in the alluvium (Adams 1981; Huot 1987a,

1987b; 1989) and the connection of the early Ubaid wares to a "transitional" Samarra type (actually Ubaid 0 ceramics at Oueili predate the Choga Mami transitional material) shows that the Ubaid was an indigenous development in the south. Indeed, the direction of cultural impact during the Ubaid seems to have been south-to-north. Ubaid 1 ceramics are not found in the north, e.g., at Yarim Tepe II, and the characteristic temple architecture of the Ubaid appears early at Eridu whereas it is found only at the end of the northern Ubaid at Gawra. The "cultural unity" of the Ubaid, thus, seems due to two interrelated processes: first, the transmission of domesticated plants and animals from the north to the south and second, a cultural expansion from the burgeoning southern villages, which were based on flourishing agricultural systems, into the north.

The leading indication of cultural unity in the northern and southern Ubaid alike is the remarkably similar temple plans at Eridu and Gawra (see Heinrich 1982 for diagrams) and the presumed commonality of beliefs and values throughout Mesopotamia. Differing from the Halaf, which was interpreted as an interregional "economic interaction sphere," the Ubaid may be posited as a "cultural interaction sphere"; indeed, in the Ubaid may be found the embryo from which the historic Mesopotamian cultural boundary uniting Assyria and Babylonia was developed.

Although the northern region (Assyria) fully shares in historic Mesopotamian cultural institutions, it is apparent that in historic times the cultural center of Mesopotamia lay in the south. The roots of this southern nucleus (too casually called "Sumerian civilization" after the language in which the earliest documents were written) can perhaps be found in the succeeding Uruk period. Although it is a key period in understanding the evolution of Mesopotamian states, only one site has yielded any significant stratigraphic sequence (the type site itself) and associated architecture and artifactual assemblage.

If the Uruk period begins with a small change in ceramic decoration (or the lack of it) and ends with the appearance of the first written documents, and if one notes a continuity in the ceramic technology and in the plans of temples, the scale of social organizational change in Uruk is in fact hardly predictable from the foregoing Ubaid. The massive temples of Uruk Levels V and IV with impressive sunken courtyards, pillared terraces, and a mighty acropolis of ceremonial buildings (some of which are as large as most Ubaid period sites) are testimony to the newly centralized political and economic activity in Uruk period urban sites.

In these Uruk conurbations massive surpluses were produced on vast estates that employed teams of workers, managers, and craft specialists. At this time writing was invented (Nissen 1988; Michalowski 1990) to keep track of this unprecedented economic differentiation and centralized direction in the great organizations of temples and palaces. This was not only a process of urbanization, but also one of ruralization, since the countryside was progressively depopulated and itself restructured according to the demands of the new city-states (Adams 1981; Nissen 1988).

The massive new division of labor in these city-states was the precondition for the establishment of southern Mesopotamian colonies up the Euphrates (Strommenger 1980; Sürenhagen 1986; Algaze 1989). This political interaction at the end of the Uruk was prefigured, of course, by the long-term knowledge of this area and its resources that can be traced into the Neolithic. The political expansion failed, however, shortly after the end of the Uruk as the colonies were abandoned. The Uruk implosion created the possibility of quick political domination of foreign regions (perhaps) by (individual) southern Mesopotamian city-states (since no political unity of southern Mesopotamia preceded that of Sargon in the twenty-fourth century B.C.). Having established hegemony over these regions, however, it was quite another matter to perpetuate control many hundreds of miles from the south, and it is no surprise that the colonies quickly disappeared.

The urban implosion of the Uruk, then, sets the pattern for interaction among the city-states in southern Mesopotamia as independent religious centers, as defensive centers against other competing city-states, and as launching pads for both local and far-flung political adventures. The division of labor and extreme social stratification in the city-states further created demand centers for foreign goods, the production of goods for exchange, and the capitalization of trade missions to distant lands. Resulting from this evolutionary "phase-transition," individual city-states became the characteristic arenas of political, social, and economic activity in southern Mesopotamia for the next three millennia.

Summary

In few areas of the world can the development—and demise—of a civilization be traced as thoroughly as it can in Mesopotamia. This is due to the ability of prehistorians to structure inquiries of social change with the consideration of the "outcomes" of such changes as historic Mesopotamian institutions. Similarly, those working in historic periods can better understand the context of social institutions by reference to the development of social and economic interactions that are depicted in the Mesopotamian prehistoric record.

In this essay into the genre of the ultra *longue durée,* I have explored the possibility that the roots of the Assyrian-Babylonian regional dichotomy characteristic of Mesopotamian historical times can be traced to the "later Neolithic" period of Mesopotamian prehistory. Bundles of northern and southern spheres of interaction, in which people in each region effected social and economic relations more intensively within, rather than across, geographic zones seem established in the sixth millennium B.C. Interaction between these zones, however, was not unimportant or infrequent and was based both on the environmental complementarity (and so the exchange of resources) and also on the long-term social interconnections of the people in those areas. In the Ubaid, such regional interaction was reflected in the formation of a belief system (clear in the common pantheon that was preserved, adapted, and transmitted over generations in the historic period) and a set of values and cultural institutions that bound them together against an external world order. In the Uruk period the process of urbanization and ruralization in the southern city-states included the invention of writing, foreign political adventures, and not least important, the development of new cultural institutions that made the south a central place in the ideological sphere that is Mesopotamian civilization.

The recent work of the Soviet expedition to northern Iraq has played an important role in understanding the nature of the early stages of this historical durée. The history of permanent agricultural villages on the northern plain has been pushed back a millennium before the Hassuna period. Now, from the recent discoveries at Nemrik and Qermez Dere, it seems that the process of agricultural development and of village life can be followed from the middle of the ninth millennium.

The nature of social organization in the Hassuna, Halaf, and northern Ubaid sites at Yarim Tepe has also been advanced by the work of the Soviet expedition. In this essay I have argued that such social and economic changes as are observed in these periods can best be evaluated within a consideration of the boundaries, scale, and meaning of the named periods. That is, trends towards social differentiation are not reflected predominantly in increased size of sites in these periods nor in any clear evidence of economic stratification (see Akkermans 1989 for a recent review of the mortuary data). It is the scale of interaction in which the sites are constituent parts that is a key factor in understanding the transformation of the moral economies of village societies into the political economies of Mesopotamian states.

The concept of interaction spheres used in this chapter is at variance with the theories of ethnic replacement that some have employed to explain the transition between cultural phases in the "later Neolithic." It must be emphasized that the concept of interaction spheres does not preclude the investigation of ethnicity in prehistory. Indeed, it advances such investigations in two distinct ways. First, the distinctive architecture, pottery, and (perhaps) burial practices of the Halaf culture appear related to the geographic extent of the Halaf that included northern Mesopotamia. Rather than seeing the Halaf as representing a large-scale ethnic replacement or demographic shift of people in northern Mesopotamia, however, it can be posited that whatever influence came into Mesopotamia from the northern Halaf region (Armenia, especially Anatolia, and northern Syria) was incorporated into a larger sphere of interregional interaction (see also Breniquet 1987). The Halaf interaction sphere thus included many ethnic groups whose elites became connected, new symbols and cultural values that were forged to promote such interactions, and new wealth and status that were created by the manipulation of such symbols and the attempts to control the flow of goods by elites.

Second, it is argued that ethnic groups are not "natural" social entities, but that ethnic groups have their origins in particular periods of time and in the context of newly emerging spheres of interaction. The concept of interaction spheres has social evolutionary significance, thus, for it specifies the conditions in which various social groups become ethnic groups by virtue of their members' self-identification (as descended from a common ancestor) and their regular interaction with other such groups that recognize that self-identification. Within interaction spheres, therefore, formerly autonomous (or nearly autonomous) local groups are transformed into ethnic groups through their participation in societies in which ethnicity, territoriality, occupation, and status provide diverse and overlapping functions for their individual members. It is not ethnic groups that promote interaction, thus, but it is through interaction spheres that ethnicity can be constructed.

Finally, the social evolutionary significance of the formation of interaction spheres in prehistory has recently been stressed by Willey in his discussion of horizonal integration in the New World and by Chang (1986) for ancient China. Willey's view is that such regional integration is the product of ideological factors and that the earliest such horizonal styles, in Mesoamerica and the Andes, the Olmec and Chavin, were not reflections of political domination but of cultural interconnection. These interaction spheres (though Willey does not use the term) speed the transmission of ideas and make possible the creation of dependencies not based on kinship, but on economic and political power.

For Chang (1986, 234–45, 409–11), the Lungshanoid horizon is an interaction sphere in that it cuts across—but does not eliminate—localized regional cultures, and within its boundary the many commonalities in material culture reflect interlinkages among these societies. From this "later Neolithic" interaction sphere, the three overlapping dynasties of the Sandai period emerge as political cores, foci of hegemonic competition, but also through which goods and ideas were circulated. Later political unification of China followed the establishment of this Chinese interaction sphere.

In Mesopotamia, as has been seen in this brief essay, the early stages in the evolution of Mesopotamian states and civilization are those in which interaction spheres are formed. Social and economic change in the "later Neolithic" and then in the Ubaid cannot be seen in increased village size, the formation of economic or political central places, or in the superfluity of wealth controlled by nascent elites. Rather, the organizational restructuring in these modest villages of the "later Neolithic" and Ubaid can be perceived in the formation of spheres of interaction in which sites, social stratification, and the creation of networks of power and new sorts of behavior are rendered possible, while others are less possible or impossible (after Wolf 1990, 586–87). If the enormity of change in the Uruk seems so unpredictable from the deceptively modest material restructurings of the preceding periods, the Uruk is also unimaginable without the development of those spheres of interaction.

Of course, the study of Mesopotamian interaction spheres does not supplant investigations of technology, subsistence, and local social systems. Knowledge of those topics alone, however, cannot lead to understanding the formation of cultural boundaries — how people define themselves, how their identities are cobbled together, and how they interact in socially constructed spaces in which spaces are also containers of power (Watts 1992, 123). That is what "Mesopotamia" is to us and what was culturally reproduced in the past.

Select Bibliography

(Note: In addition to references cited in this volume, additional references to publications of the Soviet Expedition are included. This is not, however, an exhaustive bibliography nor were exact page numbers and bibliographic details of obscure publications available to the editor).

al-Adami, K.
1968 Excavations at Tell es-Sawwan. *Sumer* 24:54–94.

al-Adami, K. A.
1968 Excavations at Tell es-Sawwan (Second Season). *Sumer* 24:57–60.

Adams, R. McC.
1981 *Heartland of Cities*. Chicago: University of Chicago Press.

Akkermans, P., M. van Loon, J. J. Roodenberg, and M. Waterbolk
1982 The 1976–1977 Excavations at Tell Bouqras. *Annales Archéologiques Arabes Syriennes* 32:45–57.

Akkermans, P.M.M.G.
1989 Halaf Mortuary Practices: A Survey. In *To the Euphrates and Beyond: Archaeological Studies in Honour of Maurits N. van Loon*, O.M.C. Haex, H. H. Curvers, and P.M.M.G. Akkermans eds., 75–88. Rotterdam: A. A. Balkema.

Akkermans, P.M.M.G. and Marie Le Mière
1992 The 1988 Excavations at Tell Sabi Abyad, a Later Neolithic Village in Northern Syria. *American Journal of Archaeology* 96:1–22.

Algaze, G.
1989 The Uruk Expansion: Cross-Cultural Exchange in Early Mesopotamian Civilization. *Current Anthropology* 30:571–608.

Antonova, E. V.
1974 *Antropomorfnaya skul'ptura drevnix zemledel'tsev Perednej i Srednej Azii*. Moscow: Nauka.

Aurenche, O., J. Evin, and F. Hours, eds.
1987 *Chronologies du Proche Orient*. Oxford: British Archaeological Reports.

Bader, N. O.
1975 Rannezemledel'cheskoe poselenie Tell' Sotto (po raskopkam 1971, 1973–1974 gg. (The early agricultural settlement of Tell Sotto according to the excavations of 1971, 1973–74, chapter 3 this volume). *Sovetskaya Arxeologiya*, 117–32.

1979 Tell Maghzaliyah: ranneneoliticheskii pamyatnik na severe Irake (Tell Maghzaliyah: an early Neolithic site in northern Iraq, chapter 2a this volume). *Sovetskaya Arxeologiya*, 99–111.

1982 Nekatorye rezul'taty rabot na rannezemledel'skom poselenii Kyultepe v severnom Irake (Some results of the work on the early agricultural settlement of Kültepe in northern Iraq, chapter 4 this volume). In *Arxeologiya Starogo i Novogo Sveta*, V. I. Gulaev ed., 50–57. Moscow: Nauka.

1983 Sources of the Hassuna Culture and the Development of Agriculture in the Sinjar Plain. Paper delivered at the First USA-USSR Archaeological Exchange, Cambridge, MA.

1984 Pervonachal'noe stanovlenie zemledeliya i skotovodsva v Severnoj Mesopotamii (The initial formation of agriculture and animal husbandry in northern Mesopotamia). *Kratkie Soobshcheniya Instituta Arxeologii* 180:60–63.

1989 *Drevenejshie Zemledel'tsy Severnoj Mesopotamii* (Earliest farmers of northern Mesopotamia). Moscow: Nauka.

Bader, N. O., N. Ya. Merpert, and R. M. Munchaev

1981 The Soviet Expedition's Surveys in the Sinjar Valley, 1977. *Sumer* 37:55–95.

Barth, F.

1969 Introduction. In *Ethnic Groups and Boundaries,* F. Barth ed., 9–38. Boston: Little, Brown.

Bar-Yosef, O.

1980 Prehistory of the Levant. *Annual Review of Anthropology* 9:101–33.

Bakhteyev, F. Kh. and Z. V. Yanushevich

1980 Discoveries of Cultivated Plants in the Early Farming Settlements of Yarim Tepe I and Yarim Tepe II in Northern Iraq. *Journal of Archaeological Science* 7:167– 178.

Bashilov, V. A. and A. V. Kuza

1972 *Drevnie Tsivilizatsii Peru/Bolivii.* Moscow. Nauka.

Bashilov, V. A., O. G. Bol'shakov, and A. V. Kuza

1980 The Earliest Strata of Yarim Tepe I. *Sumer* 36:43–64.

Bernbeck, Reinhard

1990 *Auflösungsprozesse der haüslichen Produktionsweise. Das Beispiel Mesopotamiens.* Ph.D. diss., Freie Universität Berlin.

Bibikova, V. I.

1981 Zhivotnovodstvo v severnoj Mesopotamii v V. tysyacheletii do n.e. (po materialam Xalafskogo poseleniya Yarim Tepe II). Appendix 2 (Animal husbandry in northern Mesopotamia in the fifth millennium B.C. [according to data of the Halaf settlement Yarim Tepe II]); in *Rannezemledel'cheskie Poseleniya Severnoj Mesopotamii,* R. M. Munchaev and N. Ya. Merpert eds., 299–307. Moscow: Nauka.

Bogoslovskaya, P. F.

1972 K probleme slozhniya Xalafskoj kul'tury. *Sovetskaya Arxeologiya* no. 2:3–46.

Bökönyi, S.

1973 The Fauna of Umm Dabaghiyah: A Preliminary Report. *Iraq* 35:9–11.

Braidwood, L., R. J. Braidwood, B. Howe, C. A. Reed, and P. J. Watson, eds.

1983 *Prehistoric Archaeology Along the Zagros Flanks.* Oriental Institute Publications 105. Chicago: Oriental Institute Press.

Braidwood, R.

1945 Introduction to "Tell Hassuna" by S. Lloyd and F. Safar. *Journal of Near Eastern Studies* 4:255–89.

1960 Seeking the World's First Farmers in Persian Kurdistan. *Illustrated London News,* 22 Dec. 1960:695.

1969 Prehistoric Investigations in Southeastern Turkey. *Science* 164, no. 3885: 1275–76.

1970 (Letter signed Bob Braidwood). *Archaeological Newsletter* Nov. 8:1–3. Oriental Institute, University of Chicago.

1973 The Early Village in Southwestern Asia. *Journal of Near Eastern Studies* 32:34–39.

Braidwood, R. and L. Braidwood

1960 *Excavations in the Plain of Antioch I. The Earlier Assemblages: Phases A-J.* Oriental Institute Publications 31. Chicago: Oriental Institute.

Braidwood, R., L. Braidwood, J. Smith, and C. Leslie

1952 Matarrah: A Southern Variant of the Hassuna Assemblage, excavated 1948. *Journal of Near Eastern Studies* 11:1–75.

Braidwood, R., L. Braidwood, E. Tulane, and A. Perkins

1944 New Chalcolithic Material of Samarran Type and its Implications. *Journal of Near Eastern Studies* 3:47–104.

Braidwood, R., H. Cambel, B. Lawrence, C. Redman, and R. Stewart

1974 Beginnings of Village-Farming Communities in Southeastern Turkey. *Proceedings of the National Academy of Sciences* Feb. 1974:568–72.

Braidwood, R. and B. Howe

1960 *Prehistoric Investigations in Iraqi Kurdistan.* The Oriental Institute Studies in Ancient Oriental Civilizations, vol. 31. Chicago: University of Chicago.

Breniquet, C.

1987 Nouvelle hypothèse sur la disparition de la culture de Halaf. In *Préhistoire de la Mésopotamie,* J-L. Huot ed., 231–41. Paris: CNRS.

Brinkman, J.A.

1981 Hurrians in Babylonia in the late Second Millennium B.C.: An Unexploited Minority Resource for Socio-economic and Philological Analysis. In *Studies on the Civilization and Culture of Nuzi and the Hurrians,* M. Morrison and D. Owen eds., 27-35. Winona Lake, IN: Eisenbraun's.

Bronson, B.

1975 The Earliest Farmers: Demography as Cause and Consequence. In *Population, Ecology, and Social Evolution,* S. Polgar ed., 53–78. The Hague: Mouton.

Caldwell, J. R.

1964 Interaction Spheres in Prehistory. In *Hopewellian Studies,* J. R. Caldwell and R. L. Hall eds., 133–43. Springfield: Illinois State Museum Scientific Papers, No. 12.

Campbell Thompson, R. and M.E.L. Mallowan

1933 The British Museum Excavations at Nineveh, 1931–32. *Annals of Archaeology and Anthropology, University of Liverpool* 20:71–186.

Cambel, H., and R. Braidwood

1970 An Early Farming Village in Turkey. *Scientific American* 222(3):50–56.

Cauvin, J.

1978 *Les premiers villages de Syrie-Palestine du IXe au VIIe millénaire avant J. C.* Lyon.

1988 La néolithisation de la Turquie du Sud-est dans son contexte proche-orientale. *Anatolica* 15:69–80.

Chang, K. C.

1986 *The Archaeology of Ancient China,* 4th rev. ed. New Haven: Yale University Press.

Charpin, D.

1977 L'onomastique hurrite à Dilbat et ses implications historiques. In *Methodologie et critiques I: problèmes concernant les Hurrites.* Paris: CNRS.

Childe, V. G.

1956 *Drevnejshchii Vostok v Svete Novyx Raskopok* (Russian translation of *New Light on the Most Ancient East,* 4th rev. ed., 1952).

Clarke, D.

1978 *Analytical Archaeology.* 2d ed. rev. by R. Chapman (original 1968). London: Methuen.

Coghlan, H. H.

1975 *Notes on the Prehistoric Metallurgy of Copper and Bronze in the Old World.* Oxford: Pitt-Rivers Museum.

Dabbagh, T.

1965 Hassuna Pottery. *Sumer* 21:93–112.

Davidson, T. and H. McKerrell

1976 Pottery Analysis and Halaf Trade in the Khabur Headwaters Region. *Iraq* 38:45–56.

1980 Neutron Activation Analysis of Halaf and Ubaid Pottery from Tell Arpachiyah and Tepe Gawra. *Iraq* 42:131–54.

Djaparidze, O. M. and A. I. Djavakhishvili

1971 *The Culture of the Earliest Agricultural Population in the Territory of Georgia.* Tbilisi.

Dosch, G. and K. Deller

1981 Sieben Kassitengenerationen in Temtena and Šuriniwe. In *Studies on the Civilization and Culture of Nuzi and the Hurrians,* M. Morrison and D. Owen eds., 91–113. Winona Lake: Eisenbraun's.

Eph'al, I.

1978 The Western Minorities in Babylonia in the 6th-5th Centuries B.C.: Maintenance and Cohesion. *Orientalia* 47:74–90.

Falkenstein, A.
1936 *Archaische Texts aus Uruk.* Leipzig.

Flannery, K.
1968 The Olmec and the Valley of Oaxaca: A Model for Inter-regional Interaction in Formative Times. In *Dumbarton Oaks Conference on the Olmec,* by E. Benson ed., 79–110. Washington, D.C.: Dumbarton Oaks.

Flannery, K. and J. C. Wheeler
1967 Animal Bones from Tell es-Sawwan, Level III. *Sumer* 23:179–82.

Frankel, D.
1979 *Studies on Halaf Pottery.* London: British Museum Publications.

Frifelt, K.
1989 Ubaid in the Gulf Area. In *Upon this Foundation,* E. Henrickson and I. Thuesen eds., 405–18. Copenhagen: Carsten Niebuhr Institute of Ancient Near Eastern Studies, University of Copenhagen.

Fukai, S., K. Hariuchi, and T. Matsutani
1970 *Telul eth-Thalathat. The Excavations of Tell II.* Tokyo University Iraq-Iran Archaeological Expedition, Report 2. Tokyo: University of Tokyo.

Fukai, S. and T. Matsutani
1981 *Telul eth-Thalathat. The Excavation of Tell II,* vol. 4. Tokyo: University of Tokyo.

Garstang, J.
1953 *Prehistoric Mersin.* Oxford: Clarendon Press.

Goff, B.
1963 *Symbols of Prehistoric Mesopotamia.* New Haven: Yale University Press.

Harris, D.
1989 An Evolutionary Continuum of People-Plant Interaction. In *Foraging and Farming: The Evolution of Plant Exploitation,* D. R. Harris and G. C. Hillman eds., 11–26. London: Unwin Hyman.

Heinrich, E.
1982 *Tempel und Heiligtümer im alten Mesopotamien.* Berlin: Walther de Gruyter.

Helbaek, H.
1964 Early Hassunan Vegetable Food at Tell es-Sawwan near Samarra. *Sumer* 20:45–48.
1972 Traces of Plants in the Early Ceramic Site of Umm Dabaghiyah. *Iraq* 34:17–19.

Henrickson, E. and I. Thuesen, eds.
1989 *Upon this Foundation: The Ubaid Reconsidered.* Copenhagen: Carsten Niebuhr Institute of Ancient Near Eastern Studies, University of Copenhagen.

Herzfeld, E.
1930 *Die Ausgrabungen von Samarra,* vol. 5. Die vorgeschichtlichen Töpfereien von Samarra. Berlin.

Hijara, I.
1978 Three New Graves at Arpachiyah. *World Archaeology* 10:125–28.
1980 Arpachiyah 1976. *Iraq* 42:131–54.

Hole, F.
1977 *Studies in the Archaeological History of the Deh Luran Plain: The Excavation of Chagha Sefid.* Memoirs of the Museum of Anthropology 9. Ann Arbor: University of Michigan Press.
1984 A Reassessment of the Neolithic Revolution. *Paléorient* 10:49–60.

Hole, F., K. Flannery, and J. A. Neely
1969 *Prehistory and Human Ecology of the Deh Luran Plain.* Memoirs of the Museum of Anthropology 1. Ann Arbor: University of Michigan Press.

Huot, J.
1979 Ubaidian Villages of Lower Mesopotamia: Permanence and Evolution from Ubaid 0 to Ubaid 4 as seen from Tell el 'Oueili. In *Upon this Foundation,* E. Henrickson and I. Thuesen eds., 19–42. Copenhagen: Carsten Niebuhr Institute for Ancient Near Eastern Studies, University of Copenhagen.
1987a L'apport de Tell el 'Oueili la chronologie d'Obeid. In *Chronologies du Proche Orient,* O. Aurenche, J. Even, and F. Hours eds., 465–72. Oxford: British Archaeological Reports.
1987b La phase 'Oueili de l'epoque d'Obeid. *Préhistoire de la Mésopotamie,* J-L. Huot ed., 129–51. Paris: CNRS.

Ingholt, H.
1957 The Danish Dokan Expedition. *Sumer* 13:214–15.

Jasim, S.
1985 *The Ubaid Period in Iraq.* Oxford: British Archaeological Reports.
1989 Structure and Function in an Ubaid Village. *In Upon this Foundation,* E. Henrickson and I. Thuesen eds., 79–90. Copenhagen: Carsten Niebuhr Institute of Ancient Near Eastern Studies, University of Copenhagen.

Jawad, A.J.
1965 *The Advent of the Era of Townships in Northern Mesopotamia.* Leiden: Brill.

Kamp, K. and N. Yoffee
1980 Ethnicity in Ancient Western Asia during the Early Second Millennium B.C.: Archaeological Assessments and Ethnoarchaeological Prospectives. *Bulletin of the American Schools of Oriental Research* 237:85–103.

Kaplan, J.
1960 The Relations of the Chalcolithic Pottery of Palestine to Halafian Ware. *Bulletin of the American Schools of Oriental Research* 159:32–36.

Kenyon, K.
1957 *Digging up Jericho.* New York: Praeger.

Kingery, D., P. Vandiver, and M. Prickett
1988 The Beginnings of Pyrotechnology, part II: Production and Use of Lime and Gypsum Plaster in the Pre-Pottery Neolithic Near East. *Journal of Field Archaeology* 15:219–44.

Kirkbride, D.
1968 Beidha: Early Neolithic Life South of the Dead Sea. *Antiquity* 42:263–74.
1972 Umm Dabaghiyah 1971: A Preliminary Report. An Early Ceramic Farming Settlement in Marginal North Central Jezira, Iraq. *Iraq* 34:3–15.
1973a Umm Dabaghiyah 1973: A Third Preliminary Report. *Iraq* 35:205–9.
1973b Umm Dabaghiyah 1972: A Second Preliminary Report. *Iraq* 35:1-7.
1974 Umm Dabaghiyah: A Trading Outpost? *Iraq* 36:85–92.
1975 Umm Dabaghiyah 1974: A Fourth Preliminary Report. *Iraq* 37:3–10.

Kohl, P. L.
1978 The Balance of Trade in Southwestern Asia in the Mid-Third Millennium B.C. *Current Anthropology* 19:463–92.
1987a The Ancient Economy, Transferable Technologies and the Bronze Age World-System: A View from the Northeastern Frontier of the Ancient Near East. In *Centre and Periphery in the Ancient World,* M. Rowlands, J. Gledhill, and M. T. Larsen eds., 13–24. Cambridge: Cambridge University Press.
1987b The Use and Abuse of World Systems Theory: The Case of the "Pristine" West Asian State. *Advances in Archaeological Method and Theory* 11:1–36.

Kozlowski, S. K., ed.
1990a *Nemrik 9: Pre-Pottery Neolithic Site in Iraq (General Report — Seasons 1985–86).* Warsaw: Wydawnictwa Uniwersytetu Warsaskiego.
1990b Nemrik 9, a PPN Site in Northern Iraq. In *Préhistoire du Levant: processus de Changement Culturel — Hommages à Francis Hours,* O. Aurenche, M. C. Cauvin, and P. Sanlaville eds., 29–35. Paris: CNRS.

Kushnareva, K. H. and T. N. Chubinishvili
no date *Ancient Cultures of the Southern Caucasus, V-VIII Millennia B.C.*

Lambert, W. G.
1961 Ancestors, Authors and Canonicity. *Journal of Cuneiform Studies* 11:1–14, 112.

Le Blanc, S.
1971 *Computerized, Conjunctive Archaeology and the Near Eastern Halaf.* Ph.D. diss., Washington University, St. Louis.

Le Blanc, S. and P. J. Watson
1973 A Comparative Statistical Analysis of Painted Pottery from Seven Halafian Sites. *Paléorient* 1:117–33.

Lees, G. M. and N. L. Falcon
1952 The Geographical History of the Mesopotamian Plains. *Geographical Journal* 118:24–39.

Lisitsyna, G. N.
1984 Problemy stanoveniya proizvodyashchix form xozyaystva v svete novejshschix paleoetnobotanicheskix issledonanij v perednej Azii. *Kratkie Soobshcheniya Instituta Arxeologii* 180:60–63.

Lloyd, S.
1938 Some Ancient Sites in the Jebel Sinjar District. *Iraq* 5:123–42.
1976 *Archaeology of Mesopotamia.* London: Thames and Hudson.

Lloyd, S. and F. Safar
1945 Tell Hassuna. *Journal of Near Eastern Studies* 4:255–89.

Limet, H.
1960 *Le travail du metal au pays de Sumer au temps de la IIIe dynastie d'Ur.* Paris: Société d'Edition "Les belles Lettres."

van Loon, M. and J. H. Skinner
1968 The Oriental Institute Excavations at Mueybit, Syria: Preliminary Report on the 1965 Campaign. *Journal of Near Eastern Studies* 27:265–90.

Machinist P. B.
1976 Literature as Politics: The Epic of Tukulti-Ninurta and the Bible. *Catholic Biblical Quarterly* 38:455–82.
1985 On Self-Consciousness in Mesopotamia. In *The Origins and Diversity of Axial Age Civilizations,* S.N. Eisenstadt ed., 183-202. Albany: State University of New York Press.

Mallowan, M.E.L.
1936 Excavations at Tell Chagar Bazar and the Archaeological Survey of the Habur. *Iraq* 3:1–86.
1946 Excavations in the Balikh Valley. *Iraq* 8:111–60.
1947 Excavations at Brak and Chagar Bazar. *Iraq* 9:1–259 (entire volume).
1956 *Twenty-five Years of Mesopotamian Discovery.* London: British School of Archaeology in Iraq.

Mallowan, M.E.L. and J. Cruikshank Rose
1935 Excavations at Tell Arpachiyah, 1933. *Iraq* 2:1–178.

Masry, A.
1974 *Prehistory in Northeastern Arabia: The Problem of Inter-regional Interaction.* Coconut Grove, FL: Field Research Projects.

Masson, V. M.
1964 *Srednya Aziya i Blizhnii Vostok.* Moscow-Leningrad: Nauka.

Meldgaard, J., P. Mortensen, and H. Thrane
1963 Excavations of Tepe Guran, Luristan. *Acta Archaeologica* 34:97–133.

Mellaart, J.
1965 *Earliest Civilizations of the Near East.* New York: McGraw-Hill.
1966 *The Chalcolithic and Early Bronze Ages in the Near East and Anatolia.* Beirut: Khayats.
1967 *Çatal Hüyük: A Neolithic Town in Anatolia.* London: Thames and Hudson.
1975 *The Neolithic of the Near East.* London: Thames and Hudson.

Mellink, M.
1964 Archaeology in Asia Minor. *American Journal of Archaeology* 68:149–66.

Merpert, N. Ya.
1978 Migratsii v epoxy Neolita i Eneolita. *Sovetskaya Arxeologiya,* no. 3:9–28.
1984 'Neoliticheskaya revolyutsiya' na Blizhnem Vostoke (nekotorye momenty razrabotki problemy v zarybezhnoj nauke), (The 'Neolithic Revolution' in the Near East [Some features of the treatment of this problem in foreign scholarship). In *Kratkie Soobshcheniya Instituta Arxeologii* 180:53–59.

Merpert, N. Ya. and R. M. Munchaev
1969 Excavations at Yarim Tepe. First Preliminary Report. *Sumer* 25:125–31.
1971a Excavations at Yarim Tepe 1970. The Archaeological Research in the Sinjar Valley. *Sumer* 27:23–33.
1971b Rannezemledel'cheskoe poseleniya severnoj Mesopotamii. Sovetskaya Arxeologiya (English translation in *Soviet Anthropology and Archaeology* 10 [1971–72]:203–52; see also *Earliest Agricultural Settlements in Northern Mesopotamia.* Trans. Irene Etkin Goldman; ed. Henry Field. Field Research Projects: Coconut Grove, Miami, 1972).
1971c Excavations at Yarim Tepe. Second Preliminary Report. *Sumer* 27:9–22.
1973 Early Agricultural Settlements in the Sinjar Plain, Northern Iraq. *Iraq* 35:93–113.

1977 Drevnejshaya metallurgiya Mesopotamii. *Sovetskaya Arxeologiya* 1971, no. 3 (see chapter 12 this volume).

1982a Pogrebal'nyj obryad plemyen xalafskoj kul'turj (Mesopotamiya) (Burial practices of the tribes of the Halaf culture, chapter 10 this volume). In *Arxeologiya Starogo i Novogo Sveta,* edited by V. I. Gulaev, 28–49. Moscow: Nauka.

1982b Poselenie Ubeidskoi kul'tury Yarim Tepe III v severnoi Mesopotamii (The Ubaid settlement of Yarim Tepe III in northern Mesopotamia, chapter 11 this volume). In *Sovetskaya Arxeologiya* 133–49.

1983 Peopling of the Sinjar Valley in Mesopotamia and the Structure of its Early Agricultural Settlements. Paper delivered at the First USA-USSR Archaeological Exchange, Cambridge, MA.

Merpert, N. Ya., R. M. Munchaev, and N. O. Bader

1976 The Investigations of the Soviet Expedition in Iraq 1973. *Sumer* 32:25–61.

1977 The Investigations of the Soviet Expedition in Iraq 1974. *Sumer* 33:65–104.

1978 Soviet Investigation of the Sinjar Plain. *Sumer* 34:27–71.

1981 Investigations of the Soviet Expedition in Northern Iraq, 1976. *Sumer* 37:22–54.

Michalowski, P.

1983 History as Charter: Some Observations on the Sumerian King List. *Journal of the American Oriental Society* 103:237–48.

1990 Early Mesopotamian Communicative Systems: Art, Literature, and Writing. In *Investigating Artistic Environments in the Ancient Near East,* edited by A. Gunter, 53–69. Washington, D.C.: Smithsonian Institution Press.

Moore, A.M.T.

1975 The Excavation of Tell Abu Hureyra in Syria. A Preliminary Report. *Proceedings of the Prehistoric Society* 41:50–77.

1985 The Development of Neolithic Societies in the Near East. *Advances in World Archaeology* 5:1–69.

Mortensen, P.

1970 Tell Shimshara. The Hassuna Period. *Historisk-Filisofiki Skrifter Danske* 5(2). Copenhagen.

1973 A Sequence of Samarran Flint and Obsidian Tools from Choga Mami. *Iraq* 35:37–55.

Munchaev, R. M.

1975 *Kavkaz na Zare Bronzovogo Veka.* Moscow: "Nauka"

1984 Nekatorye problemi arxeologii Mesopotamii v svete novejshix issledovanij (iz itogov rabot Sovetskoj ekspeditsii v Irake). (Some problems in the archaeology of Mesopotamia in light of the latest discoveries [from results of the Soviet expedition in Iraq, chapter 13 this volume]). In *Kratkie Soobshcheniya Instituta Arxeologii* (Brief reports of the Institute of Archaeology) 180:38–45.

Munchaev, R. M. and N. O. Bader

1969 Excavations at Yarim Tepe. *Sumer* 25:125–33.

1979 Rannezemledel'cheskoe poselenie v Severnoj Mesopotamii. *Vestnik AN SSR* 2.

Munchaev R. M. and N. Ya. Merpert

1970 Sovetskaya arxeologicheskaya ekspeditsiya v Irake. *Vestnik AN SSR,* no. 10.

1971 The Archaeological Research in the Sinjar Valley. *Sumer* 27: 23–32.

1972 Sovetskie arxeologicheskie issledovaniya v severo-zapadnom Irake. *Vestnik AN SSR,* no. 1.

1973 Excavations at Yarim Tepe 1972. Fourth Preliminary Report. *Sumer* 29:3–16.

1981 *Rannezemledel'cheskie Poseleniya Severnoj Mesopotamii.* Moscow: Nauka.

Munchaev, R. M., N. Ya. Merpert, N. O. Bader, V. A. Bashilov, O. G. Bolshakov, V. I. Gulaev, A. V. Kouza, I. G. Narimanov

1977 Issledovaniya Irakskoj Ekspeditsii. *Arxeologicheskaya Otkritiya* 1976.

Nissen, H.

1968 Survey of an Abandoned Modern Village. *Sumer* 24:107–14.

1988 *The Early History of the Near East.* Chicago: University of Chicago Press.

Oates, D.
1965 Tell al Rimah. *Iraq* 27:62–80.
1967 *Studies in the Ancient History of Northern Iraq*. Oxford: Oxford University Press (1968).

Oates, D. and J.
1976 *The Rise of Civilization.* Oxford: Elsevier.

Oates, J.
1968 Prehistoric Investigations near Mandali, Iraq. *Iraq* 30:1–20.
1969 Choga Mami 1967–68: A Preliminary Report. *Iraq* 31:115–52.
1972a A Radiocarbon Date from Choga Mami. *Iraq* 34:49–53.
1972b Prehistoric Settlement Patterns in Mesopotamia. In *Man, Settlement and Urbanism,* P. Ucko, R. Tringham, and G. Dimbleby eds., 299–310 London: Duckworth.
1973 The Background and Development of Early Farming Communities in Mesopotamia and the Zagros. *Proceedings of the Prehistoric Society* 39:147–81.
1976 Early Irrigation Agriculture in Mesopotamia. In *Problems in Economic and Social Archaeology,* G. Sieveking, I. Longworth, and K. Wilson eds., 109–36. London: Duckworth.
1978 Religion and Ritual in Sixth Millennium B.C. Mesopotamia. *World Archaeology* 10:117–24.
1980 Land use and Population in Prehistoric Mesopotamia. In *L'archéologie de l'Iraq,* M.-T. Barrelet ed., 303–34. Paris: CNRS.
1983 Ubaid Mesopotamia Reconsidered. In *The Hilly Flanks: Essays on the Prehistory of Southwest Asia presented to Robert J. Braidwood,* T. C. Young, Jr., P.E.L. Smith, and P. Mortensen eds., 251–82. Chicago: The Oriental Institute.
1987 Ubaid Chronology. In *Chronologies du Proche Orient,* O. Aurenche, J. Evin, and F. Hours eds., 473–82. Paris: CNRS.

Oates, J., T. E. Davidson, and H. McKerrell
1976 Sea-faring Merchants of Ur? *Antiquity* 51:221–34.

Oppenheim, M. Freiherr von
1943 *Tell Halaf I: Die prähistorischen Funde,* H. Schmidt ed. Berlin: Walther de Gruyter.

Perkins, A. L.
1957 *The Comparative Archaeology of Mesopotamia.* Chicago: Oriental Institute.

Porada, E.
1965 The Relative Chronology of Mesopotamia. In *Chronologies in Old World Archaeology,* R. Ehrich ed., 133–200. Chicago: University of Chicago Press.

Reade, J.
1968 Tell Taya. *Iraq* 30:234–68.
1971 Tell Taya. *Iraq* 33:87–100.

Redman, C.
1978 *The Rise of Civilization: From Early Farmers to Urban Society in the Ancient Near East.* San Francisco: W. H. Freeman.

Reilly, E.B.
1940 Test Excavations at Tilki Tepe. *Türk Tarih, Arkeoloji ve Etnografya Dergisi*:156–65.

Rubin, M. and C. Alexander
1960 U.S. Geological Survey Radiocarbon Dates V. *Radiocarbon* 2:129–85.

Sato, J, T. Sato, Y. Otomori, and H. Suzuki
1969 Univeristy of Tokyo Radiocarbon Measurements II. *Radiocarbon* 11:509–14.

Schortman, E.
1988 Interregional Interaction in Prehistory: The Need for a New Perspective. *American Antiquity* 54:52–65.

Schortman, E. and P. Urban
1987 Modeling Interregional Interaction in Prehistory. *Advances in Archaeological Method and Theory* 11:37–95.

Selimkhanov, I.R.
1970 Razgadannye sekrety drevnej bronzy. (See chap. 12, note 5, in original Russian, note 13).

Sherratt, A.
1980a Plough and Pastoralism: Aspects of the Secondary Products Revolution. In *Pattern in the Past: Studies in Memory of David L. Clarke,* N. Hammond et al. ed., 261–305. Cambridge: Cambridge University Press.

1980b Water, Soil and Seasonality in Early Cereal Cultivation. *World Archaeology* 11:313–30.

Simpson, K.
1988 Early Soundings. Qraya Modular Reports, No. 1. *Syro-Mesopotamian Studies* 4:4.

Smith, P.E.L.
1968 Ganj Dareh Tepe. *Iran* 6:158–60.
1970 Ganj Dareh Tepe. *Iran* 8:178–80.

Stein, Gil
1991 Imported Ideologies and Local Identities: North Mesopotamia in the Fifth Millennium B.C. Paper presented at fifty-sixth annual meeting of the Society of American Archaeology, New Orleans, 26 April 1991.

Strommenger, E.
1980 *Habuba Kabira.* Mainz: Verlag Philipp von Zabern.

Struever, S. and G. Houart
1972 An Analysis of the Hopewell Interaction Sphere. In *Social Exchange and Interaction,* E. Wilmsen ed., 47–78. Anthropology Papers, no. 46. Museum of Anthropology, Ann Arbor.

Stuckenrath, Jr., R. and E. K. Ralph
1965 University of Pennsylvania Radiocarbon Dates VIII. *Radiocarbon* 7:187–99.

Sürenhagen, D.
1986 The Dry Farming Belt: The Uruk Period and Subsequent Developments. In *The Origin of Cities in Dry-Farming Syria and Mesopotamia in the Third Millennium B.C.*, H. Weiss ed., 7–43. Guilford, CT: Four Quarters Publishing.

Taylor, J. P., M. V. Williams, and J. W. Waechter
1950 The Excavations at Sakce Gözü. *Iraq* 12:53–138.

Tobler, A. J.
1950 *Excavations at Tepe Gawra,* vol. 2. Philadelphia: University of Pennsylvania.

Voigt, M.
1985 Village on the Euphrates: Excavations at Neolithic Gritille in Turkey. *Expedition* 27(1):10–24.

1986 Review of *The Hilly Flanks and Beyond: Essays on the Prehistory of Southwestern Asia,* T. C. Young, Jr., P.E.L. Smith, and P. Mortensen eds. *Paléorient* 12:103–13.
1990 Reconstructing Neolithic Societies and Economies in the Middle East: An Essay. *Archaeomaterials* 4:1–14.

al-Wailly, F. and B. Abu es-Soof
1965 The Excavations at Tell es-Sawwan, First Preliminary Report (1964). *Sumer* 21:17–22.

Waterbolk, H. T.
1971 Working with Radiocarbon Dates. *Proceedings of the Prehistoric Society* 37:15–33.

Watkins, T.
1990 The Origins of House and Home? *World Archaeology* 21:336–47.
1992 Pushing Back the Frontiers of Mesopotamian History. *Biblical Archaeologist* 55 (December 4):176–181.

Watkins, T. and D. Baird
1987 *Qermez Dere: The Excavation of an Aceramic Neolithic Settlement near Tell Afar, N. Iraq.* Edinburgh: Interim Report, Project Paper No. 6, Dept. of Archaeology, University of Edinburgh.

Watkins, T., D. J. Baird, and A.V.G. Betts
1990 Qermez Dere and the Neolithic of Northern Iraq. In *Préhistoire du Levant: processus de changement culturel—Hommages à Francis Hours,* O. Aurenche, M. C. Cauvin, and P. Sanlaville eds., 23–28. Paris: CNRS.

Watkins, T. and S. Campbell
1987 The Chronology of the Halaf Culture. In *Chronologies du Proche Orient,* O. Aurenche, J. Evin, and F. Hours eds., 427–64. Oxford: British Archaeological Reports.

Watson, P. J.
1983 The Halafian Culture: A Review and Synthesis. In *The Hilly Flanks: Essays on the Prehistory of Southwestern Asia Presented to Robert J. Braidwood,* T.C. Young, Jr., P.E.L. Smith, and P. Mortensen eds., 231–50. Chicago: Oriental Institute.

Watts, Michael J.
1992 Space for Everything (A Commentary). *Cultural Anthropology* 7:115–29.

Willey, G.
1991 Horizonal Integration and Regional Diversity: An Alternating Process in the Rise of Civilization. *American Antiquity* 56:197–215.

Wolf, E.
1990 Facing Power—Old Insights, New Questions. *American Anthropologist* 92:586–96.

Yoffee, N.
1986 Sfery vzaimodestviya v Mesopotamii. In *Drevnej Tsivilizatsii Vostoka,* V. M. Masson ed., 117–28. Tashkent: Publishing House "FAN" of the Uzbek SSR.
1988a The Collapse of Ancient Mesopotamian States and Civilization. In *The Collapse of Ancient States and Civilizations,* N. Yoffee and G. L. Cowgill eds., 44–68. Tucson: University of Arizona Press.
1988b Context and Authority in Early Mesopotamian Law. In *State Formation and Political Legitimacy,* J. Toland and R. Cohen eds., 95–113. New Brunswick: Transaction Press.
1988c Aspects of Mesopotamian Land Sales. *American Anthropologist* 90:119–30.
1989 "Chuzhezemtsy" v Mesopotamii. *Vestnik Drevnej Istorii,* no. 2:95–100.

Zohary, D.
1989 Domestication of the Southwest Asian Neolithic Crop Assemblage of Cereals, Pulses, and Flax: The Evidence from the Living Plants. In *Foraging and Farming: The Evolution of Plant Exploitation,* D. R. Harris and G. C. Hillman eds., 358–73. London: Unwin Hyman.

Zohary, D. and M. Hopf
1988 *Domestication of Plants in the Old World.* Oxford: Oxford University Press.

Index

Abiboulayev, O. A., 129
Abu Hureyra, 39, 251
Adzes, 67, 69, 118, 208
Agriculture, 91. *See also* Domestication; Faunal analysis; Paleobotanical analysis
Ahmed Fottotepe, 55
Alabaster, 137, 146, 207; vessels made of, 59, 116, 145, 195, 207, 211, 216, 219, 221
Ali Aga, 7, 23, 70, 124
Ali Kosh, 85, 120, 241–42, 251
Amorites, 260–61
Amulets. See Pendants
Amuq, 125
Anthropomorphic vessels, 144, 145, 253
Arpachiyah, Tell, 4, 129–31, 144, 184, 195, 225, 233, 235, 240, 252, 254; artifacts found in, 146, 154–55, 158, 159, 178, 202–3, 205, 236, 239, 242; and burials, 207–8, 211, 217; and Halaf, 171, 178, 195, 202–3, 205, 263
Arrowheads. *See* Projectile points
Assyria, 259–61, 266–67; dichotomy with Babylonia, 267
Aswad, Tell, 132
Axes, 112. *See also* Wedge-shaped stone axes

Babylonia, 259–61, 266; dichotomy with Assyria, 267
Bader, N. O., 4, 26, 117–20, 249
Bakhteev, F. Kh., 121
Bashilov, V. A., 1, 4, 26, 41
Basketry, 34, 67
Beads, 47, 113, 146, 212, 251; alabaster, 207; bone, 187; carnelian, 178, 211; chalcedony, 240; clay, 69, 216; copper, 91, 208, 243, 245, 248, 251; diorite, 195; gypsum, 215; hematite, 240; marble, 67, 69, 91, 178, 195, 240; obsidian, 190, 215–16, 240; shell, 190, 208, 211, 219, 235, 240; slate, 207; steatite, 152, 195; stone, 37, 45, 51, 59, 69, 187, 190, 211, 216, 235; turquoise, 207, 252
Beidha, 23, 83, 95, 118
Bibikova, V. I., 122
Bitumen, 12, 34, 132, 167, 237
Blackman, J., 263
Bökönyi, S., 93, 122
Bol'shakov, O. G., 27
Bone, 31, 167, 175, 215, 263; awls, 34, 45, 59, 69, 178, 187–88, 190, 196, 208, 239, 250; blades, 87, 178, 195; needles, 34, 45, 59, 70, 187–88, 190, 196, 239, 250; polishers, 17, 34, 45; working, 263
Bouqras, 39, 70, 119–20
Bracelets, 251; lead, 245, 247–48, 252; marble, 21, 67, 69–70; stone, 25, 45, 47, 59, 250
Braidwood, R., 7, 41, 45, 88, 124
Brak, Tell, 129, 202
Bucrania motifs, 134, 155, 178, 235
Burial practices, 5, 75, 76, 82, 85, 93, 99, 104–5, 207–23, 251, 253–54; burial chambers, 212, 219, 221; ceremonies, 216, 217, 223, 254; cranial burials, 208, 210, 217, 219; cremations, 210, 212–17, 219; crematoria, 212, 215–16; Halaf

cemetery, 116, 129, 218–23; mortuary offerings, 66, 68–69, 144, 145, 168, 207–8, 211, 213, 215–17, 219, 221, 235, 243, 254; primary inhumations in domestic contexts, 39, 45, 66, 68–69, 167–68, 171, 175–76, 194, 208, 210–12, 229, 235, 243, 250–51; secondary inhumations, 208, 210

Caldwell, J. R., 257
Carbon-14 dates. *See* Radiocarbon dates
Carchemish, 125
Carnelian, 123, 178, 209
Çatal Hüyük, 46, 118, 217, 241, 242
Çayönü, 5, 39, 70, 71, 118, 131, 241–42, 251
Cemetery, Halaf. *See* Burial practices
Ceramic production: ceramic slag, 68; kilns, 76, 77, 84–85, 91, 93, 103, 132, 138–39, 142, 252, 263
Cereals. *See* Paleobotanical analysis
Chagar Bazar, 129, 155, 202, 242
Chalcedony, 123, 240
Chang, K. C., 268
Chernykh, E. N., 247, 248
Childe, V. G., 242
Choga Mami, 84, 88, 111, 125–26, 262, 266
Clarke, D., 257
Colonies, 266
Copper artifacts. *See* Metallurgy
Cranial burials. *See* Burial practices
Cremations. *See* Burial practices
Crematoria. *See* Burial practices
Cultic architecture. *See* Ritual architecture
Cultural boundaries, 269
Cyrus the Great, 260

Dabbagh, Takey, 87
Davidson, T., 263
Deh Luran Plain, 120, 125–26
Dendrochronological calibration curve, 126
Disarticulated burials. *See* Burial practices: secondary inhumations
Domestication: animal, 6, 63, 67, 69–70, 120–23, 131, 187–88, 250, 252; plant, 6, 39–40, 63, 67, 69–70, 120–23, 131–32, 250, 252, 254
Drills, 15, 146

Eridu, 265, 266
Ethnicity, 259–61; ethnic groups, 5, 258, 268; ethnic replacement, 263, 268
Evolution, social, 262, 267–68
Exchange. *See* Trade
Eynan, 83

Faunal analysis, 6, 39, 67, 69, 120–23, 133, 167, 175, 187–88, 190, 211, 221, 250, 253
Fibulae, stone, 239
Figurines, clay, 6, 17, 116, 139, 188, 202–5, 263; anthropomorphic, 22, 23, 37, 45, 47, 52–54, 67, 69, 89, 91, 132, 157–58, 167, 178, 184, 187, 202–3, 207, 239, 250–53; zoomorphic, 45, 47, 52–54, 67, 69, 178, 187, 190, 204–5, 239, 250–51, 253
Flannery, K., 263
Foundation deposits, 139, 145
Funerary structures. *See* Burial practices: burial chambers

Gadzhieva, D. V., 39
Ganj Dareh, 83, 95
Ghassul, Tell, 120
Gritille, 70–71
Gulaev, V. I., 1, 4, 41

Hacilar, 118, 241
Hagoshrim, 32
Halaf, Tell, 129, 202
Halaf cemetery. *See* Burial practices: Halaf cemetery
Halaf period and culture, 4–5, 129–61, 163–205, 251–54, 262–63; agriculture in, 45, 55, 63, 71; burial practices of, 207–23, 225; and Hassuna culture, 115–16, 120, 125–27; 207–23, 225, 235, 242–43, 246–49, 251–54, 262–63; and metallurgy, 242–43, 246–69; and Yarim Tepe, 73, 75, 83, 225, 235
Hassuna, Tell, 4, 23, 41, 45–48, 51–54, 63, 70, 115–19, 122, 124, 126, 207–8, 243, 248, 254, 262
Hassuna period and culture, 4–5, 7–8, 67–71, 252–53, 261–63, 267; agriculture in, 41, 63; artifacts of, 54–56, 58–59, 89, 91, 95, 109, 114, 152; burials in, 208–9, 217–18; and other cultures, 89, 137, 144, 158, 161, 184, 195; early sites of, 61; and metallurgy, 241–43, 246, 248–49; phases in, 43–48, 115–28; vessels of, 21; and Yarim Tepe, 75, 77, 80, 82, 84–85
Hematite, 219, 239
Hijara, I., 144, 208, 217
Hole, F., 126
Hopewell, 257
"Husking trays," 47, 89, 104, 118–19
"Huts and flowers" motifs, 157

"Interactional approach," 260
Interaction sphere, 257, 262–63, 268

International Research and Exchanges Board (IREX), 1, 6

Jafar Hüyük, 70, 71
Jarmo, 4, 5, 7, 22–23, 25, 41, 45, 48, 51–52, 54, 63, 67, 70, 73, 85, 118–20, 125–26, 251, 261
Jebel Sinjar, 3, 8, 39–41, 55, 61, 63, 118
Jemdet Nasr, 85
Jericho, 83, 118, 125, 217

Karim Shahir, 7, 41
Kassites, 261
Kilns. *See* Ceramic production
Kirkbride, D., 7, 46, 54, 123, 126, 246
Kohl, P. L., 258, 262
Korobkova, G. F., 121
Kültepe (Nakhichevan, Transcaucasia), 129
Kültepe (northern Iraq), 4–5, 23, 41, 47, 55–61, 63, 67, 69–71, 117, 250–52, 261
Kuyunjik, 87
Kuza, A. V., 26, 55

Lapis lazuli, 69
Lead artifacts, 241, 245, 247–48, 252
Le Blanc, S., 263
Lisitsyna, G. N., 121
Lloyd, S. and F. Safar, 88
Lloyd, Seton, 41, 73, 115, 124, 242
Lungshanoid horizon, 268

Mace heads, 67; alabaster, 195
Machinist, P. B., 260
McKerrell, H., 263
Maghzaliyah, Tell, 4–40, 61, 64–67, 70–71, 100, 112, 114, 118, 120, 208, 250–52, 261; defensive wall at, 25–28, 33–34, 64–66, 250, 262
Malachite, 248
Mallowan, M.E.L., 132, 144, 207, 211, 263
Masson, V. M., 7
Matarrah, 122, 124–26
Mats, 60
Medes, 259
Megiddo, 125
Mellaart, J., 7, 119, 120, 125–26, 242, 246, 263
Merpert, N. Ya., 4, 45–46, 54, 249
Mersin, 45, 125, 242
Mesopotamian civilization, 261
Metallurgy, 5, 131, 241–48; chemical analysis of metal artifacts, 247–48; copper artifacts, 37–38, 67, 69, 91, 114, 123, 139, 145, 208, 241, 243–45, 247–48, 251–52, 263; copper ore, 167, 243–45, 252; lead artifacts, 241, 245, 247–48, 252
Microliths, 13, 15, 23, 69, 139, 145
Mohammed Jaffar, 242
Moral economy, 267–68
Mortensen, P., 7, 17, 119–20, 124–25
Mortuary offerings. *See* Burial practices
Mousterian, 41
Munchaev, R. M., 1, 4
Mureybet, 119–20

Nails, clay, 239
Necropolis, Halaf, *See* Burial practices: Halaf cemetery
Nemrik, 5, 261, 267
Nineveh, 87, 263
Ninevite 1, 125
Ninevite 2, 73
Nuzi, 131

Oates, D., 7, 73
Oates, J., 4, 7, 73, 120, 262
Obsidian, 9, 28, 34, 91, 111, 117–18, 123, 131–32, 139; exchange of, 23, 69, 70, 145–46, 195, 215–16, 237, 240, 250, 263; tools, 13, 15, 17, 25, 37, 66–67, 91, 111, 117–18, 163, 175, 178, 188, 190
Oppenheim, Baron M. von, 129
Oppenheim, L., 260
Oueili, Tell, 266
Ovens, 12–13, 17, 30, 76, 81, 84–85, 95; uses of, 43–45, 65, 68, 98, 103, 132, 170, 173–75, 191–94, 231–32, 250; mangal-type, 187, 193–94; tannur-type, 56, 132, 138–39, 192, 194

Paleobotanical analysis, 6, 39, 67, 120–23; flotation, 121–22, 132–33, 208, 233, 250, 253–54
Paleozoological analysis. *See* Faunal analysis
Palettes, 99, 239, 263; flint, 188; marble, 59; obsidian, 188, 190
Palynological analysis, 39–40, 69
Pendants, 208, 212; bone, 215; clay, 199; copper 139, 145, 244–45, 252; marble, 67; obsidian, 67, 178, 195, 240; shell, 219; stone, 21, 37, 45, 67, 113, 132, 139, 146, 167, 171, 178, 195, 215, 250, 252
Political economy, 268
Population estimates, 66
Prishchepenko, L. V., 121
Projectile points, 15, 25, 66, 69–70, 117, 124, 195, 250, 252
Property, transmission of, 261

Qermez Dere, 5, 261, 267

Radiocarbon dates, 45, 52, 126–27, 135, 159, 161, 240, 246
Ramad, 95
Ras Shamra, 125
Reade, J., 73
Redman, C., 265
Rimah, Tell, 73, 249
Rings: clay, 178, 187–88, 190, 198, 239; copper, 244
Ritual architecture, 5, 132
Ruralization, 266

Sabi Abyad, Tell, 265
Safar, F., 41, 115, 124, 242
Sakçe Gözü, 46, 125
Samarra period and culture, 5, 52, 61, 70, 73, 84, 85, 87, 111, 115, 121, 125, 158, 207, 217, 246, 252, 262–63, 266
Sandai period, 268
Sculpture. *See* Figurines, clay
Seals. *See* Stamp seals
Sedentism, 71
Serpentine, 146
Shell, 190, 208, 211, 215, 219, 235, 240
Shimshara, Tell, 5, 7, 17, 23, 41, 51, 70–71, 125, 207, 251
Sickles and sickle blades, 17, 23, 34, 111, 121, 131–32, 178, 237, 250–52
Sinjar Mountains. *See* Jebel Sinjar
Sinjar Plain or Valley, 3, 9, 39, 73, 115, 119, 120, 125–26, 142, 194, 208–9, 240, 242, 249–54
"Sling bullets (pellets)," clay, 45, 69, 91, 113, 124, 131, 152, 178, 187–88, 190, 197, 239
Sotto, Tell, 4–5, 7–8, 13, 23, 25, 41–56, 58–59, 61, 63, 67–71, 95, 104, 112, 115–21, 126, 208, 243, 246, 248, 250–52, 261
Sotto period and culture, 40, 45, 47, 56, 61, 63–64, 69–71, 115, 119, 121, 124–27, 243, 246, 248, 251–52
Spalka I, 241
Spindle whorls, 69; alabaster, 146; clay, 45, 58, 152, 171, 178, 188, 190, 197, 215, 239; stone, 45, 139
Stamp seals, 146; clay, 91, 132, 199; copper, 139, 199; serpentine, 146; steatite, 146; stone, 37, 52, 67, 145, 171, 178, 195, 199, 215, 263
Statuettes. *See* Figurines, clay
Steatite, 146, 152, 195
Stein, G., 263
Stone: alabaster 59, 116, 145, 211, 216, 219, 221; tools, 17, 34, 45, 68, 81, 121, 131–32, 145–46, 178, 187–88, 190, 194, 233, 239, 254; foundations, 250; jewelry, 67, 68, 250–52; marble, 21, 34, 59, 61, 67, 69, 116, 196; vessels, 22–23, 25, 31, 45, 47–48, 51, 116, 167, 178, 190, 212, 215, 250–51
Storage: facilities, 6, 13, 17, 23, 30–32, 34, 44–45, 56–57, 65, 68–69, 75, 79–81, 121, 123, 130, 138, 142, 170, 192, 225, 231, 233, 250, 261; vessels/jars, 45, 68, 69, 105, 130, 175–76, 199, 251
Suberde, 241
Sumerian, 248

Tamerkhan, 85
Taya, 73
Tell Afar, 3, 8, 120, 163, 249, 251
Tell es-Sawwan, 23, 52, 59, 61, 70, 85, 118, 123, 125–26, 207, 241–42, 248, 262
Tell Sotto-Kültepe culture. *See* Sotto period and culture
Tell Sotto-Umm Dabighiyah complex and culture. *See* Sotto period and culture
Telul eth-Thalathat, 23, 25, 41, 45, 61, 63, 67, 70, 126, 246, 249
Tepe Gawra, 129, 131, 146, 158, 178, 184, 190, 195, 199, 202, 205, 225, 233, 235, 237, 239–40, 242, 252, 254, 265, 266
Tepe Guran, 51, 85
Tholos/tholoi, 5, 95, 130–31, 137–39, 144, 164–66, 170–76, 185–88, 190–92, 196–97, 200, 203–4, 208, 221, 235, 245, 253, 263
Tobler, A., 242
Trade, 123, 124
Transcaucasia, 129, 131, 262
"Tree of life" motif, 200
Tsalkin, V. I., 122, 133
Turquoise, 123, 207, 252, 263

Ubaid, Tell, 237, 249
Ubaid period and culture, 5, 63, 71, 73, 75, 144, 163–65, 168–70, 184–85, 194, 199, 204, 225–40, 242, 249, 253–54, 262, 265–66, 268
Ugarit. *See* Ras Shamra
Umm Dabaghiyah, 4, 7, 23, 41, 45–48, 54, 58–59, 61, 63, 70–71, 95, 104, 112, 116–23, 126, 207, 243, 246, 251, 261
Urbanization, 266
Ur III, 261
Uruk (Warka), 266
Uruk period, 85, 248, 262, 266–67, 269
Use-Wear analysis, 121

Wallerstein, I., 258
Waterbolk, H. T., 45

Wedge-shaped stone axes, 22, 34, 45, 47, 52, 67, 69, 117, 167, 178, 195, 239, 251–52
"World-system," 258

Yanushevich, Z. V., 121
Yarim Tepe I, 4–5, 23, 42, 45–48, 52–54, 73, 105, 115–27, 129–31, 142, 161, 163, 208, 218–23, 243–49, 252
Yarim Tepe II, 5, 73, 83, 115, 116, 127, 129–61, 177, 184, 190–91, 195, 197–98, 201–3, 209–18, 221, 223, 243, 245–49, 252–54, 263, 266
Yarim Tepe III, 5, 163–205, 218, 225–40, 249–50, 253–54

Zawi Chemi Shanidar, 23, 83, 120, 248
Zelikson, E. M., 39
Zoomorphic vessels, 145, 171, 177, 207, 253

About the Editors

NORMAN YOFFEE is professor and chair of the Department of Near Eastern Studies at the University of Michigan, Ann Arbor, as well as curator of archaeology in the Museum of Anthropology there. He has worked at the Institute of Archaeology in Moscow and is coeditor of *The Collapse of Ancient States and Civilizations,* published by the University of Arizona Press in 1988.

JEFFERY J. CLARK is a doctoral candidate in anthropology specializing in Near Eastern archaeology.